Annals of Mathematics Studies

Number 44

ANNALS OF MATHEMATICS STUDIES

Edited by Robert C. Gunning, John C. Moore, and Marston Morse

1. Algebraic Theory of Numbers, *by* HERMANN WEYL
3. Consistency of the Continuum Hypothesis, *by* KURT GÖDEL
11. Introduction to Nonlinear Mechanics, *by* N. KRYLOFF *and* N. BOGOLIUBOFF
16. Transcendental Numbers, *by* CARL LUDWIG SIEGEL
17. Problème Général de la Stabilité du Mouvement, *by* M. A. LIAPOUNOFF
19. Fourier Transforms, *by* S. BOCHNER *and* K. CHANDRASEKHARAN
20. Contributions to the Theory of Nonlinear Oscillations, Vol. I, *edited by* S. LEFSCHETZ
21. Functional Operators, Vol. I, *by* JOHN VON NEUMANN
22. Functional Operators, Vol. II, *by* JOHN VON NEUMANN
24. Contributions to the Theory of Games, Vol. I, *edited by* H. W. KUHN *and* A. W. TUCKER
25. Contributions to Fourier Analysis, *edited by* A. ZYGMUND, W. TRANSUE, M. MORSE, A. P. CALDERON, *and* S. BOCHNER
27. Isoperimetric Inequalities in Mathematical Physics, *by* G. POLYA *and* G. SZEGO
28. Contributions to the Theory of Games, Vol. II, *edited by* H. W. KUHN *and* A. W. TUCKER
29. Contributions to the Theory of Nonlinear Oscillations, Vol. II, *edited by* S. LEFSCHETZ
30. Contributions to the Theory of Riemann Surfaces, *edited by* L. AHLFORS *et al.*
31. Order-Preserving Maps and Integration Processes, *by* EDWARD J. McSHANE
33. Contributions to the Theory of Partial Differential Equations, *edited by* L. BERS, S. BOCHNER, *and* F. JOHN
34. Automata Studies, *edited by* C. E. SHANNON *and* J. McCARTHY
36. Contributions to the Theory of Nonlinear Oscillations, Vol. III, *edited by* S. LEFSCHETZ
37. Lectures on the Theory of Games, *by* HAROLD W. KUHN. In press
38. Linear Inequalities and Related Systems, *edited by* H. W. KUHN *and* A. W. TUCKER
39. Contributions to the Theory of Games, Vol. III, *edited by* M. DRESHER, A. W. TUCKER *and* P. WOLFE
40. Contributions to the Theory of Games, Vol. IV, *edited by* R. DUNCAN LUCE *and* A. W. TUCKER
41. Contributions to the Theory of Nonlinear Oscillations, Vol. IV, *edited by* S. LEFSCHETZ
42. Lectures on Fourier Integrals, *by* S. BOCHNER. In press
43. Ramification Theoretic Methods in Algebraic Geometry, *by* S. ABHYANKAR
44. Stationary Processes and Prediction Theory, *by* H. FURSTENBERG
45. Contributions to the Theory of Nonlinear Oscillations, Vol. V, *edited by* S. LEFSCHETZ
46. Seminar on Transformation Groups, *by* A. BOREL *et al.*
47. Theory of Formal Systems, *by* R. SMULLYAN

STATIONARY PROCESSES AND PREDICTION THEORY

BY

Harry Furstenberg

PRINCETON, NEW JERSEY

PRINCETON UNIVERSITY PRESS

1960

Copyright © 1960, by Princeton University Press

All Rights Reserved
L. C. Card 60-12226

Printed in the United States of America

To the Memory of

Jekuthiel Ginsburg

ACKNOWLEDGEMENTS

     This study was written while the author was
at Princeton University and is an elaboration of his
Princeton doctoral dissertation. During the period
of his work the author was supported by grants from
the National Science Foundation and the Office of
Ordnance Research, U. S. Army Ordnance, for which
he wishes to thank these agencies. In addition the
author is indebted to Princeton University and particularly to Professors Solomon Bochner and William
Feller for their friendship and encouragement and
for their suggestions which have been most helpful
in guiding the present work. Finally, the author
wishes to express his gratitude to Dr. Leon
Ehrenpreis without whose kind encouragement this work
would not have advanced beyond its preliminary stages.

CONTENTS

Page

CONTENTS

Page

CONTENTS

Page

CONTENTS

Page

STATIONARY PROCESSES AND PREDICTION THEORY

INTRODUCTION

The theory of linear prediction developed by Wiener and
Kolmogoroff ([15], [8]) treats the problem of extrapolating from the
known past of some variable quantity to its future. The known data is
in the form of an infinite sequence of numerical readings taken from
observations made at equally spaced intervals of time throughout the past.
The problem is to make a reliable estimate of the future values of the
variable on the basis of this data. The method of Wiener and Kolmogoroff
may, for our purposes, be described as follows. One compares each of
the past readings with a fixed linear combination of those preceding it
and asks for that particular linear combination which in some sense minimizes the discrepancy between the two. Having found the most consistent
linear combination, the predicted value for the next reading is taken as
this linear combination of the past readings.

The questions which Wiener, Kolmogoroff, and their followers have
investigated are for the most part concerned with the relationship between
this prediction procedure and the harmonic analysis of the time series in
question. The harmonic, or oscillatory, aspects of the time series are
embodied in the so-called "spectrum" of the series which analyzes the
series into its harmonic components. The spectrum of a time series turns
up in numerous questions involving "linear" properties of the time series.
For example, it was found that from the spectrum one could compute the
"expected" error of the prediction procedure described, and thereby one
could ascertain whether or not the prediction would be exact. Similar
questions may be answered for vector-valued time series as well, but the
theory in this case is considerably more complex ([5], [11], [15]).

In order to present our version of the problem of prediction it
will not be necessary to enter in great detail into the circle of problems connected with linear prediction theory. We do point out that our
description oversimplifies the actual situation in several respects. To
begin with, there need not exist an optimal linear combination and the predicted value will have to be obtained by some limiting process. Secondly,
this limiting process need not converge for an individual time series and
the procedure may be meaningful only for an ensemble of time series where
the predicted value exists as a function on the ensemble defined "almost

everywhere". As a result the customary setting for the linear prediction
problem is the theory of randomized ensembles of time series or, more
precisely, the theory of stationary stochastic processes. The known data
is then not an individual numerical sequence but a sequence of random
variables, and the problem is to find a variable defined in terms of the
past that best approximates the variable representing the next reading.
The sense in which the approximation is to be optimum is that of the
"mean square" norm so that the desired random variable always exists by
the Riesz-Fischer theorem. We emphasize that the sense in which the prediction exists here is as a random variable defined almost everywhere on
a set of time series. This is significant because a random variable defined almost everywhere on a set of points need not be well-defined at
any point and exists only as an equivalence class of well-defined functions. Consequently, the resulting prediction need not have a meaning
at any particular time series of the ensemble.

However, the existing theory still makes plausible a problem of
prediction for individual time series. Presumably if the series representing the past is sufficiently well-behaved there do not arise any convergence difficulties and the predicted value for the next reading will
be well-defined. Moreover, in principle, there is no reason to restrict
oneself to linear prediction and for a suitably restricted time series
it should be possible to speak of a non-linear prediction as well. That
is, instead of considering only linear combinations of the past as estimates of the future one can consider arbitrary functionals on the past,
and take a sequence of such functionals that minimizes the discrepancy in
the past between the estimates values and those occurring. If the series
in question is sufficiently well-behaved the sequence of functionals should
provide a sequence of values converging to an "absolutely (i.e., not necessarily linear) best estimate" of the unknown next reading.

To be more explicit, let the given time series be represented as

$$(I.1) \qquad \ldots \xi(-n), \ldots, \xi(-2), \xi(-1), \xi(0) \quad ;$$

let $f(t_1, \ldots, t_k)$ be a function of k variables, k arbitrary, and
consider as an approximation to the given series, the series ξ_f defined
by

$$(I.2) \qquad \xi_f(-n) = f(\xi(-n-1), \xi(-n-2), \ldots, \xi(-n-k)) \ .$$

Using the mean square norm to measure the discrepancy between ξ and ξ_f
we obtain for the error

$$(I.3) \qquad \varepsilon(f) = \lim_{N \to \infty} \sup \left\{ \frac{1}{N+1} \sum_{0}^{N} |\xi(-n) - \xi_f(-n)|^2 \right\}^{1/2}$$

Let ε be the g.l.b. of $\varepsilon(f)$ for all f and let $\{f_\nu\}$ be a sequence of functions with $\varepsilon(f_\nu) \longrightarrow \varepsilon$. The sequences $\{\xi_{f_\nu}(-n)\}$ then represent closer and closer approximations to $\{\xi(-n)\}$, each defined in terms of "the past", namely, in terms of the preceding values of ξ. Hence

$$(I.4) \qquad \xi^*(1) = \lim_{\nu \to \infty} f_\nu(\xi(0), \xi(-1), \ldots, \xi(-k_\nu+1)) \quad ,$$

if it exists and is independent of the sequence $\{f_\nu\}$, would be an appropriate choice for the (least-squares) estimate of the next reading $\xi(1)$.

We are now in a position to state the problem underlying this investigation. The Wiener-Kolmogoroff theory of prediction suggests that for some class of time series it is possible to affect linear and non-linear prediction of future values in a manner to be meaningful for the individual series themselves. However, the particular procedure described in the preceding paragraph, as it stands, will not take us very far. For, it may be shown that the limit in (I.4) does not exist for general minimizing sequences of functions except in the trivial case that $\{\xi(-n)\}$ is periodic. The problem that we have set ourselves here is to determine whether there is nevertheless a precise sense in which one can identify a value $\xi^*(1)$ as the optimum estimate for the unknown reading $\xi(1)$ of a time series (I.1). The problem of prediction that we pose is then no longer concerned with the computability of the future values (where for practical reasons one is restricted to linear extrapolation) but with the purely theoretical question whether the concept of a predicted value of a sequence is at all sensible. Our goal, therefore, will be to state a definition of "predictability" for time series and to demonstrate its usefulness by exhibiting as wide a class as possible of "predictable" time series.

In this description of the problem we have restricted the scope of our notion of predictability to conform to the corresponding notion for stochastic processes. However there are points of view which suggest extending this notion. Namely, suppose that $\xi(n)$ is not a numerical sequence but one that takes values in an abstract space, such as would be the case if we were to record the results of a series of coin tosses. Here, even if one identifies the abstract values with numerical values, the result of the prediction might be a number that does not correspond to one of the abstract values possible. For example, setting "heads" = 1, and "tails" = -1, the result of the Wiener-Kolmogoroff prediction procedure would be 0 (if the coin is balanced) which is an impossible value in this case.

The value 0 occurring in the foregoing situation does have a simple interpretation; namely, that both "heads" and "tails", or 1 and - 1, are equally likely to occur as the next reading. Now if one regards the concept of prediction as that of inferring _information_ about the future on the basis of the past, then the conclusion that both possible values are equally likely to occur is information that is "predicted" from the past behavior. This suggests that what we should look for is not a particular "most likely" value in the future, but a probability distribution over all the values of the future readings as conditioned by the known past. Having found this distribution, the estimate for the next value may be made for a numerical sequence in a number of ways, the most natural being, perhaps, to choose the expected value of the distribution. On the other hand, a probability distribution would be no less meaningful for non-numerical sequences - and this is the principal advantage. We shall, in fact, adopt this point of view and as a consequence, the result of prediction for a sequence, when meaningful, will not be a single value, but a probability distribution over a set of values. More generally, the desired prediction will provide a probability distribution, or measure, over the set of all possible values for all the future readings.

This approach might, incidentally, be taken in the prediction problem for stochastic processes as well; here it does not involve a radical departure from the ordinary prediction problem. Thus it is possible to define a "random prediction measure", which is a measure-valued random variable defined over the past of the process. The non-linear prediction variable of the existing theory is related to the prediction measure in the manner described above: the random predicted values are the expectation values of the random prediction measures.

It should be pointed out that the notion of prediction measures avoids another of the limitations of the Wiener-Kolmogoroff approach; namely the arbitrariness of the least squares criterion arising in the form of the error term $\varepsilon(f)$ in (I.3). As may be shown, the least squares criterion corresponds to just one of the possible choices of an estimate based on a probability distribution - namely, the expectation values, and prediction values based on other norms may be obtained as well from the prediction measure.

To sum up, our problem is to propose a definition for the predictability of a sequence; i.e., for the existence of a procedure assigning to an abstract-valued sequence a measure defined on the set of all possible futures for the sequence and representing the likelihood of the occurrence of the future events based on the known past.

Although our problem is formulated for individual time series it is not independent of the theory of stochastic processes. In fact, the class of sequences to which our analysis will be applicable will be such that to each sequence there corresponds a stationary stochastic process for which the given sequence is a "typical" sample sequence. Speaking heuristically, the stationary process plays the role of a mechanical system on which the measurements are made. This mechanical system must be of a stationary nature, for otherwise there are no grounds on which to expect any connection between past and future behavior. Our first chapter will therefore be concerned with the question of determining when a sequence can be thought of as occurring from observations on a stationary mechanism, or its abstract counterpart, a stationary process. A sequence of this kind will be said to be "regular" and it determines a "generic" point of the associated process. To make these notions precise we shall have to develop systematically the relevant theory of stationary stochastic processes. We remark that a useful tool in this part of the exposition will be the theory of commutative C^*-algebras; a stationary process will be described in an "invariant" manner as a C^*-algebra with certain special properties.

The problem of prediction for sequences turns out to be a special case of a general problem of predicting from one stationary process to another "at a point of the process". In this formulation there will be two notions of predictability that may be applicable: "continuous predictability" and "statistical predictability", and our investigation may be separated into two portions treating respectively these two notions. The more restricted notion, that of "continuous predictability" is easily defined and the problem that we treat is to exhibit a large class of regular sequences to which this form of predictability applies. We find however that this class is not as wide as one might hope, and in particular one can find continuously predictable sequences of an elementary kind such that certain sequences derived from these by ordinary algebraic operations are no longer continuously predictable. To remedy this we introduce the broader concept of "statistical predictability" and our main problem will be to show that this form of predictability does apply to the sequences in question.

In the course of our discussion the theory of stationary processes plays a double role. In the first place the definitions of predictability can be made only after imbedding the sequences into the sample spaces of appropriate stationary processes. Secondly, in exhibiting the regular

sequences that are predictable we shall not, in general, given an exact
description of the sequence (which, usually, we cannot do, but fortunately,
this will not be necessary) but rather a description of the process it
represents. Thus we shall speak of "Markoff sequences" and "multiple
Markoff sequences", these being regular sequences generic for a stationary
Markoff or multiple Markoff process.

One question, therefore, to be considered is that of describing
stationary processes. To begin with, one has the elementary processes,
the "random" and "Markoff" processes where the probability relationships
are of a simple kind. More complex processes are obtained by considering
processes "derived" from these by algebraic operations and the like.
These might be termed "sub-Markoff" processes. This gives rise to the
first non-trivial problem in our theory: to determine when a sub-Markoff
sequence is continuously predictable. In Chapter 3 we find that this
is not always the case and in Chapter 4 we derive sufficient conditions
for it to hold. In Chapter 5 we introduce the notion of a "stochastic
semigroup" by means of which one may define a still wider class of
stationary processes. Here too we shall find conditions for the predictability of the sequences arising in this way, the conditions being
in terms of the generating stochastic semigroup.

In our treatment of "statistical predictability" our main
concern will be with sequences derived from Markoff sequences and our
principal result is that these are statistically predictable. In
arriving at this result we shall come across another method of generating
stationary processes, namely, by forming "inductive functions" of known
processes. In this regard the main problem will be to see to what extent the inductive function of a regular sequence mirror the behavior
of the inductive functions of the stationary process it represents. We
shall see that this question arises in connections other than prediction
theory and we devote Chapters 7 and 8 to an exposition of these. In
particular, we shall show how a certain equidistribution theorm of
H. Weyl ([13]) is deducible in this context and we also give, by these
means another proof of a theorem of Wiener and Wintner regarding the
Fourier coefficients of a stationary process ([16]).

There is one more remark which we feel is appropriate. We
wish to point out that although prediction theory is the unifying theme of
this study, it has not always been our main concern to prove theorems regarding prediction theory. Thus our final result which may be stated
"every finitely-valued derived sequence of a finite state Markoff sequence is statistically predictable" is too special to justify the last

four chapters required for its proof. It would be more correct to say that our goal has been to show how certain concepts, most of which we present in Chapter 1, may be used to systematically analyze various stationary processes and their sample sequences. Prediction theory has therefore often been the excuse rather than the reason for carrying through a certain analysis illustrating the concepts and methods that we introduce.

CHAPTER 1. STOCHASTIC PROCESSES AND STOCHASTIC SEQUENCES

§1. Preliminaries: Commutative C^*-Algebras ([10]).

1.1. <u>C^*-Algebras</u>. We denote by $C(X)$ the algebra of all
complex valued continuous functions on a compact Hausdorff space X.   When
provided with the norm

$$\|f\| = \sup_{x \in X} |f(x)| \quad ,$$

and the involution $f \longrightarrow f^*$, $f^*(x) = \overline{f(x)}$, $C(X)$ becomes a commutative
C^*-algebra. In other words it is a Banach algebra:

(a) $\|\lambda f\| = |\lambda|\, \|f\|$, $\|f + g\| \leq \|f\| + \|g\|$, $\|fg\| \leq \|f\|\, \|g\|$,
     $C(X)$ is complete in the metric induced by $\|\cdot\|$, and

(b) the involution satisfies $f^{**} = f$, $(\lambda f)^* = \bar{\lambda} f^*$,
     $(f + g)^* = f^* + g^*$, $(fg)^* = f^* g^*$, $\|ff^*\| = \|f\|^2$.

These are the conditions for a C^*-algebra. To an extent the converse is
true as well. If A is any commutative normed algebra with an identity
element and an involution satisfying (a) and (b) then there is a compact
Hausdorff space X, and an isomorphism of A with $C(X)$ preserving
norms and involutions. The class of algebras $C(X)$ is thus identical
with the class of commutative C^*-algebras with identity. This will be of
considerable importance for us, for we shall frequently recognize an algebra as a C^*-algebra and it will be useful to know that it is the algebra
of continuous functions on a space X.

To study more closely the nature of this equivalence let A be
a commutative C^*-algebra with identity. We consider the set of all algebraic homomorphisms of A onto the complex numbers. These form a set
X_A and the elements $x \in A$ induce functions $\hat{x}$ on X_A by defining
$\hat{x}(h) = h(x)$ for $h \in X_A$. X_A may then be given the weakest topology that
renders all these functions continuous. With this topology X_A is a
compact Hausdorff space and the mapping $x \longrightarrow \hat{x}$ sends A onto a subalgebra of $C(X_A)$. Ordinarily, the mapping $x \longrightarrow \hat{x}$ would reduce norms;
however when A is a commutative C^*-algebra the norm of x in A agrees
exactly with the supremum norm of $\hat{x}$ on X_A. This implies that our

representation of A in $C(X_A)$ is 1 - 1 so that A may be identified with a subalgebra of $C(X_A)$, which, by the completeness of A, is uniformly closed. Since two points of X_A that give the same value to each $\hat{x}$ must correspond to the same homomorphism of A, hence must be identical, it follows that the subalgebra of $C(X_A)$ corresponding to A separates points in X_A. Moreover it may be shown that involutions go over into conjugates in this representation, i.e., that $(\hat{x}^*) = \overline{\hat{x}}$. This implies that the subalgebra in question contains with every function on X_A its complex conjugate.

At this point we invoke the Stone-Weierstress theorem to the effect that a subalgebra of $C(X)$ separating points of X and containing with each function its complex conjugate is dense either in $C(X)$ or in the subalgebra of functions vanishing at some fixed point. Since the image of A in $C(X_A)$ satisfies the hypotheses of the theorem and moreover is uniformly closed and contains the function 1, it follows that this algebra is $C(X_A)$ itself. In this way, the identity of A and $C(X_A)$ is established.

We note that the compact Hausdorff space X_A is, moreover, uniquely determined (up to a homeomorphism) by the condition that $A \cong C(X_A)$. For if $A = C(Y)$ for some space Y then since each point of Y provides a homomorphism of A onto the complex numbers we have a mapping of Y into X_A. Since $C(Y)$ separates points in Y it follows that this mapping is 1 - 1 and it will be continuous by the definition of the topology of X_A. The image of Y in X_A will be compact; hence if the two were not identical there would be a non-zero function on X_A vanishing on the image of Y. But then the 0 element of $A = C(Y)$ would be identified with a nonzero element of $A = C(X_A)$ which is impossible. Hence $Y \approx X_A$. A corollary to this is that the only homomorphisms of $C(X)$ onto the complex numbers are those given by the points of X — namely, $X \approx X_{C(X)}$.

We shall, as a rule, identify the algebras A and $C(X_A)$. So an element $g \in A$ may be interchangeably thought of as an abstract element of A or as a function on the space X_A. In the former case we shall write $x(g)$ for the pairing of an $x \in X_A$ and $g \in A$ while in the latter we shall write $g(x)$, the meaning being the same in both cases.

Now let A and B be two C*-algebras (henceforth it will be understood that our C*-algebras are commutative and possess an identity) and suppose $B \subset A$. Clearly every homomorphism (onto the complex numbers) of A induces a homomorphism of B so that if $A \cong C(X_A)$ and $B \cong C(X_B)$, we shall have a map (the term "mapping" will always be meant to imply continuity) $\beta : X_A \longrightarrow X_B$. We can show that β has to be onto: $\beta(X_A) = X_B$.

For, in any case, since β is continuous $\beta(X_A)$ will be a compact subset of X_B. If it is a proper subset then there exists a non-zero element of B vanishing on $\beta(X_A)$. This means that every homomorphism on A applied to this element gives 0, so that the element must be 0 which is a contradiction. Hence β is onto. An important consequence of this should be pointed out. Namely, since β is onto, every homomorphism of B is the restriction to B of some homomorphism of A, i.e., <u>every homomorphism of a C^*-subalgebra of</u> A <u>may be extended to</u> A.

Consider next the converse situation where we are given two compact Hausdorff spaces X and Y and a map β of X onto Y. To every $g \in C(Y)$ we may associate the composite function $g \circ \beta$ on X. If $g \circ \beta$ vanishes identically then since β is onto it follows that g is identically zero. Hence the correspondence $g \longrightarrow g \circ \beta$ imbeds $C(Y)$ in a 1 – 1 manner into $C(X)$ and we may write $C(Y) \subset C(X)$, identifying g with $g \circ \beta$. We thus find that the C^*-subalgebras of a C^*-algebra A are all obtained by taking a mapping β of the homomorphism space X_A onto a compact Hausdorff space Y and identifying those functions on X_A of the form $g \circ \beta$ with $g \in C(Y)$.

Whenever we have a mapping β of a compact Hausdorff space X onto a space Y we may refer to Y as an <u>identification space</u> of X; for with an appropriate topology, Y is the space obtained by identifying the points of X that have the same image under β. The mapping β then becomes the <u>canonical map</u> taking a point into the class of equivalent points to which it belongs. Moreover the algebra $C(Y)$ when identified with a subalgebra of $C(X)$ consists of just those functions on X that take on equal values at points with the same image under β. Namely, if $f = g \circ \beta$ with $g \in C(Y)$, then obviously $\beta(x_1) = \beta(x_2)$ implies $f(x_1) = f(x_2)$. The converse is also true. For a function taking equal values at equivalent points has the form $f = g \circ \beta$ for some function g on Y and it need only be shown that g is continuous. But if F is a closed subset of the complex plane then $g^{-1}(F) = \beta(f^{-1}(F))$ and since $f^{-1}(F)$ is closed and hence compact so is $\beta(f^{-1}(F))$. Hence $g^{-1}(F)$ is closed so that g is continuous.

Returning to the two C^*-algebras A and B with $B \subset A$, let us call two homomorphisms of A equivalent if they agree on B. This means that two points of X_A are considered equivalent if they have the same image in X_B under the canonical map $\beta : X_A \longrightarrow X_B$. It follows from the foregoing discussion that the subalgebra B corresponds to just those functions on X_A taking equal values at equivalent points. As a result of this the elements of B may be thought of in several ways. To begin with they are elements of B, they are also the continuous functions on X_B, and finally, they are the continuous functions on X_A taking equal values at equivalent points. We shall not distinguish between these three except where necessary so that $\hat{g} \in B$, $g \in C(X_B)$ and $g \circ \beta \in C(X_A)$ are all to be identified. In particular $g(x)$ will make sense for x either in X_A or in X_B.

Let T be a homomorphism of a C^*-algebra B into a C^*-algebra A. Every homomorphism h of A onto the complex numbers determines a homomorphism $h \circ T$ of A and so there is induced a mapping of X_A onto X_B. We shall find it convenient to denote this mapping again by T so that one may write either $x(Tf)$ or $Tx(f)$ (or also $Tf(x)$ or $f(Tx)$). It will be noted that although T may be given only as an algebraic homomorphism it will necessarily preserve involutions as well: $(Tf)^* = T(f^*)$. This is seen by considering the induced map T of X_A onto X_B. For the isomorphism of A with $C(X_A)$ preserves the involution so that for $x \in X_A$, $f^*(x) = \overline{f(x)}$. Hence for $x \in X_A$, $(Tf)^*(x) = \overline{Tf(x)} = \overline{f(Tx)} = f^*(Tx) = Tf^*(x)$.

If $T : B \longrightarrow A$ is $1 - 1$ then we are in the situation previously considered, $B \subset A$. Here the induced map $T = \beta$ takes X_A onto X_B. An incidental consequence of this is that the map $T : B \longrightarrow A$ is norm preserving. For if $f \in B$,

$$\|f\| = \sup_{x \in X_B} |f(x)|$$

and since T is onto, the latter is

$$\sup_{x \in X_A} |f(Tx)| = \sup_{x \in X_A} |Tf(x)| = \|Tf\|$$

and $\|f\| = \|Tf\|$.

If $T : B \longrightarrow A$ is onto then T on X_A to X_B is $1 - 1$. Namely if $Tx_1 = Tx_2$, then $f(Tx_1) = f(Tx_2)$ for all f in B, or $Tf(x_1) = Tf(x_2)$ for all such f. But since T takes B onto A, this implies that as homomorphisms of A, $x_1 = x_2$ so that T is $1 - 1$.

Finally we note that a mapping T of X_A into X_B induces an algebraic homomorphism, again denoted by T, of B into A. Again the induced map will be $1 - 1$ or onto if the original map was onto or $1 - 1$ respectively. These statements follow just as readily as the previous ones.

1.2. <u>E-algebras</u>. If an algebra A is given as the algebra of continuous functions on a space X, then the borel measures on X determine linear functionals on the algebra A:

$$E_\mu(f) = \int_X f(x)d\mu(x) \quad .$$

In the following definition we describe in terms of an abstract C^*-algebra the situation that arises by choosing μ to be a probability measure. (Compare Segal's "probability algebra", [12].)

DEFINITION 1.1. An E-algebra is a C^*-algebra with a distinguished functional E satisfying

(i) $E(ff^*) \geq 0$ with equality possible only if $f = 0$,

(ii) $E(1) = 1$.

When viewed as functions on X_A, the elements ff^* are precisely the positive continuous functions and (i) requires that the functional E be positive, i.e., take on positive values at positive functions. It is clear that if we have a probability measure μ on X_A, it will induce a functional E_μ satisfying (i) and (ii) if μ takes on positive values on non-empty open sets. Conversely by the Riesz-Markoff representation theorem any functional on $C(X_A)$ satisfying (i) and (ii) is induced by such a probability measure. In general we shall denote by P the probability measure on X_A corresponding to the functional E of an E-algebra A.

Suppose that T is an algebraic homomorphism of B with A which furthermore preserves the functional E: $E(Tf) = E(f)$, $E(\cdot)$ denoting the distinguished functional on both algebras. In particular, $E(T(ff^*)) = E(ff^*)$ so that if $f \neq 0$, $E(T(ff^*)) > 0$. But $T(ff^*) = (Tf)(Tf^*)$ so that it follows from $f \neq 0$ that $E((Tf)(Tf^*)) \neq 0$ and $Tf \neq 0$. Thus we conclude that a <u>homomorphism of E-algebras preserving</u> E <u>is</u> $1 - 1$ <u>and hence</u> (§1.1) also <u>norm-preserving</u>. As a result, an algebraic homomorphism of an E-algebra preserving the functional E, is an isomorphism preserving the entire E-algebra structure. As regards the homomorphism spaces, we note that T is a measure preserving map of X_A onto X_B. Namely from

$$\int_{X_A} f(Tx)dP_A(x) = \int_{X_B} f(x)dP_B(x)$$

we deduce that $P_A(T^{-1}(\Delta)) = P_B(\Delta)$ for any borel set Δ in X_B; this is the sense in which T is to be thought of as measure preserving. (In general, if T maps X into Y and μ is a measure on X there is induced a measure ν on Y by

$$(1.1) \qquad\qquad \nu(\Delta) = \mu(T^{-1}(\Delta)) \quad,$$

for borel sets Δ in Y. T is measure preserving if the induced measure agrees with the existing measure. Note that (1.1) does not imply

$$(1.2) \qquad\qquad \mu(\Delta) = \nu(T(\Delta))$$

for borel subsets Δ of X.)

By means of the measure P on the homomorphism space of an E-algebra A we may define the Lebesgue spaces of A:

DEFINITION 1.2. If A is an E-algebra, $L^p(A)$ $(1 \leq p < \infty)$ will denote the Banach spaces of measurable functions on X_A satisfying

$$\int_{X_A} |f^p(x)|\,dP(x) < \infty$$

(the elements of $L^p(A)$ are equivalence classes of functions modulo the functions vanishing almost everywhere with respect to P). L(A) will denote the E-algebra of all bounded measurable functions on X_A.

One easily establishes that if $B \subset A$ then $L^p(B) \subset L^p(A)$ and $L(B) \subset L(A)$. Moreover A itself is contained in all the $L^p(A)$ and in L(A). Finally if $A \subset B \subset L(A)$ then $L^p(B) = L^p(A)$ for all p and L(B) = L(A).

If for two E-algebras we have $B \subset A$, then there is an important transformation from A to L(B) given by

DEFINITION 1.3. For every $z \in A$, the <u>conditional expectation of</u> z <u>with respect to</u> B, written E(z|B), is the uniquely determined element $\bar{z}$ of L(B) satisfying

$$(1.3) \qquad\qquad\qquad E(zw) = E(\bar{z}w)$$

for all $w \in B$ ([3]).

The uniqueness of the solution to (1.3) in L(B) is easily established since if there were two solutions the difference would be orthogonal to all of B, hence to L(B) and therefore must be 0. The existence follows from the fact that by the Hilbert space structure of $L^2(A)$ which contains $L^2(B)$ as a subspace, we know that there is a solution $\bar{z}$ to (1.3) belonging to $L^2(B)$. It remains to show that $\bar{z}$ is in $L(B) \subset L^2(B)$. To do so we point out that one easily shows that $z \geq 0$ implies $\bar{z} > 0$ and that $\bar{1} = 1$. It follows then from $- \|z\| \leq z \leq \|z\|$ that $- \|z\| \leq \bar{z} \leq \|z\|$ so that $\bar{z} \in L(B)$.

By the uniqueness of the solution to (1.3) it will be seen that

$$
\begin{aligned}
&\text{(a)} \quad B_1 \subset B_2 \subset L(B_1) \ \text{ implies } \ E(z|B_1) = E(z|B_2) \ , \\
(1.4) \quad &\text{(b)} \quad B_1 \subset B_2 \ \text{ implies } \ E(E(z|B_2)|B_1) = E(z|B_1), \ \text{ and} \\
&\text{(c)} \quad u \in B \ \text{ implies } \ E(uz|B) = uE(z|B) \ .
\end{aligned}
$$

§2. Preliminaries: Stationary Stochastic Processes

2.1. <u>One-sided and Two-sided Processes</u>. Conventionally, a
stochastic process with discrete parameter is defined as a sequence of
measurable functions on a probability space and it is stationary if there
is a measure preserving transformation of the probability space which
transforms each function into the next. More precisely, a stationary
stochastic process consists of a measure space Ω, a probability measure
P on Ω, a measure preserving transformation T of Ω and a sequence
$\{x_n\}$ of measurable functions on Ω satisfying $x_n(T\omega) = x_{n-1}(\omega)$. If
the x_n are essentially bounded (i.e., $|x_n(\omega)| \subset M$, but for a set of
ω of measure 0) they will generate an E-subalgebra of the algebra of
all bounded measurable functions on Ω. As elements in this E-algebra
they may be identified with the continuous functions on a compact
Hausdorff space. This enables us to choose as a particular version of the
sample space, a compact Hausdorff space — the random variables of the
process will then be continuous functions on this space. For our purposes
it will be convenient to assume from the start that the sample space is
a compact Hausdorff space and moreover to allow the random variables to
take their values in an abstract topological space.

DEFINITION 2.1. A <u>one-sided process</u> X consists of a separable
compact Hausdorff space Ω_X, a sequence $\{x_n\}_{n \leq 0}$ of continuous functions
from Ω_X to a compact metric space Λ, a probability measure P on Ω_X
and a map T of Ω_X into Ω_X satisfying

  (i) $x_n(T\omega) = x_{n-1}(\omega)$
  (ii) the x_n separate points in Ω_X
  (iii) for every borel set $\Delta \subset \Omega_X$, $P(T^{-1}(\Delta)) = P(\Delta)$
  (iv) $P(\Delta) > 0$ for every non-empty open set $\Delta \subset \Omega_X$.

X is a <u>two-sided process</u> if the index n ranges from $-\infty$ to ∞ and T
is a homeomorphism of the corresponding space Ω_X. Ω_X will be referred
to as the <u>sample space</u> of the process X.

In this definition we have suppressed the adjective "stationary"
because we shall deal almost exclusively with stationary processes so that
a non-stationary process will be explicitly referred to as such.

We note that conditions (iii) and (iv) together imply that the
transformation T is onto. For $T(\Omega_X)$ is in any case compact so that
$\Omega_X - T(\Omega_X)$ is an open set and if it is non-empty $P(\Omega_X - T(\Omega_X)) > 0$ and
$P(T(\Omega_X)) < 1$. However T is measure preserving by (iv) so that
$P(T(\Omega_X)) = P(T^{-1}T(\Omega_X)) \geq P(\Omega_X) = 1$ which gives a contradiction so T is
necessarily onto. This shows that not only is the function x_{n-1} determined by x_n according to (1) but also conversely so that all the x_n

are determined by any one of them.

Condition (ii) is not an essential one in the sense that it can always be satisfied by modifying the space Ω_X. Namely, if (ii) is not satisfied we pass to the identification space of Ω_X obtained by identifying pairs of points ω_1 and ω_2 if $x_n(\omega_1) = x_n(\omega_2)$ for all $n \leq 0$. If $\omega_1 \sim \omega_2$ then we have, in particular, $x_{n-1}(\omega_1) = x_{n-1}(\omega_2)$ for $n \leq 0$, or by (1), $x_n(T\omega_1) = x_n(T\omega_2)$ so that $T\omega_1 \sim T\omega_2$. Therefore T will be defined on the identification space and it follows that with the new space, the x_n determine a process.

In particular suppose that X is a two-sided process with variables x_n defined for all integers n. We may form from this a one-sided process by considering only the x_n with $n \leq 0$, reducing the space Ω_X so that (ii) will be satisfied for these variables. The resulting process we shall denote by X^- and the corresponding space (properly Ω_{X^-}) we denote by Ω_X^-. Ω_X^- is then an identification space of Ω_X and we denote the canonical map of Ω_X onto Ω_X^- by β. We shall think of X^- and X as the one-sided and two-sided versions of a single process. This will be justified when we show later that X may also be determined uniquely from X^-.

In certain connections the variables x_n play a secondary role and only the space Ω_X, the measure P, and the transformation T are of significance. It will be convenient to consider such a set-up in its own right and we give it a formal definition in terms of the algebra $C(\Omega_X)$:

DEFINITION 2.2. An <u>abstract (one-sided) process</u> X is a separable E-algebra A_X with an endomorphism T of A_X preserving the functional E. X is two-sided if T is an automorphism of A_X.

We recall from §1.2 that an algebraic homomorphism of an E-algebra preserving the functional E must also preserve norms. Hence, the operator T provides an E-algebra isomorphism of A_X with $T(A_X)$.

If we start with an ordinary process X (one- or two-sided) we obtain an abstract process by taking $A_X = C(\Omega_X)$ with E the functional induced by P. E satisfies (i) and (ii) of Definition 1.1 because P is a probability measure satisfying (iv) of Definition 2.1. If X is two-sided then T, being a homeomorphism of Ω_X, induces an automorphism of A_X so that A_X will be two-sided. If X^- is the one-sided version of X then A_{X^-}, which we shall write as A_X^-, is the algebra of continuous functions on an identification space of Ω_X, hence may be identified with a subalgebra of A_X. These algebras will be referred to as the E-algebras of the process X.

The E-algebras of a process X may be obtained directly as

follows. Let ψ range over $C(\Lambda)$ where Λ is the range of the x_n. Since the x_n, $-\infty < n < \infty$, separate points in Ω_X it follows that so do the set of all $\psi(x_n)$. Hence the set of these functions generates the algebra $A_X = C(\Omega_X)$. Similarly A_X^- is generated by the subset of the $\psi(x_n)$ for which $n \leq 0$.

When dealing with an ordinary two-sided process we are able to associate with it a one-sided process in a natural way. This is not the case for an abstract two-sided process. To see this let X be an ordinary process with variables x_n and let X^1 be the process obtained by setting $x_n^1 = x_{n-1}$. Then clearly $A_{X^1} = A_X$; however $A_{X^1}^- \subsetneq A_X^-$ in general. On the other hand to a one-sided abstract process it is possible to associate a canonical two-sided abstract process. Namely, we have

> THEOREM 2.1. If Y is a one-sided abstract process
> with algebra A_Y, there exists a unique two-sided
> process X with $A_X \supset A_Y$ such that the set of
> $x \in A_X$ with $T^m x \in A_Y$ for some m is dense in
> A_X. (Here, as always, "unique" means unique up to
> an isomorphism preserving the relevant structure.)

PROOF. Let $A_Y^{(n)}$, $n \geq 0$, be denumerably many copies of A_Y and suppose each $A_Y^{(n)}$ imbedded in $A_Y^{(n+1)}$ by identifying the element z of $A_Y^{(n)}$ with the element Tz in $A_Y^{(n+1)}$. This is an imbedding since (§1.2) the map T preserves E and is therefore $1-1$; moreover the image of $A_Y^{(n)}$ in $A_Y^{(n+1)}$ is isomorphic (as an E-algebra) to $A_Y^{(n)}$ for the same reason. We now have $A_Y^{(0)} \subset A_Y^{(1)} \subset \ldots \subset A_Y^{(n)} \subset \ldots$ and we may form the union of these algebras A_Y^∞. In A_Y^∞, the functional E is defined as is the transformation T, and T still preserves E. Moreover $T(A_Y^{(n+1)}) = A_Y^{(n)}$. Set A_X equal to the completion of A_Y^∞ in the norm determined upon it by the norms of the algebras $A_Y^{(n)}$. Since T is norm preserving (§1.1) it extends to A_X and so does E; since T takes A_Y^∞ onto A_Y^∞ it also follows that T takes A_X onto A_X. Since T is $1-1$ and onto it is an automorphism of A_X so that A_X defines a two-sided process X. Identifying $A_Y^{(0)}$ with A_Y we have $A_Y \subset A_X$. Moreover every x in A_Y^∞ belongs to some $A_Y^{(m)}$ and so $T^m x \in A_Y^{(0)} = A_Y$ for some m so that, since A_Y^∞ is dense in A_X, A_X will have the properties required of it. Now the latter property shows that it is unique for if A_Z satisfies the conditions of the theorem, then since T is invertible in A_Z and $A_Y \subset A_Z$ it follows that $A_Y^\infty \subset A_Z$ and since A_Y^∞ must be dense in A_Z it follows that $X = Z$.

We now turn to the case of an ordinary one-sided process Y with variables y_n, $n \leq 0$, and we wish to find a two-sided (ordinary) process X with $Y = X^-$. To do so let Y' be the abstract process corresponding

to Y so that $A_{Y'}$ is the algebra of functions of Ω_Y. By Theorem 2.1
there exists a unique process X' with $A_{Y'} \subset A_{X'}$ and X' two-sided,
such that furthermore for a dense subset of $z \in A_{X'}$, $T^m z \in A_{Y'}$, for
some m. We define an ordinary two-sided process X by taking Ω_X to
be the homomorphism space of $A_{X'}$, P the corresponding probability
measure, and T the homeomorphism induced on Ω_X by the automorphism
T of $A_{X'}$. There remains only to define the variables x_n. Since
$A_{Y'} \subset A_{X'}$ there is induced a canonical map β from Ω_X to the homomorphism space of $A_{Y'}$ which is Ω_Y. We then define the x_n by setting
$x_n = y_n \circ \beta$ for $n \le 0$ and $x_n(\omega) = x_0(T^{-n}\omega)$ for $n > 0$, and $\omega \in \Omega_X$.
To show that these functions define a process it is only necessary to
show that the x_n separate points in Ω_X, the other requirements being
clearly satisfied. Suppose then that there exist ω_1 and ω_2 in Ω_X
with $x_n(\omega_1) = x_n(\omega_2)$ for all n. For each m we will then have
$x_n(T^{-m}\omega_1) = x_n(T^{-m}\omega_2)$ for all n, certainly for $n \le 0$, and therefore
$y_n(\beta T^{-m}\omega_1) = y_n(\beta T^{-m}\omega_2)$. Since the y_n separate points in Ω_Y,
$\beta T^{-m}\omega_1 = \beta T^{-m}\omega_2$. Now for a dense set of z in $A_{X'}$ $T^m z \in A_{Y'}$ for some
m, and so $z(\omega_1) = z(T^m T^{-m}\omega_1) = T^m z(T^{-m}\omega_1) = T^m z)\beta T^{-m}\omega_1) = T^m z(\beta T^{-m}\omega_2) = z(\omega_2)$. Hence for all z in $A_{X'}$ $z(\omega_1) = z(\omega_2)$ and therefore $\omega_1 = \omega_2$.
Therefore X as defined here is a two-sided process. Finally since
$x_n = y_n \circ \beta$ for $n \le 0$ it follows that $A_X^- = A_Y$.

Now suppose that W and X are two two-sided processes satisfying $W^- = X^- = Y$. We shall show that W and X are essentially identical.
Namely consider the subalgebra of A_W consisting of functions on Ω_W of
the form $\psi(w_{i_1}, w_{i_2}, \ldots, w_{i_k})$ and the corresponding algebra for X;
denote these two by A_W^O and A_X^O. It is easy to show that the mapping
$\psi(w_{i_1}, w_{i_2}, \ldots, w_{i_k}) \longrightarrow \psi(x_{i_1}, x_{i_2}, \ldots, x_{i_k})$ gives a homomorphism of
A_W^O onto A_X^O preserving the functional E and preserving norms. First
one shows that E is preserved, i.e., that $E(\psi(w_{i_1}, w_{i_2}, \ldots, w_{i_k})) = E(\psi(x_{i_1}, x_{i_2}, \ldots, x_{i_k}))$; this obviously holds if all the $i_\nu \le 0$ and
since $E(\psi(w_{i_1}, w_{i_2}, \ldots, w_{i_k})) = E(\psi(w_{i_1}-r, w_{i_2}-r, \ldots, w_{i_k}-r))$ it follows
for all sets of i_ν. Similarly the norm must be preserved. From either of
these we deduce that if $\psi(w_{i_1}, \ldots, w_{i_k}) = 0$ so is $\psi(x_{i_1}, \ldots, x_{i_k})$ so
that our mapping is well defined and similarly the mapping can be shown to
be 1 - 1. It follows that $A_W^O \cong A_X^O$ and since these are dense in A_W
and A_X we have $A_W \cong A_X$, the isomorphism relating to the algebraic
structure, the functional E and the transformations T; moreover the
variables w_n are carried into the x_n. This shows that the spaces Ω_X

and Ω_W are homeomorphic in a manner preserving the structure of the processes and sending the w_n into the x_n. This completes the proof of

THEOREM 22. There is a one-to-one correspondence between one-sided and two-sided processes defined by $X \longrightarrow X^-$.

2.2. <u>The Sample Paths of a Process</u>. In our definition of a process the points of the space Ω_X were elements of an abstract compact Hausdorff space and devoid of any further significance. We shall show that these points may be viewed as sample paths of the process X, i.e., as functions from the integers to Λ. To make this precise we state

DEFINITION 2.3. If Λ is a compact metric space, Λ_∞ will denote the set of all functions on the integers with values in Λ. Λ_∞^- will denote the set of functions from the integers $n \leq 0$ to Λ and Λ_∞^+, the set of functions from the integers $n > 0$ to Λ. Each of these spaces is given the weak topology induced by the coordinate functions, i.e., the topology of pointwise convergence. The elements of Λ_∞, Λ_∞^-, and Λ_∞^+ will be referred to as <u>doubly-infinite</u>, <u>left-infinite</u>, and <u>right-infinite</u> Λ-<u>sequences</u> respectively.

There are natural maps from Λ^∞ to both Λ_∞^- and Λ_∞^+; we shall be concerned primarily with the first of these, denoted β, defined by $\beta(\omega)(n) = \omega(n)$ for $\omega \in \Lambda_\infty$, $\beta(\omega) \in \Lambda_\infty^-$ and $n \leq 0$. We shall also make use of the shift operator in Λ_∞ and in Λ_∞^- defined by

$$(2.1) \qquad\qquad T\omega(n) = \omega(n-1) \ .$$

(The shift in the opposite direction could be defined in Λ_∞^+, but we shall not need it.) T then represents an operator either in Λ_∞ or in Λ_∞^- and the two operators are related by

$$(2.2) \qquad\qquad T\beta = \beta T \ .$$

Now let X be a two-sided process with variables x_n taking their values in Λ and let X^- be the one-sided version of X. To every $\omega \in \Omega_X$ we may associate the doubly-infinite Λ-sequence $\gamma(\omega)$ where

$$(2.3) \qquad \gamma(\omega)(n) = x_n(\omega) \qquad\qquad -\infty < n < \infty \ ,$$

and for every $\omega' \in \Omega_X^-$ we obtain a left-infinite Λ-sequence $\gamma'(\omega)$ defined in the analogous way. We thus have $\gamma : \Omega_X \longrightarrow \Lambda_\infty$ and $\gamma' : \Omega_X^- \longrightarrow \Lambda_\infty^-$.

THEOREM 2.3. γ and γ' are $1-1$ (continuous) maps and

$$(2.4) \quad
\begin{aligned}
&\text{(a) } T\gamma(\omega) = \gamma T(\omega) \text{ for } \omega \in \Omega_X \\
&\text{(b) } T\gamma'(\omega) = \gamma'T(\omega) \text{ for } \omega \in \Omega_X^- \\
&\text{(c) } \gamma'\beta(\omega) = \beta\gamma(\omega) \text{ for } \omega \in \Omega_X \ .
\end{aligned}$$

The ranges of γ and γ' are closed T-invariant
subsets of Λ_∞ and Λ_∞^- respectively.

PROOF. In order that γ and γ' be continuous functions it
suffices that the coordinates of $\gamma(\omega)$ and $\gamma'(\omega')$ depend continuously
on ω and ω' and this follows from the continuity of the x_n. Since
the x_n in each case separate the points of the sample space Ω_X or
Ω_X^-, it follows that γ and γ' are 1 - 1. (a), (b) and (c) of (2.4)
follow by direct substitution using the definitions of β, γ, γ', and T.
(a) and (b) imply moreover that the ranges of γ and γ' are invariant
under T and these ranges are necessarily closed as images of compact
sets.

The import of the first part of this theorem is that Ω_X and
Ω_X^- may be identified with subsets of Λ_∞ and Λ_∞^-; the equations in
(2.4) ensure that under this identification the operators T and β de-
fined for Ω_X and Ω_X^- go over into the T and β defined for Λ_∞ and
Λ_∞^-. As a consequence of this we may suppose that from the start Ω_X and
Ω_X^- were taken as subsets of Λ_X and Λ_X^- respectively so that the points
of Ω_X and Ω_X^- are Λ-sequences.

Ω_X and Ω_X^- are now sequence spaces and the relationship between them is given in the next theorem.

THEOREM 2.4. A left-infinite Λ-sequence is in Ω_X^- if
and only if it is the image under β of a sequence in
Ω_X. A doubly-infinite Λ-sequence ω is in Ω_X if and
only if each of the left-infinite Λ-sequences $\beta(T^{-n}\omega)$
is in Ω_X^- where $n \geq 0$.

PROOF. The first assertion follows from the fact that Ω_X^- is
an identification space of Ω_X; hence β takes Ω_X onto Ω_X^-. The
"only if" portion of the second assertion follows from the same fact since
if $\omega \in \Omega_X$ so do each of the $T^{-n}\omega$. To prove the converse, assume that
$\omega \in \Lambda_\infty$ and that $\beta(T^{-n}\omega) \in \Omega_X^-$ for all $n \geq 0$; we shall show that $\omega \in \Omega_X$.
Since β takes Ω_X onto Ω_X^- there is for each n a point $\omega_n \in \Omega_X$ with
$\beta(\omega_n) = \beta(T^{-n}\omega)$. Consider the points $\{T^k\omega_k\}_{k>0}$ in Ω_X and let
$\omega' \in \Omega_X$ be a limit point of some subsequence of this sequence (Ω_X is
separable and compact — hence, sequentially compact), say $T^{k_m}\omega_{k_m} \longrightarrow \omega'$.
We have then for each n, $\omega_{k_m}(n - k_m) \longrightarrow \omega'(n)$. Since $k_m \longrightarrow \infty$,
$n - k_m$ will eventually be negative so that ω_{k_m} may be replaced by
$\beta(\omega_{k_m}) = \beta(T^{-k_m}\omega)$ and we find $\omega(n) = T^{-k_m}\omega(n - k_m) \longrightarrow \omega'(n)$. Hence
$\omega'(n) = \omega(n)$, $\omega' = \omega$, so that $\omega \in \Omega_X$.

2.3. _Subprocesses_. If X is a Λ-valued process with variables x_n and φ is a continuous function from Λ_∞^- to some other compact space Λ', we obtain a Λ'-valued process Y by setting

$$(2.5) \qquad y_n = \varphi(\ldots, x_{n-2}, x_{n-1}, x_n) \ .$$

The index n may range over all the integers or only over the negative ones depending upon whether X is two-sided or one-sided. We suppose that the function φ is independent of n so that we may write

$$y_n(T\omega) = \varphi(\ldots, x_{n-2}(T\omega), x_{n-1}(T\omega), x_n(T\omega)) =$$

$$\varphi(\ldots, x_{n-3}(\omega), x_{n-2}(\omega), x_{n-1}(\omega)) = y_{n-1}(\omega) \ .$$

The sample space of Y can then be taken to be the identification space of Ω_X induced by the functions y_n on Ω_X, the probability measure P and shift transformation T for Y being obtained in the natural way from those for X.

DEFINITION 2.4. The process Y related to X by (2.5) is said to be _the subprocess of_ X _determined by_ φ and X is said to be an _extension_ of Y.

The algebra A_Y of the subprocess Y is the algebra of continuous functions on an identification space of Ω_X, hence is a subalgebra of A_X. Alternatively, we recall that A_X is generated by the functions $\psi(x_n)$, $\psi \in C(\Lambda)$ and A_Y is generated by the functions $\psi'(y_n)$, $\psi' \in C(\Lambda')$, so that (2.5) shows again that $A_Y \subset A_X$. We are therefore led to make the following definition for abstract processes:

DEFINITION 2.5. An abstract process Y is a _subprocess_ of the abstract process X if there is a $1-1$ homomorphism Γ of A_Y into A_X such that $E(\Gamma z) = E(z)$, $\Gamma T(z) = T\Gamma(z)$ for $z \in A_Y$. X is then said to be an _extension_ of Y.

Note that if Y^- is a subprocess of X^-, where Y^- and X^- are abstract one-sided processes, then it follows from Theorem 2.1 that the two-sided version Y is a subprocess of the two-sided process X. The converse is not true since even if $X = Y$, X^- and Y^- may be distinct.

For ordinary processes, the map φ defining a subprocess of X induces maps Φ^- and Φ from Λ_∞^- to $\Lambda_\infty'^-$ and from Λ_∞ to Λ_∞' respectively. These are defined by

$$(2.6) \quad \begin{array}{ll} \Phi^-(\omega)(n) = \varphi(\ldots, \omega(n-2), \omega(n-1), \varphi(n)) \ , & n \leq 0 \\[2mm] \Phi(\omega)(n) = \varphi(\ldots, \omega(n-2), \omega(n-1), \omega(n)), & -\infty < n < \infty \ . \end{array}$$

It follows directly from (2.5) that the sample paths of Y are determined as follows:

> THEOREM 2.5. The subsets $\Omega_Y^- \subset \Lambda_\infty'^-$ and $\Omega_Y \subset \Lambda_\infty'$ are given by $\Omega_Y^- = \Phi^-(\Omega_X^-)$, $\Omega_Y = \Phi(\Omega_X)$.

It will often occur that the variables of a process Y may not be represented as continuous functions on the sample space of X as is the case when Y is a subprocess of X, but may still be represented as measurable functions on Ω_X. It is therefore convenient to have the following definition:

DEFINITION 2.6. An abstract process Y is an <u>L-extension</u> of an abstract process X if there is an isomorphism Γ of A_Y into $L(A_X)$ satisfying $E(\Gamma z) = E(z)$, $\Gamma T(z) = T\Gamma(z)$ and such that $A_X \subset \Gamma(A_Y) \subset L(A_X)$.

It should be pointed out that both E and T are defined on $L(A_X)$, the former because $L(A_X)$ is an E-algebra (Definition 1.2), and the latter because T is induced by a transformation of the compact Hausdorff space of the algebra A_X. In fact, $L(A_X)$ itself would define an abstract process except for the fact that it is in general not separable.

2.4. <u>Ergodicity</u>. In what follows we shall have occasion to use the ordinary notion of ergodicity (or metric transitivity) for a stationary process, as well as several variants of this notion.

DEFINITION 2.7. A (one- or two-sided) process X is <u>topologically ergodic</u> if any closed set $\Delta \subset \Omega_X$ satisfying $T\Delta \subset \Delta$ must have measure 0 or 1.

Note that since P assigns positive measures to non-empty open sets it follows that a closed set with measure 1 is identical to Ω_X.

For completeness we include

DEFINITION 2.8. X is ergodic if any measurable set $\Delta \subset \Omega_X$ satisfying $T\Delta \subset \Delta$ must have measure 0 or 1.

Clearly an ergodic process is topologically ergodic.

DEFINITION 2.9. X is λ-ergodic, $0 < \lambda < 1$, if the only solution in $L(A_X)$ to the equation $Tz = e^{2\pi i \lambda} z$ is $z = 0$.

There are a number of analogies that may be drawn between the more familiar notion of ergodicity and the more elementary one of topological ergodicity. In particular the connection between averages of functional values and ergodicity is very similar to one between maxima of functional values and topological ergodicity. For example, the following is the analogue of Birkhoff's ergodic theorem:

THEOREM 2.6. X is topologically ergodic if and only
if for any continuous function f on Ω_X we have

$$\lim_{N \to \infty} \max(|f(\omega)|, |f(T\omega)|, \ldots, |f(T^N\omega)|) = \|f\|$$

for almost all ω in Ω_X, where

$$\|f\| = \sup_{\omega \in \Omega_X} |f(\omega)| \quad .$$

This is equivalent to the statement that the trans-
lates of ω are dense in Ω_X for almost all ω if
and only if X is topologically ergodic.

The proof is straightforward and we omit it.
Regarding these ergodicity properties we may state

THEOREM 2.7. A one-sided process X^- is ergodic in one
of the preceding senses if and only if the corresponding
two-sided process X is ergodic in the same sense.

PROOF. Suppose first that X^- is topologically ergodic and that
$\Delta \subset \Omega_X$ is closed and satisfies $T\Delta \subset \Delta$. $\beta(\Delta) \subset \Omega_X^-$ is closed and satisfies $T(\beta\Delta)) = \beta(T(\Delta)) \subset \beta(\Delta)$ so that either $\beta(\Delta) = \Omega_X^-$ or $\beta(\Delta)$
has measure 0. In the latter case $\Delta \subset \beta^{-1}(\beta(\Delta))$ must have measure 0
since β is measure preserving, so we need only consider the case that
$\beta(\Delta) = \Omega_X^-$. Since β need not be 1 - 1 the latter condition by itself
does not imply that $\Delta = \Omega_X$; however we shall see that this together
with the condition $T\Delta \subset \Delta$ does imply $\Delta = \Omega_X$. Namely, let $\omega \in \Omega_X$ and
consider $\{T^{-k}\omega\}_{k \geq 0}$. For each k we have $\beta(T^{-k}\omega) \in \beta(\Delta)$, so there
is an ω_k in Δ with $\beta(\omega_k) = \beta(T^{-k}\omega)$ which means that, as points in
Λ_∞^-, $\omega_k(n) = T^{-k}\omega(n) = \omega(n + k)$ for $n \leq 0$, or $\omega_k(n - k) = \omega(n)$ for
$n \leq k$, or finally, $T^k\omega_k(n) = \omega(n)$ for $n \leq k$. It follows directly from
the definition of the topology in Λ_∞ that $T^k\omega_k \longrightarrow \omega$. Now $\omega_k \in \Delta$ so
that $T^k\omega_k \in \Delta$; since Δ is moreover closed, we conclude that $\omega \in \Delta$.
Since ω was arbitrary it follows that $\Delta = \Omega_X$. Hence X is topologically
ergodic. If, conversely, X is topologically ergodic, then, if $\Delta \subset \Omega_X^-$
is closed and T-invariant, it follows easily that $\beta^{-1}(\Delta) \subset \Omega_X$ is closed
and T-invariant and $P(\beta^{-1}(\Delta)) = P(\Delta)$ is either 0 or 1.

Consider next the question of ergodicity for X^- and X. The
ergodic theorem implies that this is equivalent to the assertion that

$$(2.7) \qquad \lim_{N \to \infty} \frac{1}{N+1} \sum_{0}^{N} T^n u = E(u)$$

almost everywhere for all u in A_X^- or in A_X. Now if (2.7) holds for
all u in A_X it obviously holds for all n in A_X^-. But the converse
is also true. For if (2.7) holds for all n in A_X^- it also holds for
all n with some translate $T^k u$ in A_X^-. Since by Theorem 2.1 these u
are dense it A_X it follows that (2.7) holds for all u in A_X.

A similar argument carries over to the case of λ-ergodicity.
In fact, $X(X^-)$ is λ-ergodic if and only if

$$(2.8) \qquad \lim_{N \to \infty} \frac{1}{N+1} \sum_{0}^{N} e^{-2\pi i n \lambda} T^n u = 0$$

for all u in $L^2(A_X) (L^2(A_X^-))$ in the weak topology of $L^2(A_X) (L^2(A_X^-))$.
Namely, if X is λ-ergodic, there is no non-trivial solution in $L(A_X)$
to $Tz = e^{2\pi i \lambda} z$. It follows that there is no non-trivial solution to the
latter equation in $L^2(A_X)$ for if z is a solution, so is $z/_{|z|}$ (where
$0/_{|0|} = 0$). But any weak limit of a subsequence of the sequence in (2.8)
would be such a solution, hence must vanish identically, hence (2.8) holds.
Conversely, if (2.8) holds, applying it to a solution of $Tz = e^{2\pi i \lambda} z$ we
find that $z = 0$. The same argument applies to λ-ergodicity in X^-. The
assertion of the theorem now follows for λ-ergodicity in exactly the
same way as for ergodicity since (2.8) must be true either for both A_X
and A_X^- or for neither.

§3.   Stochastic Sequences

3.1. <u>Definitions</u>. In this section we define the class of time
series that may be said to exhibit "definite statistical behavior". This
will constitute the first of two conditions that will restrict the class
of sequences to which our analysis is applicable.

DEFINITION 3.1. If $\xi(n)$ is a numerical left-infinite sequence,
the <u>average</u> of ξ, denoted $E(\xi)$, is defined by

$$E(\xi) = \lim_{N \to \infty} \frac{1}{N+1} \sum_{n=0}^{N} \xi(-n)$$

when this limit exists. If S is a subset of the negative integers, the
<u>upper</u> and <u>lower density</u> of S refer to

$$\bar{D}(S) = \lim_{N \to \infty} \sup \frac{1}{N+1} \sum_{n=0}^{N} \chi_S(-n), \quad \underline{D}(S) = \lim_{N \to \infty} \inf \frac{1}{N+1} \sum_{n=0}^{N} \chi_S(-n)$$

where $x_S(-n) = 1$ or 0 according as $-n \in S$ or $-n \notin S$. When $\bar{D}(S) = \underline{D}(S)$ we refer to the common value as $D(S)$, the _density_ of S. (Analogous definitions may be made for right-infinite and doubly-infinite sequences and sets but we shall not need them.)

DEFINITION 3.2. A left-infinite Λ-sequence (Definition 2.3) is _stochastic_ if for any k-tuple $(\psi_1, \ldots, \psi_k)$ of functions in $C(\Lambda)$, the sequence

$$(3.1) \qquad \zeta(n) = \psi_1(\xi(n))\psi_2(\xi(n-1)) \ldots \psi_k(\xi(n-k+1))$$

possesses an average.

The sequence $\zeta(n)$ of (3.1) is a _derived sequence_ of $\xi(n)$ in accordance with

DEFINITION 3.3. A Λ'-sequence $\zeta(n)$ is a _derived sequence_ of $\xi(n)$ if there exists a continuous function φ from Λ_∞^- to Λ' such that

$$(3.2) \qquad \zeta(n) = \varphi(\ldots, \xi(n-2), \xi(n-1), \xi(n)) \quad .$$

By Definition 3.2, a sequence $\xi(n)$ is stochastic if certain numerical valued derived sequences possess averages. Since the functions on Λ_∞^- occurring in (3.1) span a dense subset of $C(\Lambda_\infty^-)$ it follows that if $\zeta(n)$ is stochastic, any numerical derived sequence possesses an average.

A derived sequence of a derived sequence of $\xi(n)$ is evidently again a derived sequence of $\xi(n)$. Hence

THEOREM 3.1. A derived sequence of a stochastic sequence is again a stochastic sequence.

If $\xi(n)$ is a uniform limit (in the metric of Λ) of a sequence of stochastic sequences $\xi^k(n)$, any Λ'-sequence derived from $\xi(n)$ will be a uniform limit (in the metric of Λ') of the corresponding derived sequences of $\xi^k(n)$. It follows that

THEOREM 3.2. A uniform limit of stochastic sequences is a stochastic sequence.

3.2. _Distribution Functions of Stochastic Sequences_. To illustrate the sense in which a stochastic sequence exhibits definite statistical behavior we shall prove

THEOREM 3.3.   If $\xi(n)$ is a real-valued stochastic sequence, there exists a distribution function $F(\lambda)$ on the real line such that at each point of continuity λ_o of F, the set

$$S(\lambda_o) = \{n : \xi(n) < \lambda_o\}$$

possesses a density $D(S(\lambda_o)) = F(\lambda_o)$.

PROOF. The sequence $\xi(n)$ takes its values in an interval [a, b] of the real line and we shall determine the distribution function $F(\lambda)$ in terms of a measure on [a, b]. To obtain this measure define a linear functional L on C([a, b]) by setting L(f) equal to the average of the sequence $\zeta(n) = f(\xi(n))$. Since L is linear and when applied to non-negative functions it gives non-negative results and since L(1) = 1 it follows that

$$L(f) = \int_a^b f(\lambda)dF(\lambda)$$

for a distribution function $F(\lambda)$ satisfying $F(b + 0) - F(a) = 1$.

Now suppose that λ_o is a point of continuity of F and let $X_{-\varepsilon}$, X, X_ε denote respectively the characteristic functions of the half-open intervals $[a, \lambda_o - \varepsilon)$, $[a, \lambda_o)$, $[a, \lambda_o + \varepsilon)$. Choose f_1 and f_2 continuous on [a, b] such that

$$X_{-\varepsilon} \leq f_1 \leq X \leq f_2 \leq X_\varepsilon \quad .$$

For any prescribed $\delta > 0$, ε can be chosen so small that $F(\lambda_o + \varepsilon) - F(\lambda_o - \varepsilon) < \delta$. However

$$F(\lambda_o - \varepsilon) = \int_a^b X_{-\varepsilon}(\lambda)dF(\lambda) \leq \int_a^b f_1(\lambda)dF(\lambda)$$

$$= L(f_1) = \lim_{N \to \infty} \frac{1}{N+1} \sum_{n=0}^{N} f_1(\xi(-n)) \leq \liminf_{N \to \infty} \frac{1}{N+1} \sum_{n=0}^{N} X(\xi(-n))$$

$$\leq \limsup_{N \to \infty} \frac{1}{N+1} \sum_{n=0}^{N} X(\xi(-n)) \leq \lim_{N \to \infty} \frac{1}{N+1} \sum_{n=0}^{N} f_2(\xi(-n))$$

$$= L(f_2) = \int_a^b f_2(\lambda)dF(\lambda) \leq \int_a^b X_\varepsilon(\lambda)dF(\lambda) = F(\lambda_o + \varepsilon) \quad .$$

Hence

$$\limsup_{N \to \infty} \frac{1}{N+1} \sum_{n=0}^{N} x(\xi(-n)) - \liminf_{n \to \infty} \frac{1}{N+1} \sum_{n=0}^{N} x(\xi(-n)) < \delta$$

and since δ is arbitrary it follows that the average of $x(\xi(-n))$ exists. This average is clearly the density of $S(\lambda_o)$ and by the continuity of $F(\lambda)$ at λ_o, the limit in question agrees with $F(\lambda_o)$.

Proceeding along these lines we could show that there is not only one random variable associated with $\xi(n)$ (i.e., the random variable with distribution function $F(\lambda)$) but an entire stationary process may be associated with the sequence. We do not enter into the details now as we shall arrive at this result in the next section from a different point of view.

3.3. <u>Examples</u>. The simplest examples of stochastic sequences are periodic sequences or those that are eventually periodic. Less trivial is the case of almost periodic sequences (the analogue for the integers of almost periodic functions on the line) restricted to the negative integers. That these are stochastic follows from two facts: a sequence derived from an almost periodic sequence is almost periodic, and the Bohr Mean Value existence theorem for almost periodic sequences restricted to the negative integers, so that the averages of such sequences exist. We might remark that the case of almost periodic sequences already shows that in Theorem 3.3 we must restrict ourselves to points of continuity of $F(\lambda)$. For Bohr has shown ([1]) that almost periodic functions exist for which $S(\lambda)$ has no density and the same argument goes through for almost periodic sequences.

The widest class of stochastic sequences are obtained through the following theorem.

THEOREM 3.4. Almost all sample sequences of a one-sided stationary process are stochastic sequences.

PROOF. The existence almost everywhere of the averages of any particular derived sequence

$$(3.3) \qquad z_n(\omega) = \varphi(\ldots, x_{n-2}(\omega), x_{n-1}(\omega), x_n(\omega))$$

follows from Birkhoff's ergodic theorem. It follows that the averages in question exist simultaneously for a denumerable set of φ at almost every ω. Since Ω_X^- is separable it follows that a denumerable set of φ may be found such that the resulting z lie dense in $C(\Omega_X^-)$ and hence all the averages of sequences (3.3) exist simultaneously with probability 1.

If we take X to be the process in which the variables x_n are independent and take on the values 0, 1 each with probability 1/2, then our theorem is implied by the statement that almost all numbers are "normal" to the base 2. Here normality is taken in the strong sense that every k-tuple of 0's and 1's occurs in the binary expansion of the number with frequency 2^{-k}. For a particular block of 0's and 1's, $a_{i_1}, \ldots, a_{i_k}$, the frequency of occurrence of this block in a left-infinite $\{0, 1\}$-sequence $\xi(n)$ is the same as the average of the sequence

$$(3.4) \qquad \zeta(n) = \varphi(\ldots, \xi(n - 2), \xi(n - 1), \xi(n))$$

where φ is the function equal to 0 unless $(\xi(n - k + 1), \ldots, \xi(n - 1), \xi(n)) = (a_{i_1}, \ldots, a_{i_{k-1}}, a_{i_k})$ in which case $\varphi = 1$. Thus normality implies the existence of averages for divided sequences of $\xi(n)$ of the type in (3.4) Since linear combinations of these are dense in the set of all numerical derived sequences of $\xi(n)$, it follows that $\xi(n)$ is stochastic.

§4. The Process Associated with a Stochastic Sequence

4.1. <u>The Abstract Process of a Stochastic Sequence</u>. In this section we shall find that the converse to Theorem 3.4 holds in the sense that any stochastic sequence arises as a "typical" sample sequence of a uniquely determined stationary process. The stationary process in question will be found by first obtaining the abstract process which involves associating to a stochastic sequence an E-algebra. The algebra turns out to be essentially the algebra of derived sequences of the given sequence.

DEFINITION 4.1. If $\zeta(n)$ is a numerical left-infinite sequence, the essential supremum of ζ, denoted by $\|\zeta\|_\infty$, is given by

$$(4.1) \qquad \|\zeta\|_\infty = \inf \{\lambda : \bar{D}\{n : |\zeta(n)| > \lambda\} = 0\}$$

(i.e., the analogue for sequences of the L^∞ norm for measurable functions).

DEFINITION 4.2. $\zeta(n)$ is a <u>null sequence</u> if $\|\zeta\|_\infty = 0$.

Thus the sequence $\zeta(n) = 1/(|n| + 1)$ is a null sequence. Note that if $T\zeta$ represents the shifted sequence (Equation (2.1)) then $\|T\zeta\|_\infty = \|\zeta\|_\infty$; in particular, the translate of a null sequence is a null sequence.

With any left-infinite Λ-sequence $\xi(n)$ we may associate the algebra of all numerical valued derived sequences of $\xi(n)$, the operations of this algebra being pointwise addition and multiplications of the sequences. Denote this algebra by $A_O(\xi)$. If $\zeta \in A_O(\xi)$ then

$$\zeta(n) = \varphi(\ldots, \xi(n - 2), \xi(n - 1), \xi(n))$$

where $\varphi \in C(\Lambda^-)$ and if the shift transformation T is given by (2.1), then

$$(4.2) \quad \begin{aligned} T\zeta(n) = \zeta(n - 1) &= \varphi(\ldots, \xi(n - 3), \xi(n - 2), \xi(n - 1)) \\ &= T\varphi(\ldots, \xi(n - 2), \xi(n - 1), \xi(n)) \end{aligned}$$

where $T\varphi$ is the translate of φ under the transformation T of Λ_∞^-. It follows that $A_O(\xi)$ is invariant under translations T.

Consider the set $N \subset A_O(\xi)$ of null sequences. It is easily seen that N is an ideal in $A_O(\xi)$ and that moreover $TN \subset N$. Hence we may define an algebra $A_1(\xi)$ by setting $A_1(\xi) = A_O(\xi)/N$, and the translation T on $A_O(\xi)$ induces an operator T on $A_1(\xi)$. The functional $\|\zeta\|_\infty$ which defined a semi-norm on $A_O(\xi)$ defines an actual norm, denoted again $\|\zeta\|_\infty$, on $A_1(\xi)$. We can easily verify

THEOREM 4.1. The completion $A^-(\xi)$ of $A_1(\xi)$ is a C^*-algebra and the operator T in $A_1(\xi)$ induces a norm preserving endomorphism of $A^-(\xi)$.

Now suppose that $\xi(n)$ is a stochastic sequence. Every numerical derived sequence of $\xi(n)$, hence every sequence in $A_O(\xi)$, possesses an average which defines a functional $E(\zeta)$ in $A_O(\xi)$. E carries over to $A_1(\xi)$ since E clearly vanishes in N, and since $|E(z)| \leq \|z\|_\infty$ for $z \in A_1(\xi)$ it follows that E has a unique extension to $A^-(\xi)$ as a continuous linear functional. Clearly $E(1) = 1$ and $E(zz^*) \geq 0$; to show that $A^-(\xi)$ is an E-algebra with this functional it need only be shown that $E(zz^*) = 0$ implies $z = 0$. Suppose then that $z \in A^-(\xi)$ and $E(zz^*) = 0$. If we take

$$z = \lim_{k \to \infty} z_k$$

with $z_k \in A_1(\xi)$ it will suffice to prove that $\|z_k\|_\infty \longrightarrow 0$. Let the sequences ζ_k in $A_O(\xi)$ represent the $z_k \in A_1(\xi)$. Since $\|z_k - z_\ell\|_\infty \longrightarrow 0$ as $k, \ell \longrightarrow \infty$, for arbitrary $\varepsilon > 0$ there will exist a k_ε such that the set of n for which $|\zeta_k(n) - \zeta_\ell(n)| > \varepsilon$ has density 0 for $k, \ell > k_\varepsilon$. It follows that if for some $k > k_\varepsilon$, $|\zeta_k(n)| > 2\varepsilon$ on a set of non-zero upper density, then for all $\ell > k_\varepsilon$,

$|\zeta_\ell(n)| > \varepsilon$ on a set with the same upper density. In particular, $E(|\zeta_\ell|^2)$ would be bounded away from 0. But since $E(zz^*) = 0$, $E(|\zeta_\ell|^2) = E(z_\ell z_\ell^*) \longrightarrow 0$. Hence as $k \longrightarrow \infty$, $\|\zeta_k\|_\infty \longrightarrow 0$ as was to be shown.

Thus $A^-(\xi)$ is an E-algebra. To see that it defines an abstract process we remark that the transformation T in $A_0(\xi)$ obviously preserves the functional E, hence so does the induced shift transformation in $A^-(\xi) : E(Tz) = E(z)$. Since T is an endomorphism of $A^-(\xi)$, the latter will be the algebra of a one-sided process. The corresponding two-sided process is determined as in Theorem 2.1 and the corresponding algebra is denoted $A(\xi)$.

4.2. <u>The Process $X(\xi)$</u>. We may also associate with a stochastic sequence $\xi(n)$ an ordinary process $X(\xi)$. In order to define the variables x_n of this process let $\Omega^-(\xi)$ be the homomorphism space of $A^-(\xi) : A^-(\xi) = C(\Omega^-(\xi))$. The transformation T and functional E in $A^-(\xi)$ induce a mapping T and a probability measure P on $\Omega^-(\xi)$ such that P is preserved by T. We shall obtain a one-sided process by defining a set of variables x_n on $\Omega^-(\xi)$ that separate the points of that space and satisfying $x_n(T\omega) = x_{n-1}(\omega)$, $\omega \in \Omega^-(\xi)$.

To do so suppose that $\xi(n)$ is a Λ-sequence, let ψ range over $C(\Lambda)$ and consider the sequences in $A_0(\xi)$ of the form

$$(4.3) \qquad \zeta_\psi^{(k)}(n) = \psi(\xi(k + n)), \qquad\qquad n \leq 0,\, k \leq 0 \ .$$

Clearly the mapping $\psi \longrightarrow \zeta_\psi^{(k)}$ is an algebraic homomorphism of $C(\Lambda)$ into $A_0(\xi)$ and hence induces a homomorphism of $C(\Lambda)$ into $A^-(\xi) : \psi \longrightarrow z_\psi^{(k)}$. Hence if $\omega \in \Omega^-(\xi)$, $\psi \longrightarrow \omega(z_\psi^{(k)})$ is a homomorphism of $C(\Lambda)$ onto the complex numbers so that we may write

$$(4.4) \qquad \omega(z_\psi^{(k)}) = \psi(\lambda) \ , \qquad\qquad \lambda \in \Lambda \ .$$

By the preceding equation each $\psi(\lambda)$ depends continuously on ω; it follows that λ itself depends continuously on ω so that we may write $\lambda = x_k(\omega)$ where x_k is a map of $\Omega^-(\xi)$ to Λ. The functions x_k are thus defined by the relationship

$$(4.5) \qquad \psi(x_k(\omega)) = \omega(z_\psi^{(k)}) \ .$$

Consider next $T\omega(z_\psi^{(k)})$; by definition this is $\omega(Tz_\psi^{(k)})$ where $Tz_\psi^{(k)}$ is the element of $A^-(\xi)$ corresponding to the translate of the sequence $\zeta_\psi^{(k)}$, i.e., to the sequence

$$T\zeta_\psi^{(k)}(N) = \psi(\xi(n + k - 1)) \quad .$$

It follows that $T\zeta_\psi^{(k)} = \zeta_\psi^{(k-1)}$ or $Tz_\psi^{(k)} = z_\psi^{(k-1)}$ so that

$$(4.6) \qquad \psi(x_k(T\omega)) = \omega(Tz_\psi^{(k)}) = \omega(z_\psi^{(k-1)}) = \psi(x_{k-1}(\omega)) \quad .$$

This being true for all ψ in $C(\Lambda)$ we have

$$(4.7) \qquad\qquad x_k(T\omega) = x_{k-1}(\omega) \quad .$$

Finally we wish to show that the variables x_k separate the points of $\Omega^-(\xi)$. To do so we observe that the sequences $\zeta_\psi^{(k)}$ generate a dense subset of $A_0(\xi)$ and hence of $A^-(\xi)$. This is so because the sequence $\zeta_\psi^{(k)}$ corresponds to the function $\psi(\lambda(k))$ on Λ_∞^-. ($\lambda(k)$ is the kth component of a sequence in Λ_∞^-) and the functions of this form separate points in Λ_∞^-, hence they generate $C(\Lambda_\infty^-)$. It follows that the z_ψ^k separate points in $\Omega^-(\xi)$. Now if $x_k(\omega_1) = x_k(\omega_2)$ for all k then by (4.5) $\omega_1(z_\psi^{(k)}) = \omega_2(z_\psi^{(k)})$ for all k and all ψ. But this implies that $\omega_1 = \omega_2$ so that the x_k separate points in $\Omega^-(\xi)$.

DEFINITION 4.3. The process $X^-(\xi)$ defined by a stochastic sequence $\xi(n)$ is the process with variables x_n, space $\Omega^-(\xi)$, transformation T and probability measure P described in the foregoing paragraphs. The corresponding two-sided version will be denoted $X(\xi)$, with space $\Omega(\xi)$.

Consider now the sequences of $A_0(\xi)$ which, after identifying sequences that differ by a null sequence, form a dense subset of $A^-(\xi)$. Each of these corresponds to a function on the space $\Omega^-(\xi)$ and we seek an exact relationship between the derived sequences of $\xi(n)$ in $A_0(\xi)$ and the functions on $\Omega^-(\xi)$. Consider, first of all, the functions of the form $\psi(x_k)$ where $\psi \in C(\Lambda)$. By (4.5) and (4.3), $\psi(x_k)$ corresponds to the sequence

$$\zeta_\psi^k(n) = \psi(\xi(k + n)) \quad .$$

Since the correspondence between functions on $\Omega^-(\xi)$ and sequences in $A_0(\xi)$ preserves algebraic operations, it follows that a function of the form $\psi(x_{j_1}, x_{j_2}, \ldots, x_{j_k})$ corresponds to the sequence $\zeta(n) = \psi(\xi(n + j_1), \xi(n + j_2), \ldots, \xi(n + j_k))$. Finally taking uniform limits in the sequence algebra and in $A^-(\xi)$ we deduce that for any $\psi \in C(\Lambda_\infty^-)$, the function

$$(4.8) \qquad\qquad z = \psi(\ldots, x_{-2}, x_{-1}, x_0)$$

corresponds to the sequence $\zeta(n)$ where

$$(4.9) \qquad \zeta(n) = \psi(\ldots, \xi(n - 2), \xi(n - 1), \xi(n)) \ .$$

From this we may draw several consequences. First of all since
the x_k are Λ-valued, $\Omega^-(\xi)$ may be imbedded in Λ_∞^- and hence (§1.1)
there is induced a restriction map of $C(\Lambda_\infty^-)$ onto $C(\Omega^-(\xi)) = A^-(\xi)$.
Since the x_k are the coordinate functions imbedding $\Omega^-(\xi)$ into Λ_∞^-
it follows that every continuous function on $\Omega^-(\xi)$ when considered the
restriction of a function on Λ_∞^- has the form (4.8). But then it corresponds to the derived function (4.9). In other words we have shown that

$$(4.10) \qquad\qquad A^-(\xi) = A_1(\xi) \ ,$$

i.e., that every element of $A^-(\xi)$ corresponds to a derived sequence of
$\xi(n)$.

The points of Λ_∞^- are left-infinite Λ-sequences and, in particular, $\xi \in \Lambda_\infty^-$ as do its translates $T^m\xi$. Thus (4.9) states that the
sequence $\zeta(n)$ corresponding to the function z of (4.8) may be obtained by evaluating the function ψ on Λ_∞^- at the points $T^n\xi$. From
this we deduce

THEOREM 4.2. Every point in $\Omega^-(\xi)$ is a limit of translates $T^m\xi$ of the original sequence $\xi(n)$.

PROOF. If not, there would be a non-zero function on Λ_∞^-
vanishing at each $T^m\xi$ but non-zero on $\Omega^-(\xi)$. If this function is denoted by ψ then according to (4.9) the corresponding sequence in $A_0(\xi)$
vanishes identically whereas the corresponding function in $A^-(\xi)$ would
not be 0. This is a contradiction.

We shall later need the following.

THEOREM 4.3. Let $\varphi : \Lambda_\infty^- \longrightarrow \Lambda'$ and suppose that
$\eta(n)$ is the derived sequence of $\xi(n)$ defined by

$$(4.11) \qquad \eta(n) = \varphi(\ldots, \xi(n - 2), \xi(n - 1), \xi(n)) \ .$$

If $\xi(n)$ is stochastic and $X(\xi)$ is its associated
process then the process $X(\eta)$ associated to the
stochastic sequence η is the subprocess of $X(\xi)$ determined by φ. (Cf. Definition 2.4 and Theorem 3.1.)

PROOF. The numerical derived sequences of $\eta(n)$ are evidently derived sequences of $\xi(n)$; hence $A_o(\eta) \subset A_o(\xi)$ and $A^-(\eta) \subset A^-(\xi)$ so that $X(\eta)$ is a subprocess of $X(\xi)$. If the variables of $X(\eta)$ are denoted y_k then y_k is defined by the correspondence of $\psi(y_k)$ to the sequence

$$\zeta_\psi^k(n) = \psi(\eta(k + n))$$

where $\psi \in C(\Lambda')$. Using (4.11) we have $\zeta_\psi^k(n) = \psi\varphi(\ldots, \xi(k + n - 2), \xi(k + n - 1), \xi(k + n))$ so that by (4.8) and (4.9) we find that ζ_ψ^k corresponds to the function $\psi\varphi(\ldots, x_{k-2}, x_{k-1}, x_k)$. Hence we may write $y_k = \varphi(\ldots, x_{k-2}, x_{k-1}, x_k)$ and this proves our assertion that $X(\eta)$ is the subprocess of $X(\xi)$ determined by φ.

The algebra $A^-(\xi)$ is, according to (4.10), essentially a sequence algebra, namely, it is the algebra of sequences derived from $\xi(n)$, identifying sequences that differ by a null sequence. The two-sided algebra $A(\xi)$ has so far been defined abstractly but it is possible to relate it to a sequence algebra as well. For the algebra $A(\xi)$ is characterized by the property that for a dense subset of $z \in A(\xi)$ there exists an m with $T^m z \in A^-(\xi)$, or $z = T^{-m}w$ with $w \in A^-(\xi)$. Since the operator T in $A^-(\xi)$ corresponds to translation to the right in $A_o(\xi)$ it follows that a dense subset of $A(\xi)$ will be obtained if we consider the translates to the left of sequences in $A_o(\xi)$. More precisely, let $B_o(\xi)$ be the set of all sequences defined on <u>all but finitely many negative integers</u> with the property that some translate to the right is a derived sequence of $\xi(n)$. Since in identifying sequences that differ by a null sequence, any finite set of values of the sequences are insignificant, this equivalence relation may be extended to $B_o(\xi)$ so that the resulting algebra $B_1(\xi)$ contains $A_1(\xi)$. The completion of $B_1(\xi)$ is then the algebra $A(\xi)$.

4.3. <u>Examples</u>. By way of illustration we shall describe the processes $X^-(\xi)$ and $X(\xi)$ when $\xi(n)$ is a periodic sequence. We note that if $\xi(n)$ has the period ν then so does every derived sequence of $\xi(n)$ so that $A_o(\xi)$ will be at most ν-dimensional. Moreover if ν is the minimal period of $\xi(n)$ then ν is exactly the dimension of $A_o(\xi)$. Now a periodic sequence cannot be a null sequence without being identically 0 so that $A_o(\xi) = A_1(\xi)$ and by (4.10), or by the fact that a finite dimensional normed space is always complete, $A_o(\xi) = A^-(\xi)$. Since every sequence of $A_o(\xi)$ is determined by its last ν entries it follows that $A^-(\xi)$ is isomorphic to the algebra of functions on a space with ν elements: $\Omega^-(\xi) \approx \{1, 2, \ldots, \nu\}$. Moreover $T^\nu = $ identity so that $T^{-1} = T^{\nu-1}$. It follows that T is an automorphism of $A^-(\xi)$ so that

$X^-(\xi)$ is already two-sided: $X^-(\xi) = X(\xi)$. Since T is measure preserving, it follows that the probability measure on $\Omega^-(\xi) = \Omega(\xi)$ is obtained by assigning probability $1/\nu$ to each of the ν points of the
space.

If the sequence $\xi(n)$ is not periodic but it is "eventually"
periodic, say with period ν, then each translate $T^\nu \zeta(n)$ of a derived
sequence $\zeta(n)$ of $\xi(n)$ will agree with $\zeta(n)$ except at a finite number
of values of n. Hence $\eta(n) = T^\nu \zeta(n) - \zeta(n)$ would be a null sequence
that would not necessarily be identically 0. Thus $A_1(\xi) \neq A_0(\xi)$.
However we would still have $T^\nu =$ identity in $A_1(\xi)$ so that $A_1(\xi)$ defines the same process as before.

The property $A^-(\xi) = A(\xi)$ just observed for periodic sequences
holds for almost periodic sequences as well. In fact, by the definition
of almost periodicity, there exists a sequence of integers $n_k \longrightarrow \infty$,
for a given almost periodic sequence $\xi(n)$, such that $T^{n_k}\xi \longrightarrow \xi$ uniformly. This is equivalent to $x_{-n_k} \longrightarrow x_0$ in $A^-(\xi)$. Hence
$x_{j-n_k} \longrightarrow x_j$ for all j so that each x_j is a limit of functions in
$A^-(\xi)$ and hence $A(\xi) = A^-(\xi)$.

$$\S 5. \quad \text{Regular Sequences}$$

5.1. <u>Definitions</u>. For the purposes of prediction theory, the
class of left-infinite sequences that permit satisfactory analysis is
still further restricted. The condition to be imposed in this section will
be such as to exclude sequences that exhibit "sudden" changes of behavior.
An example of such a sequence would be one that is periodic up to a point
and then violates its periodicity, e.g.,

$$(5.1) \qquad \xi(n) = (-1)^n, \quad n \leq -2, \quad \xi(-1) = \xi(0) = 1 \ .$$

We first prove

THEOREM 5.1. Let $\xi(n)$ be a stochastic left-infinite
Λ-sequence. The following statements regarding $\xi(n)$
are equivalent:
> (a) If $\Delta_1, \Delta_2, \ldots, \Delta_k$ are open subsets of
> Δ, then the set of n for which $\xi(n) \in \Delta_1$,
> $\xi(n-1) \in \Delta_2, \ldots, \xi(n-k+1) \in \Delta_k$ is
> either empty or has strictly positive upper
> density.
> (b) If $\zeta(n)$ is a non-negative derived sequence

of $\xi(n)$ that is not identically 0, then $\zeta(n) > 0$ on a set with strictly positive upper density.

(c)   If $\zeta(n)$ is a non-negative derived sequence of $\xi(n)$ that is not identically 0 then $\zeta(n) > 0$ on a set with strictly positive <u>lower</u> density.

(d)   There are no null sequences in $A_o(\xi)$ but for the sequence identically 0.

(e)   As a point of Λ_∞^-, $\xi \in \Omega^-(\xi)$.

PROOF.   The implication from (a) to (b) is made in several steps. First we note that (a) implies that if Γ is any open subset of Λ^k (the k-fold cartesian product of Λ with itself) then the set of n for which $\xi(n)$, $\xi(n - 1)$, $\ldots$, $\xi(n - k + 1)) \in \Gamma$ is either empty or has positive upper density. This is so because Γ is the union of sets of the form $\Delta_1 \times \Delta_2 \times \ldots \times \Delta_k$. From this we conclude that if ψ is a non-negative function on Λ^k then $\psi(\xi(n), \xi(n - 1), \ldots, \xi(n - k + 1))$ is either identically 0 or positive on a set of positive upper density. Now suppose $\zeta(n)$ is a derived sequence of $\xi(n)$

$$\zeta(n) = \varphi(\ldots, \xi(n - 2), \xi(n - 1), \xi(n))$$

that is non-negative and not identically 0. We may assume that φ is non-negative and that φ is the uniform limit of non-negative functions on Λ_∞^- that depend upon finitely many coordinates. Suppose that for some n, $\zeta(n) = \alpha$ where $\alpha > 0$; there exists a non-negative ψ on Λ^k for some k such that

$$(5.2) \qquad |\zeta(n) - \psi(\xi(n), \xi(n - 1), \ldots, \xi(n - k + 1))| < \frac{\alpha}{3}$$

for all n. But then

$$\psi(\xi(n), \xi(n - 1), \ldots, \xi(n - k + 1)) > \frac{2}{3}\alpha$$

when $\zeta(n) = \alpha$ so that the function $\psi'(\xi(n), \xi(n - 1), \ldots, \xi(n - k + 1))$ where $\psi' = \max(\psi - 2/3\alpha, 0)$ is not identically 0. Hence it is strictly positive on a set of positive upper density; therefore $\psi(\xi(n), \xi(n - 1), \ldots, \xi(n - k + 1) > 2/3\alpha$ on a set of positive upper density and by (5.2) $\zeta(n) > \alpha/3$ on this set of n. This shows that (a) implies (b).

To see that (b) implies (c) we remark that if $\zeta(n) > 0$ on a non-empty set of n, then also $\zeta(n) - \delta > 0$ on a non-empty set of n

for sufficiently small δ . Since this is again a derived sequence of
$\xi(n)$, by (b) $\zeta(n) - \delta > 0$ on a set of positive upper density. Now let
$F(\lambda)$ be the distribution function of the stochastic sequence $- \zeta(n)$
(Theorem 3.3). $F(\lambda)$ will have a point of continuity between $- \delta$ and
0 and hence $D \{n : - \zeta(n) < - \delta'\}$ exists for some δ' , $0 < \delta' < \delta$.
But $0 < \bar{D}\{n : - \zeta(n) < - \delta\} \leq D\{n : - \zeta(n) < - \delta'\} \leq \underline{D}\{n : - \zeta(n) < 0\} =$
$\underline{D}\{n : \zeta(n) > 0\}$ and this proves our assertion.

Condition (d) is contained in the stronger statement:

(d') The essential supremum (Definition 4.1) of any
derived sequence of ξ is given by

$$\|\zeta\|_\infty = \sup_{n \leq 0} |\zeta(n)| \quad .$$

That (b) implies (d') is easily shown. For by Definition 4.1,
the sequence $\zeta'(n) = \max (|\zeta(n)| - \|\zeta\|_\infty - \varepsilon, 0)$ can be positive only on
a set of n with 0 density and hence by (b) $\zeta'(n) = 0$ and
$|\zeta(n)| \leq \|\zeta\|_\infty + \varepsilon$. This being true for every $\varepsilon > 0$ $|\zeta(n)| \leq \|\zeta\|_\infty$ and
(d') follows.

To see that (d) implies (e) suppose that $\xi(n)$ satisfies (d)
and that $\xi \notin \Omega^-(\xi)$. There then exists a non-negative function φ on
Λ_∞^- vanishing on $\Omega^-(\xi)$ but non-zero at ξ . Consider the derived sequence

$$\zeta(n) = \varphi(\ldots, \xi(n - 2), \xi(n - 1), \xi(n)) \quad ;$$

we have $\zeta(0) > 0$. By (4.8) and (4.9) the element z in $A^-(\xi)$ corresponding to $\zeta(n)$ is given by

$$z = \varphi(\ldots, x_{-2}, x_{-1}, x_0) \quad .$$

Now the x_n are the variables of $X^-(\xi)$ and as such the sequences
$(\ldots, x_{-2}, x_{-1}, x_0)$ lie in $\Omega^-(\xi)$ by the definition of the imbedding of
the sample space into a sequence space (§2.2). Hence $z \equiv 0$. It follows
that $\zeta(n)$ is a null sequence and by (d) $\zeta(n) \equiv 0$. This contradicts
$\zeta(0) > 0$ and therefore we must have $\xi \in \Omega^-(\xi)$.

Finally suppose that $\xi \in \Omega^-(\xi)$ and let $\Delta_1, \Delta_2, \ldots, \Delta_k$ be
open subsets of Λ . There exist continuous functions $\psi_1, \psi_2, \ldots, \psi_k$
on Λ with ψ_j positive in Δ_j and vanishing on $\Lambda - \Delta_j$. Consider
the element $z \in A^-(\xi)$ defined by

$$(5.3) \qquad\qquad z = \psi_1(x_0)\psi_2(x_{-1})\ldots\psi_k(x_{-k+1}) \quad .$$

By (4.8) and (4.9) this corresponds to the sequence

$$(5.4) \qquad \zeta(n) = \psi_1(\xi(n))\psi_2(\xi(n-1))\ldots\psi_k(\xi(n-k+1)) \quad .$$

We see that $\zeta(n) > 0$ if and only if the conditions $\xi(n) \in \Delta_1$, $\xi(n-1) \in \Delta_2$, ..., $\xi(n-k+1) \in \Delta_k$ are fulfilled. Now since $\xi \in \Omega^-(\xi)$ all its translates $T^m\xi \in \Omega^-(\xi)$ (Theorem 2.3). Hence we may rewrite (5.4) as

$$(5.5) \qquad \zeta(n) = z(T^{-n}\xi) \quad ,$$

z being defined by (5.3). It follows that if the set of n for which $\xi(n) \in \Delta_1$, $\xi(n-1) \in \Delta_2$, ..., $\xi(n-k+1) \in \Delta_k$ is non-empty then $z \neq 0$. However since $z \geq 0$ this implies $E(z) > 0$ and by the definition of E on $A^-(\xi)$ this means that the average of the sequence $\zeta(n)$ in (5.4) is strictly positive. Finally this implies that the upper density of the set of points for which $(n) > 0$ is strictly positive and this is equivalent to (a). This completes the proof of the theorem.

DEFINITION 5.1. A left-infinite sequence $\xi(n)$ is <u>regular</u> if it is stochastic and it satisfies the five equivalent conditions of Theorem 5.1.

Thus in the example of (5.1) although $\xi(n)$ is stochastic it is not regular since the pair of values $(1, 1)$ occurs once in $\xi(n)$ but not with positive upper density, thus violating (a).

We remark that if $\xi(n)$ is regular then by (d) the ideal N of $A_0(\xi)$ consists only of the 0 element and hence $A_1(\xi) \overset{\sim}{=} A_0(\xi)$. Combining this with (4.10) we find that $A^-(\xi) \overset{\sim}{=} A_0(\xi)$ or that $A^-(\xi)$ is simply the algebra of numerical valued derived sequences of $\xi(n)$. The correspondence between the elements of $A^-(\xi)$ as functions on $\Omega^-(\xi)$ and the sequences in $A_0(\xi)$ is obtained by way of (e) which asserts that $\xi \in \Omega^-(\xi)$. Namely, if $z \in A^-(\xi)$ is a function on $\Omega^-(\xi)$ it may be evaluated at ξ and each of its translates $T^m\xi$. The resulting sequence is the sequence in $A_0(\xi)$ corresponding to z. More precisely we may write

$$(5.6) \qquad \zeta(-n) = z(T^n\xi) \quad .$$

This is a direct consequence of equations (4.8) and (4.9) relating the functions on $\Omega^-(\xi)$ (or on Λ_∞^-) with sequences in $A_0(\xi)$.

The following theorem is easily verified:

THEOREM 5.2. A derived sequence of a regular sequence

is regular, and a uniform limit of regular sequences
is again a regular sequence.

5.2. **Existence of Regular Sequences.** Theorem 3.4, which states
that almost all sample sequences of a stationary process are stochastic
and thereby ensures a plentiful supply of stochastic sequences may be
sharpened to give

THEOREM 5.3. Almost all sample sequences $\{x_n(\omega)\}$ of
a one-sided stationary process are regular.

PROOF. By Theorem 3.4 and (b) of Theorem 5.1 what remains to
be shown is that for almost all ω in the sample space and for any non-
negative φ on Λ_∞^- (Λ is the range of the variables x_n of the process
in question), the derived sequence of $\{x_n(\omega)\}$ corresponding to φ is
either identically 0 or it is non-zero on a set of positive upper density.
To do so let us call the pair (φ, ω) regular, where $\varphi \in C(\Lambda_\infty^-)$ and
It would certainly suffice to show that for each rational r, the de-
rived sequence is either never $> r$ or is $> r$ on a set of positive
upper density. Let us call the pair (φ, ω) regular if $\varphi \in C(\Lambda_\infty^-)$,
$\omega \in \Omega_X^-$ and if the derived sequence $\varphi(\ldots, x_{n-2}(\omega); x_{n-1}(\omega), x_n(\omega))$
has this property for all rationals. Observe that if $\varphi_k \longrightarrow \varphi$ uniformly and the (φ_k, ω) are regular then (φ, ω) is also regular. For
if the derived sequence for φ is somewhere $> r$, it is also
$> r + 2\epsilon$ for a rational $\epsilon > 0$. Hence if $\|\varphi - \varphi_k\| < \epsilon$, the sequence
for φ_k is $> r + \epsilon$ on a set of positive upper density and so the
original sequence for φ is $> r$ on a set of positive upper density.
Hence (φ, ω) is regular.

It follows that it is only necessary to show that for almost all
ω and a denumerable dense set of $\varphi \in C(\Lambda_\infty^-)$ the pair (φ, ω) is regular.
Since the condition for regularity involves the denumerable set of
rationals it suffices to show that for a fixed φ and r, the derived
sequence for (φ, ω) is either never $> r$ or is so on a set of positive
upper density. Let $\theta = \max(\varphi - r, 0)$ and let

$$\Delta = \left\{ \omega \in \Omega_X^- : \lim_{N \to \infty} \frac{1}{N+1} \sum_{n=0}^{N} \right.$$

(5.7)

$$\left. \theta(\ldots, x_{n-2}(\omega), x_{n-1}(\omega), x_n(\omega)) = 0 \right\} \quad .$$

Our theorem will follow if we show that for almost all $\omega \in \Delta$, $\Phi_n(\omega) = \theta(\ldots, x_{n-2}(\omega), x_{n-1}(\omega), x_n(\omega)) = 0$ for all $n \leq 0$. From $x_{n-1}(\omega) = x_n(T\omega)$ it follows that $T\Delta \subset \Delta$ so that if we show that for almost all ω in Δ, $\Phi_0(\omega) = 0$ then for almost all ω in Δ, $\Phi_0(T^n(\omega)) = 0$ for all $n \geq 0$ and hence $\Phi_n(\omega) = 0$ for $n \leq 0$. To show that Φ_0 vanishes almost everywhere in Δ, let X_Δ be the characteristic function of Δ; $X_\Delta(\omega) = $ or 0 according as $\omega \in \Delta$ or $\omega \notin \Delta$. By the ergodic theorem

$$(5.8) \qquad \lim_{N \to \infty} \frac{1}{N+1} \sum_{n=0}^{N} X_\Delta(T^n\omega)\Phi_0(T^n\omega) = \psi(\omega)$$

exists almost everywhere and satisfies

$$(5.9) \qquad \int_{\Omega_X^-} \psi(\omega)dP(\omega) = \int_{\Omega_X^-} X_\Delta(\omega)\Phi_0(\omega)dP(\omega) \quad .$$

Comparing (5.8) with (5.7) we see that $\psi(\omega) \equiv 0$. For if $X_\Delta(T^m\omega) \neq 0$ for some m, then $T^m\omega \in \Delta$ and then all successive $T^{m+j}\omega \in \Delta$ and the average of $\Phi_0(T^{m+j}\omega) = \Phi_j(T^m\omega)$ vanishes. It follows from (5.9) that $X_\Delta\Phi_0 = 0$ almost everywhere and hence Φ_0 vanishes almost everywhere in Δ. This completes the proof of the theorem.

5.3. <u>Generic Points</u>. If $\xi(n)$ is a regular sequence then $T^n\xi \in \Omega^-(\xi)$ for $n \geq 0$, by (e) of Theorem 5.1. Hence if $z \in A^-(\xi) \cong C(\Omega^-(\xi))$ we may speak of $z(T^n\xi)$. By (5.6), this is $\zeta(-n)$ where ζ is the sequence corresponding to z. Since by the definition of E in $A^-(\xi)$, $E(z)$ is the average of the sequence $\zeta(-n)$ we may also write

$$(5.10) \qquad E(z) = \lim_{N \to \infty} \frac{1}{N+1} \sum_{n=0}^{N} z(T^n\xi) \quad .$$

The point ξ is thereby distinguished by the fact that the functional E is determined entirely from the values of a function at ξ and its translates $T^n\xi$. In the following definition we consider the analogous situation for an arbitrary abstract process.

DEFINITION 5.2. If X is an abstract process, $A_X^- = C(\Omega_X^-)$, the algebra of a one-sided version of X (for example, X itself), and $\omega \in \Omega_X^-$, then ω is a <u>generic</u> point of X if for every $z \in A_X^-$, the sequence $\{z(T^n\omega)\}$ has an average equal to $E(z)$.

Thus if $\xi(n)$ is a regular sequence then ξ is a generic point of $X(\xi)$.

One immediate consequence of this definition is that the translates $T^n\omega$ of a generic point of X lie dense in Ω_X^-. For if this were not the case we could find a non-negative z, vanishing at all the $T^m\omega$ but not identically 0. Then on the one hand $E(z) > 0$ since $z \neq 0$ and on the other

$$\lim_{N \to \infty} \frac{1}{N+1} \sum_{n=0}^{N} z(T^n\omega) = 0$$

so that ω could not be a generic point.

For an ordinary process X we have the following theorem.

THEOREM 5.4. ω_0 is a generic point of X if and only if the sequence $\{\xi(n) = x_n(\omega_0)\}_{n \leq 0}$ is a regular sequence with $X(\xi)$ identical to X.

PROOF. Note first that when Ω_X^- is identified with a subset of Λ_∞^- as in §2.2 where Λ is the range space of the x_n, then the point ω_0 is identified with the sequence ξ. It follows then that if $\xi(n)$ is regular, ω_0 is generic for $X(\xi)$, and therefore for X if the two processes are identical. Suppose conversely that ω_0 is generic for X. To show that $\xi(n)$ is stochastic let φ be a function in $C(\Lambda_\infty^-)$ and consider the derived sequence $\zeta(n)$ of $\xi(n)$ defined by φ. Clearly $\zeta(n) = z_n(\omega_0)$ where

(5.11) $$z_n = \varphi(\ldots, x_{n-2}, x_{n-1}, x_n)$$

and since $z_n = T^{-n}z_0$, $\zeta(n) = z_0(T^{-n}\omega_0)$, $n \leq 0$ so that the average of $\zeta(n)$ exists because ω_0 is generic. In the same way we can show that $\xi(n)$ is regular, using (d) of Theorem 5.1. For if $\zeta(n)$ is a derived sequence of $\xi(n)$ corresponding to φ, and $\zeta(n)$ is a null sequence then the average of $\{|\zeta(n)|\}$ vani hes; hence so does $E(|z_0|)$ which is the average of $\{|z_0(T^{-n}\omega_0)|\}$ where z_n is defined by (5.11). But then $|z_0| = 0$, $z_0 = 0$ and therefore $z_0(T^{-n}\omega_0) = 0$ for all n, so that $\zeta(n)$ is identically 0.

To show that $X(\xi) = X$ we first show that $\Omega_X^- = \Omega^-(\xi)$ where both are taken as subsets of Λ_∞^-. For this recall that by Theorem 4.2 the translates of ξ are dense in $\Omega^-(\xi)$ and that by the remark following Definition 5.2 the translates of ω_0 and hence of ξ are dense in Ω_X^-. Hence $\Omega_X^- = \Omega^-(\xi)$. Since the transformation T has a common definition in Ω_X^- and Ω_ξ^-, we shall have $X = X(\xi)$ once it is shown that the probability measures are the same and that the variables of the processes are the same.

Since the probability measures are determined by the distinguished functional E which is determined by the values of a function at the translates of a generic point, it follows that the probability measures are the same for X and $X(\xi)$. As regards the variables of the two processes recall that the variables of $X(\xi)$ are defined by Equation (4.5): $\psi(x_k(\omega)) = \omega(z_\psi^k)$ where z_ψ^k is the element of $A^-(\xi)$ corresponding to the sequence $\zeta_\psi^k(n) = \psi(\xi(n + k))$. It suffices therefore to verify (4.5) for the x_k of X and all ω in Ω_X^-. However for this it suffices to take a dense set of ω; in particular it suffices to verify

$$(5.12) \qquad\qquad \psi(x_k(T^n\omega_0)) = T^n\omega_0(z_\psi^k)$$

for all $n \geq 0$. By (5.6) it follows that the sequence on the right side of (5.12) is just the sequence $\zeta_\psi^k(-n) = \psi(\xi(-n + k))$ since ω_0 is identified with ξ. Since $\xi(n) = x_n(\omega_0)$, $\psi(\xi(-n + k)) = \psi(x_{-n+k}(\omega_0)) = \psi(x_k(T^n\omega_0))$ which is the left side of (5.12). This completes the proof of the theorem.

As to whether generic points of a process are the rule or the exception we have:

> THEOREM 5.5.   If X is ergodic then almost all points
> of Ω_X^- are generic for X and conversely, if almost
> all points of Ω_X^- are generic for X then X is
> ergodic.

PROOF.   In one direction the assertion follows from Birkhoff's ergodic theorem. In the converse we use the following criterion for ergodicity which is considerably more elementary than the ergodic theorem: X is ergodic if and only if for all $u \in L^2(A_X^-)$ the weak limit in $L^2(A_X^-)$ of

$$u_N = \frac{1}{N+1} \sum_{n=0}^{N} T^n u$$

exists and is a constant. Clearly it suffices to know this for a dense subset of $L^2(A_X^-)$, in particular for $u \in A_X^-$ and for these it is an immediate consequence of the assumption that almost all points of Ω_X^- are generic for X.

Finally we prove

THEOREM 5.6. If ω is a generic point for X and
Y is a subprocess of X, and $\Phi : \Omega_X \longrightarrow \Omega_Y$ is
the induced map, then $\Phi(\omega)$ is generic for Y.

PROOF. This follows directly from Definition 5.2 since $\Phi(\omega)$
is generic for Y if (5.10) holds for $z \in A_Y^-$ and $\xi = \Phi(\omega)$. But
$= z \circ \Phi(T^n\omega)$ where $z \circ \Phi \in A_X^-$ so that the desired equality holds
because ω is generic for X.

§6. Examples

6.1. <u>The Almost Periodic Case</u>. Periodic and almost periodic
sequences have already been shown to be stochastic. The former are also
obviously regular. It is also not difficult to see that almost periodic
sequences are regular. For by (d) of Theorem 5.1 this will be the case
if no derived sequence can be a null sequence without vanishing identically.
Now a derived sequence of an almost periodic sequence is again an almost
periodic sequence so we have only to show that a null sequence that is
almost periodic vanishes identically. This however follows from the fact
that the Fourier coefficients of such a sequence must all necessarily
vanish.

Take as a particular example the almost periodic sequence
$\xi(n) = \cos n\alpha$ with α/π irrational; we shall determine $A^-(\xi)$, $A(\xi)$,
$\Omega^-(\xi)$ and $\Omega(\xi)$ for this particular sequence. To do so we first con-
sider the process X defined as follows. Let Ω_X^- be the circle, $|z| = 1$,
with normalized Lebesgue measure as the probability measure. Define T
on Ω_X^- by setting $T\omega = e^{-i\alpha}\omega$ for $\omega \in \Omega_X^-$. T is measure preserving and
so we have defined an abstract process. Now define the variables x_n,
$n \leq 0$, on Ω_X^- by setting $x_0(\omega) = R(\omega)$, the real part of the complex
number ω. We might note that $x_1(\omega) = R(e^{i\alpha}\omega)$ and $x_0(\omega) = R(\omega)$ together determine ω uniquely and hence the x_n for $n \leq 0$ certainly
separate points in Ω_X^- and so the x_n are variables for the abstract
process X.

Now it is known that the transformation T in this case is
ergodic, i.e., leaves no non-trivial sets invariant in Ω_X^- ([7]). It
follows that almost all points of Ω_X^- are generic for X. However it is
also evident that if ω is a generic point of X then so is $e^{i\beta}\omega$ for
any β. For the latter point induces upon Ω_X^- the measure induced by
ω but rotated by $e^{i\beta}$. Since this measure is still Lebesgue measure
$e^{i\beta}\omega$ must again be generic. Hence every point of Ω_X^- is generic for
X. In particular $\omega = 1$ is generic and so $x_n(1) = x_0(T^{-n}1) =$
$x_0(e^{in\alpha}) = R(e^{in\alpha}) = \cos n\alpha$ is a regular sequence and is generic for X

(Theorem 5.4). Also $X^-(\xi) = X$ so $\Omega^-(\xi)$ is the circle and $A^-(\xi)$ is the set of continuous functions on the circle. Next we notice that $A^-(\xi)$ is closed under T^{-1}, since the map T on $\Omega^-(\xi)$ is $1 - 1$. Hence $X^-(\xi)$ is already two-sided and $X^-(\xi) = X(\xi) = X$.

6.2. <u>The Random Case</u>. In the foregoing paragraph a process X was constructed with the property that every sample point is generic for it. This will be the case whenever the process corresponds to an almost periodic sequence $\xi(n)$. To see this we note that ξ itself and its translates are generic for $X(\xi)$ and that therefore a uniform limit of translates of ξ will be generic for $X(\xi)$, since evidently a uniform limit of generic sequences is again generic. Now an almost periodic sequence may be characterized by the fact that weak (i.e., pointwise) limits of translates and uniform limits are identical so that every sequence in $\Omega^-(\xi)$ is a uniform limit of translates of ξ and hence is generic for $X(\xi)$.

In contrast to this consider the case when X is a process with independent variables; we shall refer to such a process as a random process. In particular suppose that the x_n take on the values a_1, ..., a_r with probabilities p_1, ..., p_r, $\Sigma p_i = 1$. Setting $\Lambda = \{a_1, \ldots, a_r\}$ we have $\Omega_X^- = \Lambda_\infty^-$ since $P\{x_0 = a_{i_0}, x_{-1} = a_{i_1}, \ldots, x_{-k} = a_{i_k}\} > 0$ for any sequence $a_{i_0}, a_{i_1}, \ldots, a_{i_k}$; hence there must exist in Ω_X^- a sequence terminating in $a_{i_k}, \ldots, a_{i_1}, a_{i_0}$, and this implies $\Omega_X^- = \Lambda_\infty^-$. We denote the process just described by $X(p_1, \ldots, p_r)$. Allowing the p_i to vary (but still satisfying $p_i > 0$, $\Sigma p_i = 1$) we obtain distinct processes, all of which however have the same sample space, Λ_∞^-. Each of these processes is ergodic ([3]) and so for each p_1, ..., p_r almost every point in Λ_∞^- (with respect to the measure induced by $X(p_1, \ldots, p_r)$) is generic for $X(p_1, \ldots, p_r)$. It follows that the sample space of any one of these processes contains points generic for every other one of these processes as well; in particular, not all the sample paths can be generic for the original process.

6.3. <u>Operations on Regular Sequences</u>. We have mentioned (Theorems 3.1 and 5.1) that a sequence derived from a stochastic or regular sequence is again stochastic or regular. It is natural to inquire what can be said of sequences derived from two stochastic or regular sequences, e.g., $\xi(n) = \xi_1(n) + \xi_2(n)$, or $\xi(n) = \xi_1(n)\xi_2(n)$ where $\xi_1(n)$ and $\xi_2(n)$ are stochastic or regular. The answer is that in general nothing may be said and the sum or product of two regular sequences need not even be stochastic.

A counterexample may be constructed as follows. Take $\Omega = \Lambda_\infty^-$ where $\Lambda = \{-1, 1\}$ and take P as the measure on Ω corresponding to

the process $X(1/2, 1/2)$ of §6.2. We shall obtain two processes on Ω
by defining two transformations T_1 and T_2 of Ω both of which preserve P. First we define X_1 as the process obtained by taking T_1 to
be the ordinary shift on Λ_∞^-: $T_1\omega(n) = \omega(n - 1)$. The variables of X_1
are defined by the coordinate functions on Λ_∞^- : $x_n'(\omega) = \omega(n)$.

To define X_2 choose a sequence of positive integers $\{n_k\}$
satisfying $n_{k+1} > 2(n_1 + n_2 + \ldots + n_k)$. Define T_2 on Ω by setting

$$(6.1) \qquad T_2\omega(-n) = \begin{cases} \omega(-n-1) & \text{if} \quad n \neq n_1 + n_2 + \ldots + n_k \\[2ex] -\omega(-n-1) & \text{if} \quad n = n_1 + n_2 + \ldots + n_k \ . \end{cases}$$

It is easy to verify that T_2 preserves the measure P. We define the
variables of X_2 by letting $x_0'' = x_0'$ and then taking $x_{-n}'' = T_2^n x_0''$. We
thus obtain two processes X_1 and X_2 defined on a common sample space
with a common measure. (Note that although Ω is a sequence space it
does not give the standard representation of $\Omega_{\overline{X_2}}$ as a sequence space
since T_2 is not the ordinary shift operation.) From (6.1) we find that
for all n, $x_n'' = \pm x_n'$ so that since the x_n' are mutually independent it
follows that the x_n'' are also independent. Hence both X_1 and X_2 are
ergodic processes so that almost all points of Ω are generic for either
of them and hence almost all points of Ω are generic for both simultaneously. If ω is such a point it follows from Theorem 5.4 that the
two sequences $\{x_{-n}'(\omega)\}, \{x_{-n}''(\omega)\}$ are both regular. The first of these
is by definition $\{\omega(-n)\}$. For the second we have by (6.1) $x_{-n}''(\omega) =$
$\pm \omega(-n)$ depending upon the number of $n_1 + n_2 + \ldots + n_k$ lying between
o and n. More precisely we have

$$(6.2) \qquad x_{-n}''(\omega) = \begin{cases} \omega(-n) & \text{if} \quad n_1 + \ldots + n_{2k} < n \leq n_1 + \ldots + n_{2k+1} \\[2ex] -\omega(-n) & \text{if} \quad n_1 + \ldots + n_{2k-1} < n \leq n_1 + \ldots + n_{2k} \ . \end{cases}$$

It follows from this that the product of the two regular sequences
$x_{-n}'(\omega)x_{-n}''(\omega)$ is given by

$$(6.3) \quad \xi(n) = x_{-n}'(\omega)x_{-n}''(\omega) = \begin{cases} 1 & \text{if} \quad n_1 + \ldots + n_{2k} < n \leq n_1 + \ldots + n_{2k+1} \\[2ex] -1 & \text{if} \quad n_1 + \ldots + n_{2k-1} < n \leq n_1 + \ldots + n_{2k}. \end{cases}$$

The rapidness with which the n_k increase implies that $\xi(n)$ will retain the same value for longer and longer intervals in such a way that it cannot possess an average. Namely since $n_{k+1} > 2(n_1 + \ldots + n_k)$, the average of $\xi(n)$ from $-(n_1 + \ldots + n_{2k})$ to 0 is

$$\frac{1}{n_1 + \ldots + n_{2k} + 1}(-n_{2k} + n_{2k-1} - + \ldots + n_1)$$

$$< \frac{1}{n_1 + \ldots + n_{2k} + 1}(-n_{2k} + (n_1 + \ldots + n_{2k-1}))$$

$$< \frac{-n_{2k}}{2(n_1 + \ldots + n_{2k}) + 2} < \frac{-n_{2k}}{3n_{2k} + 2} \longrightarrow -\frac{1}{3}$$

and similarly the average from $-(n_1 + \ldots + n_{2k-1})$ to 0 is greater than

$$\frac{n_{2k-1}}{3n_{2k-1} + 2} \longrightarrow \frac{1}{3} \quad .$$

Thus the averages of $\xi(n)$ oscillate between $\pm 1/3$ and $\xi(n)$ cannot be stochastic. We also deduce that

$$\eta(n) = x'_{-n}(\omega) + x''_{-n}(\omega)$$

is not stochastic; for $\eta(n)^2 = 2 + 2\xi(n)$. Hence neither sums nor products of regular sequences need even be stochastic.

§7. Ergodic Properties of Regular Sequences

7.1. <u>Criteria for Ergodicity</u>. In this section we examine how the ergodicity properties of a process (§2.4) are reflected in the behavior of its generic points.

DEFINITION 7.1. A regular sequence $\xi(n)$ is <u>topologically ergodic, ergodic, or λ-ergodic</u> if $X(\xi)$ (Definition 4.3) is respectively <u>topologically ergodic, ergodic, or λ-ergodic</u> (Definitions 2.7, 2.8, and 2.9).

We consider now the question of characterizing these properties directly in terms of the individual sequences without passing to the corresponding process $X(\xi)$ first. This is done in the four theorems of this section.

DEFINITION 7.2. Let X be an abstract process and $\omega \in \bar{\Omega}_X$, Δ

any subset of Ω_X^-. The ω-_density_ of Δ is defined by $D_\omega(\Delta) = \bar{D}(S(\omega, \Delta))$
where $S(\omega, \Delta)$ is the set of integers n for which $T^n\omega \in \Delta$.

We now have:

LEMMA 7.1. If P is the probability measure on Ω_X^- and
ξ is a generic point of X then for Δ closed
$D_\xi(\Delta) \leq P(\Delta)$ and for Δ open $D_\xi(\Delta) \geq P(\Delta)$.

PROOF. If Δ is closed, there is a sequence of continuous
functions f_n on Ω_X^- with the f_n decreasing monotonically to x_Δ, the
characteristic function of Δ. We shall then have $E(f_n) \longrightarrow E(x_\Delta) = P(\Delta)$.
On the other hand

$$D_\xi(\Delta) = \lim_{N \to \infty} \sup \frac{1}{N+1} \sum_{n=0}^{N} x_\Delta(T^n\xi) \leq \lim_{N \to \infty} \frac{1}{N+1} \sum_{n=0}^{N} f_k(T^n\xi) = E(f_k)$$

so that

$$D_\xi(\Delta) \leq \lim_{k \to \infty} E(f_k) = P(\Delta) \quad .$$

If Δ is open then $D_\xi(\Delta \cup (\Omega_X^- - \Delta)) = 1$ where $\Omega_X^- - \Delta$ is closed. On
the other hand, it is clear that for disjoint sets Δ_1 and Δ_2,
$D_\xi(\Delta_1 \cup \Delta_2) \leq D_\xi(\Delta_1) + D_\xi(\Delta_2)$ so that $D_\xi(\Delta) + D_\xi(\Omega_X^- - \Delta) \geq 1$ and since
$D_\xi(\Omega_X^- - \Delta) \leq P(\Omega_X^- - \Delta) = 1 - P(\Delta)$, we conclude that $D_\xi(\Delta) \geq P(\Delta)$.

THEOREM 7.1. A regular sequence $\xi(n)$ is topologically
ergodic if and only if for every numerical derived sequence $\zeta(n)$ of $\xi(n)$ and every $\epsilon > 0$ there is an
integer N such that

(7.1) $\max (|\zeta(\ell)|, |\zeta(\ell - 1)|, \ldots, |\zeta(\ell - N)|) > \|\zeta\|_\infty - \epsilon$

for all ℓ but for a set of upper density $< \epsilon$.

PROOF. Suppose first that $\xi(n)$ is topologically ergodic; let
$\zeta(n)$ be a derived sequence and let $\epsilon > 0$. Let z denote the function
on $\Omega^-(\xi)$ corresponding to the sequence $\zeta(n)$ so that $\zeta(-n) = z(T^n\xi)$
(Equation (5.6)). For each $n > 0$ define

$\Delta_n = \{\omega \in \Omega^-(\xi) : \max (|z(\omega)|, |z(T\omega)|, \ldots, |z(T^n\omega)|) \leq \|\zeta\|_\infty - \epsilon\}$

and let

$$\Delta = \bigcap_{n > 0} \Delta_n \ .$$

Each Δ_n is closed so that Δ will also be closed. Moreover if $\omega \in \Delta_{n+1}$ then $T\omega \in \Delta_n$ so that $T\Delta_{n+1} \subset \Delta_n$ and $T\Delta \subset \Delta$. Now since the points $\{T^m\xi\}$ are dense in $\Omega^-(\xi)$ (Theorem 4.2), it is clear that none of the $T^m\xi$ belong to Δ; hence $\Delta \neq \Omega^-(\xi)$. If $X(\xi)$ is topologically ergodic, then $T\Delta \subset \Delta$ and $\Delta \neq \Omega^-(\xi)$ imply that Δ have measure 0. Now Δ is a monotone limit of sets $\Delta_{n+1} \subset \Delta_n$; hence there is an N with $P(\Delta_n) < \varepsilon$. By Lemma 7.1, $D_\xi(\Delta_N) \leq P(\Delta_N) < \varepsilon$ since Δ_N is closed. This N satisfies the requirements of the theorem since (7.1) will be violated only when $T^\ell\xi \in \Delta_N$ which can happen only for a set of ℓ with upper density $< \varepsilon$.

Now suppose that the condition of the theorem is fulfilled and that $\Delta \subset \Omega^-(\xi)$ is closed with $T\Delta \subset \Delta$. If $\Delta \neq \Omega^-(\xi)$ there exists a continuous function z on $\Omega^-(\xi)$ with maximum value 1 and vanishing on Δ. Suppose moreover that Δ does not have measure 0 so that $P(\Delta) = \alpha > 0$; from this we shall obtain a contradiction.

First of all, z corresponds to a sequence in $A_0(\xi)$ (§5.1) $z(T^n\xi) = \zeta(-n)$. Take $\varepsilon < \alpha$ and set

$$\Gamma = \{\omega \in \Omega^-(\xi) : \max (|z(\omega)|, |z(T\omega)|, \ldots, |z(T^N(\omega)| < 1 - \varepsilon\}$$

where N is such that $\max (|\zeta(\ell)|, |\zeta(\ell - 1)|, \ldots, |\zeta(\ell - N)|) > 1 - \varepsilon$ holds for all ℓ but for a set with upper density $< \varepsilon$. Since $z = 0$ on Δ and $T\Delta \subset \Delta$ it follows that $\Delta \subset \Gamma$ and hence $P(\Gamma) \geq \alpha$. But Γ is open so $D_\xi(\Gamma) \geq P(\Gamma) \geq \alpha$. But this would mean that the set of ℓ for which $\max (|\zeta(\ell)|, |\zeta(\ell - 1)|, \ldots, |\zeta(\ell - N)| < 1 - \varepsilon$ has upper density $\geq \alpha > \varepsilon$ which contradicts our choice of N.

THEOREM 7.2. A regular sequence $\xi(n)$ is ergodic if and only if for every numerical derived sequence $\zeta(n)$ of $\xi(n)$ and every $\varepsilon > 0$ there is an N such that

$$(7.2) \qquad \left| \frac{1}{N + 1} \sum_{j=0}^{N} \zeta(\ell - j) - E(\zeta) \right| < \varepsilon$$

for all ℓ outside of a set of upper density $< \varepsilon$.
($E(\zeta)$ is the average of the sequence $\zeta(n)$.)

PROOF. Suppose $\xi(n)$ is ergodic, $\zeta(n)$ a numerical derived sequence of $\xi(n)$ and let $z \in A^-(\xi)$ be the function on $\Omega^-(\xi)$ corresponding to $\zeta(n)$. Define

$$\Delta_n = \left\{ \omega \in \Omega^-(\xi) \; : \; \left| \frac{1}{n+1} \sum_{j=0}^{n} z(T^j \omega) - E(z) \right| \geq \varepsilon \right\}$$

and set

$$\Delta = \limsup_{n \to \infty} \Delta_n = \{ \omega : \omega \in \text{infinitely many } \Delta_n \} \quad .$$

If $\omega \in \Delta$ then ω is clearly not a generic point for the process; hence Δ has measure 0, assuming $X(\xi)$ is ergodic. Consequently $P(\Delta_n) \longrightarrow 0$ so that there exists an N for which $P(\Delta_N) < \varepsilon$, for a preassigned $\varepsilon > 0$. By Lemma 7.1 since Δ_N is closed, $D_\xi(\Delta_N) < \varepsilon$ and this means that the set of ℓ for which (7.2) does not hold has upper density $< \varepsilon$ as was to be shown.

Now assume that the condition of the theorem is fulfilled for all $\zeta(n)$ derived from $\xi(n)$. From this we may deduce that if $z \in A^-(\xi)$ there is a sequence $\{N_k\}$, $N_k \longrightarrow \infty$, such that

$$(7.3) \qquad \frac{1}{N_k+1} \sum_{j=0}^{N_k} T^j z \longrightarrow E(z)$$

in the norm of $L^2(A^-(\xi))$. For this choose N_k such that

$$\left| \frac{1}{N_k+1} \sum_{j=0}^{N_k} \zeta(\ell+j) - E(z) \right| < \frac{1}{k}$$

for ℓ outside a set of upper density $< \frac{1}{k}$ where $\zeta(n)$ is the sequence corresponding to z. Then

$$E\left(\left| \frac{1}{N_k+1} \sum_{j=0}^{N_k} T^j z - E(z) \right|^2 \right) = \lim_{N \to \infty} \frac{1}{N+1} \sum_{n=0}^{N} T^n \left\{ \left| \frac{1}{N_k+1} \sum_{j=0}^{N_k} T^j z - E(z) \right|^2 \right\} (\xi)$$

$$= \lim_{N \to \infty} \frac{1}{N+1} \sum_{n=0}^{\infty} \left\{ \left| \frac{1}{N_k+1} \sum_{j=0}^{N_k} \zeta(-n-j) - E(\zeta) \right|^2 \right\}$$

since $z(T^{n+j}\xi) = \zeta(-n-j)$. Now since

$$\left| \frac{1}{N_k + 1} \sum_{j=0}^{N_k} \zeta(-n-j) - E(\zeta) \right|^2 < \frac{1}{k^2}$$

but for a set of density $1/k$ at which it takes on at most the value $4\|z\|_\infty^2$ it follows that

$$E\left(\left| \frac{1}{N_k + 1} \sum_{j=0}^{N_k} T^j z - E(z) \right|^2 \right) \leq \frac{1}{k^2} + \frac{4}{k} \|z\|_\infty^2$$

so that as $k \longrightarrow \infty$, (7.3) holds. It is moreover easy to show that if (7.3) holds for a dense set of $z \in L^2(A^-(\xi))$ it holds for all of $L^2(A^-(\xi))$. Now suppose $X(\xi)$ were not ergodic so that there would exist a non-trivial measurable set Δ in $\Omega^-(\xi)$ with $T\Delta \subset \Delta$. The characteristic function of this set would satisfy $Tx_\Delta = x_\Delta$ almost everywhere. Taking $z = x_\Delta$ in (7.3) would give $x_\Delta = E(z)$, a constant which shows Δ must have measure 0 or 1. This completes the proof.

In Chapter 9 we shall need the following stronger form of the "only if" portion of the preceding theorem. We note that in the case of Theorem 7.1 the corresponding sharper form would have been trivial.

THEOREM 7.3. If $\xi(n)$ is an ergodic sequence and $\zeta(n)$ a numerical derived sequence of $\xi(n)$ then for any $\varepsilon > 0$ there exists an N_0 such that

$$\left| \frac{1}{N + 1} \sum_{j=0}^{N} \zeta(\ell - j) - E(\zeta) \right| < \varepsilon$$

for all $N > N_0$ and all ℓ outside a set of upper density $< \varepsilon$ and independent of N.

PROOF. We may suppose that ζ is real valued; by addition of an appropriate constant, to the terms $\zeta(n)$, the assertion of the theorem may be reduced to the following. If $E(\zeta) < 0$, there exists an N_0 such that for all $N > N_0$,

$$\sum_{j=0}^{N} \zeta(\ell - j) < \delta N$$

for all ℓ outside a set of upper density $< \varepsilon$. It will be convenient to consider this over the positive integers so that the desired inequality takes the form

$$\sum_{j=1}^{N} \zeta(\ell + j) < \delta N$$

for $\ell \geq 0$ and outside a set of upper density $< \varepsilon$. Suppose that we have $E(\zeta) = -\alpha < 0$ and that $\|\zeta\|_{\infty} = \beta$. We decompose the positive integers into two sets R and S defined by

$$R = \left\{ n : \sum_{j=0}^{k} \zeta(n + j) \geq 0 \quad \text{for some} \quad k \right\}$$

$$S = \left\{ n : \sum_{j=0}^{k} \zeta(n + j) < 0 \quad \text{for all} \quad k \right\} .$$

We observe that S must be infinite for if it were finite, then for sufficiently large n,

$$\sum_{j=n}^{n'} \zeta(j) \geq 0$$

for appropriate n' and we could not have

$$\lim_{N \to \infty} \frac{1}{N} \sum_{1}^{N} \zeta(j) = -\alpha \quad .$$

We may also assume R is infinite for otherwise there is nothing to prove. Now apply Theorem 7.2 to the sequence $\zeta(n)$; then for arbitrary δ_1, $\delta_2 > 0$ we can find an N_1 with

$$\frac{1}{N_1 + 1} \sum_{j=0}^{N_1} \zeta(\ell + j) > - (\alpha + \delta_1)$$

for all ℓ outside a set U of upper density $< \delta_2$. We now proceed to define a set $\bar{S}$ consisting of blocks of $N_1 + 1$ consecutive integers defined as follows. Let n_1 be the smallest element of S not in U (if no such n_1 exists then $\bar{S}$ will be empty); the block of integers $(n_1, n_1 + 1, \ldots, n_1 + N_1)$ is then taken to belong to $\bar{S}$. If n_2 is the least element of $S \geq n_1 + N_1 + 1$ but not in U then $(n_2, n_2 + 1, \ldots, n_2 + N_1)$ also belongs to $\bar{S}$. Proceeding in this manner we obtain the set $\bar{S}$ with the property that if $n \in S$ but $n \notin U$ then $n \in \bar{S}$. Now let N_k be a sequence of integers with $N_k \in S$ and $N_k \longrightarrow \infty$. This is possible because R is infinite. We also suppose that N_k is so large that the number of integers in U which are $< N_k$ is $< 2\delta_2 N_k$. Letting σ_k denote the number of integers of $\bar{S}$ that are $< N_k$ we find by decomposing these integers into blocks that

$$\sum_{\substack{n \in \bar{S} \\ n < N_k}} \zeta(n) > - (\alpha + \delta_1)\sigma_k - N_1(\beta - \alpha - \delta_1) \ .$$

Now the number of elements of S less than N_k that are not in $\bar{S}$ must be $\leq$ the number of elements of U that are $< N_k$ and hence

$$\sum_{\substack{n \in S \cup \bar{S} \\ n < N_k}} \zeta(n) > - (\alpha + \delta_1)\sigma_k - N_1(\beta - \alpha - \delta_1) - 2\delta_2 N_k \beta \ .$$

Now consider an integer $n < N_k$ that does not belong to $S \cup \bar{S}$. It is clear by the definition of $\bar{S}$ that the first integer to the right of n belonging to $S \cup \bar{S}$ is in S. Say this integer is n'. We contend that

$$\sum_{j=n}^{n'-1} \zeta(j) \geq 0 \ .$$

For suppose that the sum in question were < 0; there will exist a largest $n'' > n$ and $n'' < n'$ with

$$\sum_{j=n}^{n'-1} \zeta(j) \geq 0 \ .$$

Consider

$$\sum_{j=n''}^{m} \zeta(j) \ .$$

If $m < n'$ then this sum must be negative for otherwise n'' could have been chosen as m. If $m > n'$ the sum may be decomposed into

$$\sum_{j=n''}^{n'-1} \zeta(n) + \sum_{j=n'}^{m} \zeta(n)$$

the first of which has already been shown to be negative and the second of which is negative by the definition of S. Thus

$$\sum_{j=n''}^{m} \zeta(j)$$

is always negative so that $n'' \in S$ which contradicts $n'' < n'$. As a result we must have

$$\sum_{j=n}^{n'-1} \zeta(j) \geq 0 \quad .$$

It follows that the $\zeta(n)$ for $n \notin S \cup \bar{S}$ and $n < N_k$ are decomposed into blocks of consecutive terms whose sums are ≥ 0. Hence

$$\sum_{n=1}^{N_k-1} \zeta(n) > -(\alpha + \delta_1)\sigma_k - N_1(\beta - \alpha - \delta_1) - 2\delta_2 N_k \beta \quad .$$

Dividing by N_k and letting $k \longrightarrow \infty$ we find that

$$-\alpha \geq -(\alpha + \delta_1) \liminf_{k \to \infty} \frac{\sigma_k}{N_k} - 2\delta_2 \beta \quad .$$

Since δ_1 and δ_2 could be chosen arbitrarily small we find that there is an N_1 and a corresponding $\bar{S}$ with

$$\liminf_{k \to \infty} \frac{\sigma_k}{N_k} > 1 - \epsilon$$

so that the upper density (along the sequence N_k) of the integers not in $\bar{S}$ is $< \epsilon$. To show that this implies that the ordinary upper density (i.e., not only for N_k) of the complement of $\bar{S}$ is $< \epsilon$ we must show that we can choose N_k in S with $N_{k+1}/N_k \longrightarrow 1$. Now if this were not the case it would follow that $\zeta(n)$ could not be stochastic. For suppose that for a large N_k, $N_{k+1} > (1 + \delta')N_k$ where N_{k+1} is the next integer in S after N_k. As we have shown

$$\sum_{n=N_k+1}^{N_{k+1}-1} \zeta(n) \geq 0$$

so that

$$\sum_{n=1}^{N_{k+1}-1} \zeta(n) \geq \sum_{n=1}^{N_k} \zeta(n) \sim -\alpha N_k > \frac{-\alpha}{1+\delta'} N_{k+1}$$

which could not be possible for large k. We conclude then that the complement of $\bar{S}$ has ordinary upper density $< \varepsilon$. To complete the proof of the theorem we shall show that there is an N_o such that for $\ell \in \bar{S}$,

$$\sum_{j=1}^{N} \zeta(\ell + j) < \delta N \quad .$$

For this we recall that by the definition of $\bar{S}$, if $\ell \in \bar{S}$ then $\ell - m \in S$ for some $m \leq N_1$. But then

$$\sum_{j=-m}^{N} \zeta(\ell + j) < 0$$

and

$$\sum_{j=1}^{N} \zeta(\ell + j) < N_1 \beta$$

so that for the desired inequality we need only take $N_o > N_1 \beta / \delta$.

Finally we consider λ-ergodicity.

THEOREM 7.4. A regular sequence $\xi(n)$ is λ-ergodic if and only if for every numerical derived sequence $\zeta(n)$ of $\xi(n)$ and $\varepsilon > 0$ there is an N such that

$$(7.4) \qquad \left| \frac{1}{N'+1} \sum_{j=0}^{N} e^{-2\pi i j \lambda} \zeta(\ell - j) \right| < \varepsilon$$

for all ℓ outside of a set of upper density $< \varepsilon$.

PROOF. The proof of this theorem proceeds exactly as the proof of Theorem 7.2 once two facts are established. The first is that if $X(\xi)$ is λ-ergodic then for almost all $\omega \in \Omega^-(\xi)$

$$(7.5) \qquad \lim_{N \to \infty} \frac{1}{N+1} \sum_{j=0}^{N} e^{-2\pi i j \lambda} T^j z(\omega) = 0 \quad .$$

The second fact is that, conversely, if for every $z \in L^2(A^-(\xi))$

$$(7.6) \qquad \lim \frac{1}{N_k + 1} \sum_{j=0}^{N_k} e^{-2\pi i j \lambda} T^j z = 0$$

in the sense of weak convergence in $L^2(A^-(\xi))$ for an appropriate sequence N_k, then $X(\xi)$ is λ-ergodic. The first follows from the fact ([3], p. 469) that the limit in (7.5) does exist almost everywhere and if this limit is w we have $Tw = e^{2\pi i \lambda} w$. By λ-ergodicity $w = 0$. The second assertion is more elementary since a function satisfying $Tw = e^{2\pi i \lambda} w$ has to be 0 if (7.6) holds with $z = w$. The details of the proof now follow as for ordinary ergodicity.

It might be pointed out that the criterion of Theorem 7.2 for ergodicity is, in a sense, a refinement of the condition that a sequence be stochastic. For whereas the latter merely requires the existence of an average for every derived sequence; ergodicity requires some uniformity in the approach to this average. This observation may help clarify the role of ergodicity in some of our results.

Finally, we remark that the theorems just proven have the following corollary:

> COROLLARY. A uniform limit of topologically ergodic, ergodic or λ-ergodic sequences is again respectively topologically ergodic, ergodic or λ-ergodic.

7.2. _Examples_. Let us call a sequence a "random sequence" if it is generic for a random process, i.e., a process with independent variables. A random sequence is both ergodic, (and hence topologically ergodic) and λ-ergodic for $0 < \lambda < 1$. Both of these follow from the fact that if X is a random process, then for a dense subset of A_X^-, the variables z and $T^n z$ will be independent for sufficiently large n. On the other hand if $Tn = e^{2\pi i \lambda} u$, $0 \leq \lambda < 1$, then all translates of n are proportional to u, and this implies that u is constant if $\lambda = 0$ and $u = 0$ for $0 < \lambda < 1$, so that X is both ergodic and λ-ergodic.

If $\xi(n)$ is an almost periodic sequence then $\xi(n)$ is ergodic. To see this we apply the criterion of 7.2 and observe that a derived sequence of an almost periodic sequence is again almost periodic. Now if $\zeta(n)$ is almost periodic, the proof of the existence of an average for $\zeta(n)$ shows that this average is approached uniformly, i.e.,

$$\frac{1}{N + 1} \sum_{j=0}^{N} \zeta(\ell - j) \longrightarrow E(\zeta)$$

uniformly in ℓ . It follows that for N sufficiently large the left hand side of (7.2) may be made arbitrarily small, for all ℓ without any exceptional set.

On the other hand a non-constant almost periodic sequence cannot be λ-ergodic for every λ . For it follows readily from the criterion of Theorem 7.4 that if $\xi(n)$ is λ-ergodic then for any ℓ

$$(7.7) \qquad \lim_{N \to \infty} \frac{1}{N + 1} \sum_{j=0}^{N} e^{-2\pi i j \lambda} \zeta(\ell - j) = 0$$

where $\zeta(n)$ is a derived sequence of $\xi(n)$. However if we consider (7.7) for $\zeta(n) = \xi(n)$ this implies that the Bohr-Fourier coefficient of $\xi(n)$ corresponding to $e^{2\pi i \lambda}$ vanishes. Since this cannot happen for every $\lambda \neq 0$ unless $\xi(n)$ is a constant our assertion follows.

It is possible to construct sequences that are not ergodic by a method of superposition of two sequences that are generic for different processes. To make this precise let us consider a partition of the negative integers into infinitely many blocks of consecutive integers: ..., I_{-n}, ..., I_{-2}, I_{-1}, I_0 where the intervals I_n are arranged according to the natural order of the integers. Suppose that I_{-n} has k_n integers and that

$$(i) \qquad\qquad k_n \longrightarrow \infty$$

$$(ii) \qquad\qquad \frac{k_{2n}}{k_{2n+1}} \longrightarrow \frac{\alpha}{\beta}$$

$$(iii) \qquad\qquad \frac{k_{2n}}{k_{2n-1}} \longrightarrow \frac{\alpha}{\beta}$$

where $\alpha + \beta = 1$, α, $\beta \neq 0$.

Now let $\xi_1(n)$ and $\xi_2(n)$ be two regular sequences; we shall form a new sequence $\xi(n)$ from these by means of the partition $\{I_{-n}\}$. Namely, the last k_0 terms of $\xi_2(n)$ are placed in order in the positions of I_0, the next k_2 terms of $\xi_2(n)$ are placed in the positions of I_{-2}, the next k_4 in I_{-4} and so on. The last k_1 terms of $\xi_1(n)$ are placed in order in the positions of I_{-1}, the next k_3 in the positions of I_{-3} and so on till all the I_{-2n} are occupied by the entries of $\xi_2(n)$, and the I_{-2n+1} by the entries of $\xi_1(n)$. Conditions (ii) and (iii) now imply that the entries of $\xi_2(n)$ and of $\xi_1(n)$ occur in the ratio $\alpha : \beta$. From this one can show the following: If $\xi_1(n)$ and $\xi_2(n)$ are both stochastic Λ-sequences then the new sequence, $\xi(n)$, will again be stochastic. If $\varphi \in C(\Lambda_\infty^-)$ and $\zeta(n)$, $\zeta_1(n)$, $\zeta_2(n)$ are the derived sequences of $\xi(n)$, $\xi_1(n)$, $\xi_2(n)$ corresponding to φ then the averages of ζ, ζ_1, ζ_2 are related by

$$(7.8) \qquad\qquad E(\zeta) = \alpha E(\zeta_2) + \beta E(\zeta_1) \quad .$$

The sequence $\xi(n)$ defined here will be termed an (α, β) superposition of $\xi_1(n)$ and $\xi_2(n)$.

Suppose that $\xi_1(n)$ is a random $\{-1,1\}$-sequence, i.e., a regular sequence generic for a random process $X(p_1, q_1)$ (§6.2), and suppose that $\xi_2(n)$ is also a random $\{-1, 1\}$-sequence, but generic for $X(p_2, q_2)$ with $p_1 \neq p_2$. Let $\xi(n)$ be an (α, β) superposition of $\xi_1(n)$ and $\xi_2(n)$ and consider the process $X(\xi)$. The sample space of $X^-(\xi)$ may be imbedded in $\{-1, 1\}_\infty^-$ while for $\xi_1(n)$ and $\xi_2(n)$ we shall have $\Omega^-(\xi_1) = \Omega^-(\xi_2) = \{-1, 1\}_\infty^-$ (§6.2). If P_1 and P_2 represent the probability measures for $X(\xi_1)$ and $X(\xi_2)$ respectively and P is the probability measure for $X(\xi)$, we find from (7.8) that

$$(7.9) \qquad\qquad P(\Delta) = \alpha P_2(\Delta) + \beta P_1(\Delta)$$

for any borel set $\Delta \subset \{-1, 1\}_\infty^-$. In particular $\Omega^-(\xi)$ must be all of $\{-1, 1\}_\infty^-$. From (7.9) it is clear that $X(\xi)$ is not ergodic. For P_1 and P_2 are not identical and hence the points in $\{-1, 1\}_\infty^-$ generic for $X(\xi_1)$ will be disjoint from those generic for $X(\xi_2)$. Referring to these two sets as Δ_1 and Δ_2 we have from (7.9), $P(\Delta_1) = \beta$, $P(\Delta_2) = \alpha$. But clearly Δ_1 and Δ_2 are invariant under T so that there exist non-trivial invariant subsets for $X(\xi)$, and $X(\xi)$ will not be an ergodic process. As regards the sequence $\xi(n)$ itself we can deduce that $\xi(n)$ is regular from the fact that it is stochastic and that $\Omega^-(\xi) = \{-1, 1\}_\infty^-$ whence $\xi \in \Omega^-(\xi)$. We thus have constructed a regular sequence that is not ergodic.

The sequence constructed above is however still topologically ergodic, as may easily be verified. A non-topologically-ergodic sequence may also be found by forming an (α, β) superposition of a random $\{-1, 1\}$ sequence with a random $\{-1, 0, 1\}$-sequence. That such a sequence cannot be topologically-ergodic may be verified either directly from the sequence, using Theorem 7.1 or by an analysis of the corresponding process and its sample space.

CHAPTER 2. THE PREDICTION PROBLEMS FOR SEQUENCES

The problem of prediction to be treated here, as we have stated
in the Introduction, is that of assigning to a sequence representing the
past behavior of some variables, a probability measure on the set of all
possible futures of the variable. Our procedure at first will be to set
down the basic properties that such a probability measure would be expected to possess. It will appear that in certain cases these properties
uniquely determine the unknown measure; for such sequences the problem of
prediction becomes trivial, in a sense. These "trivially" predictable
sequences will be the object of our study in this and the succeeding three
chapters. Subsequently we shall consider how a measure may be chosen for
prediction when more than one such measure is admissible.

§8. The Support of the Prediction Measure

8.1. <u>L-sequences</u>. Let $\xi(n)$ denote the given left-infinite
sequence. Each possible future of $\xi(n)$ is a right-infinite sequence;
however we shall think of it rather as a doubly-infinite sequence extending $\xi(n)$. Thus the set on which the prediction measure will have to
be defined is the set of all doubly-infinite sequences extending $\xi(n)$.

Our first step will be to make an assumption restricting the set
of doubly-infinite sequences that will be admitted to extend $\xi(n)$. Such
an assumption amounts to restricting the support of an allowable prediction measure. The purpose of the present restriction is to exclude
doubly-infinite sequences in which the events of the future cannot be
approximated by events of the past. In particular it guarantees that
the future of a Λ-sequence will be a Λ-sequence.

DEFINITION 8.1. A doubly-infinite Λ-sequence $\xi(n)$ is an
L-<u>sequence</u> if for every numerical derived sequence $\zeta(n)$ of $\xi(n)$
(Definition 3.3)

$$(8.1) \qquad \sup_{n \leq o} \{|\zeta(n)|\} = \sup_{-\infty < n < \infty} \{|\zeta(n)|\} \quad .$$

The initial "L" in this definition is intended to call to mind the theorem of Littlewood ([9]) to the effect that an n-body system confined to a fixed region in space throughout the past will, with probability 1, remain in that region in the future. The property of being an L-sequence is an extension of this and may be described as the requirement that whatever has been true throughout the entire past remains true in the future. We shall see presently that Littlewood's theorem extends to the statement that almost all sample sequences of a two-sided process are L-sequences. As a result it appears plausible to allow only L-sequences as the doubly-infinite sequences extending $\xi(n)$.

If the left-infinite sequence $\xi(n)$ is a regular sequence then restricting its extensions to L-sequences has a particular significance:

> THEOREM 8.1. Let $\xi(n)$ be a regular left-infinite
> Λ-sequence and suppose that $\tilde{\xi}(n)$ in Λ_∞ extends
> $\xi(n) : \beta(\tilde{\xi}) = \xi$. Then $\tilde{\xi}$ is an L-sequence if and only
> if $\tilde{\xi} \in \Omega(\xi)$.

PROOF. Suppose first that $\tilde{\xi} \in \Omega(\xi)$. By Theorem 2.4 this implies that $\beta(T^{-m}\tilde{\xi}) \in \Omega^-(\xi)$ for all m. Now let $\zeta(n)$ be a derived sequence of $\tilde{\xi}(n)$. For each m, $T^{-m}\zeta$ will be a derived sequence of $T^{-m}\tilde{\xi}$ and $\beta(T^{-m}\zeta)$ will be a derived sequence of $\beta(T^{-m}\tilde{\xi}) \in \Omega^-(\xi)$. In each case the derived sequence corresponds to the same function φ on Λ_∞^-. It follows then that the entries of the sequence $\beta(T^{-m}\zeta)$ are values of φ at points of $\Omega^-(\xi)$ and hence the supremum of the absolute values of these entries is dominated by the supremum of $|\varphi|$ on $\Omega^-(\xi)$. Since the translates $T^n\xi$ are dense in $\Omega^-(\xi)$ (Theorem 4.2), the latter agrees with the supremum of $\{|\varphi(T^n\xi)|\} = \{|\zeta(-n)|\}$. What we have just shown is that

$$(8.2) \qquad \sup_{n \leq 0} \{|\beta(T^{-m}\zeta)(n)|\} \leq \sup_{n \leq 0} \{|\zeta(n)|\}$$

or

$$(8.3) \qquad \sup_{n \leq m} \{|\zeta(n)|\} \leq \sup_{n \leq 0} \{|\zeta(n)|\}$$

and since the right hand side is evidently dominated by the left hand side if $m \geq 0$, we must have equality. Letting $m \longrightarrow \infty$ we obtain (8.1).

Conversely suppose that $\tilde{\xi}(n)$ is an L-sequence, extending $\xi(n)$. To show that $\tilde{\xi} \in \Omega(\xi)$ we apply the criterion of Theorem 2.4, namely, we show that for all $m \geq 0$, $\beta(T^{-m}\tilde{\xi}) \in \Omega^-(\xi)$. In doing so we shall make use of the metric on Λ which we denote by $d(., .)$. In order for

$\beta(T^{-m}\xi)$ to belong to $\Omega^{-}(\xi)$ it suffices that the translates $T^{\ell}\xi$ come arbitrarily close to $\beta(T^{-m}\tilde{\xi})$. For this we will show that the distances $d(T^{\ell}\xi(j),\ \beta(T^{-m}\tilde{\xi})(j))$ can be made arbitrarily small for any finite set of j, by choosing ℓ appropriately. Namely, let $\zeta(n)$ be the derived sequence of $\tilde{\xi}(n)$ defined by

$$(8.4) \qquad \zeta(n) = \left\{ 1 + \sum_{j=-k}^{0} d(\tilde{\xi}(j+n),\ \beta(T^{-m}\tilde{\xi})(j)) \right\}^{-1} .$$

Since $\beta(T^{-m}\tilde{\xi})(j) = \tilde{\xi}(j+m)$ it follows that when $n = m$, $\zeta(n) = 1$. In general $\zeta(n) \leq 1$ so that

$$\sup_{-\infty < n < \infty} |\zeta(n)| = 1 .$$

Since $\tilde{\xi}(n)$ is by assumption an L-sequence we must also have

$$\sup_{n \leq 0} |\zeta(n)| = 1$$

and this is equivalent to

$$\inf_{n \leq 0} \sum_{j=-k}^{0} d(\tilde{\xi}(j+n),\ \beta(T^{-m}\tilde{\xi})(j)) = 0$$

or

$$\inf_{\ell \geq 0} \sum_{j=-k}^{0} d(T^{\ell}\xi(j),\ \beta(T^{-m}\tilde{\xi})(j)) = 0$$

which implies that the $T^{\ell}\xi(j)$ for $j = -k,\ -k+1,\ \ldots,\ -1,\ 0$ are simultaneously close to $\beta(T^{-m}\tilde{\xi})(j)$, as was to be shown. This proves the theorem.

Since we shall be concerned only with the prediction problem for regular sequences, the theorem just proved shows that in studying the "possible" extensions of the given sequence, we need only consider those that lie in the two-sided sample space, $\Omega(\xi)$, determined by the given left-infinite sequence $\xi(n)$. Another way of looking at the matter is to regard the points of $\Omega^{-}(\xi)$ as homomorphisms of $A^{-}(\xi)$ onto the complex numbers and the points of $\Omega(\xi)$ as homomorphisms of the larger algebra $A(\xi)$. The "possible" futures of the given regular sequence $\xi(n)$ correspond to the extensions to $A(\xi)$ of a given homomorphism of $A^{-}(\xi)$. Our problem will be to determine a suitable measure on the set of extensions of this homomorphism. (Recall that by §1.1 such extensions always exist.)

In this form the sequential nature of the spaces $\Omega^-(\xi)$, $\Omega(\xi)$ may be ignored and, in particular, the problem of prediction applies to abstract processes as well, the known data being, in this case, in the form of a homomorphism of the one-sided algebra onto the complex numbers.

8.2. <u>Littlewood's Theorem</u>. We now proceed to prove the analogue of Littlewood's Theorem regarding the n-body problem.

THEOREM 8.2. Almost all sample sequences of a two-sided process are L-sequences.

PROOF. Let X be the process in question and $\{x_n\}$ its variables. In order that $\{x_n(\omega)\}$ be an L-sequence (8.1) must hold for any derived sequence $\zeta(n)$ of $\xi(n) = x_n(\omega)$. Suppose then that $\zeta(n) = \varphi(\ldots, x_{n-2}(\omega), x_{n-1}(\omega), x_n(\omega))$; we may therefore write $\zeta(-n) = z(T^n\omega)$ where $z = \varphi(\ldots, x_{-2}, x_{-1}, x_0)$ and $\varphi \in C(\Lambda_\infty^-)$, and Λ is the range of the x_n. Equation (8.1) then reads

$$(8.5) \qquad \sup_{n \geq 0} |z(T^n\omega)| = \sup_{-\infty < n < \infty} |z(T^n\omega)| \quad .$$

For a particular ω, if (8.5) has been verified for a dense set of $z \in A_X^-$, it will follow for all of A_X^-. Hence to show that for almost all ω, (8.5) holds for all $z \in A_X^-$, it suffices to show that for a particular z, (8.5) holds for almost all ω, A_X^- being separable.

To prove this define the sets $C_{r,\varepsilon}$ for $\varepsilon > 0$ by

$$(8.6) \quad C_{r,\varepsilon} = \{\omega : \sup_{-r+1 \leq n < \infty} |z(T^n\omega)| < \sup_{-\infty < n < \infty} |z(T^n\omega)| - \varepsilon < |z(T^{-r}\omega)|\}$$

It is clear that $TC_{r,\varepsilon} = C_{r+1,\varepsilon}$ from which it follows that $P(C_{r,\varepsilon}) = P(C_{r+1,\varepsilon})$. On the other hand the $C_{r,\varepsilon}$ are disjoint for different r; hence $P(C_{r,\varepsilon}) = 0$ and also

$$P\left(\bigcup_{r=-\infty}^{\infty} C_{r,\varepsilon}\right) = 0 \quad .$$

Now if

$$\sup_{n \geq 0} |z(T^n\omega)| < \sup_{-\infty < n < \infty} |z(T^n\omega)| - \varepsilon \quad ,$$

then since

$$|z(T^{-r}\omega)| > \sup_{-\infty < n < \infty} |z(T^n\omega)| - \varepsilon$$

for some r, it follows that ω lies in some $C_{r,\varepsilon}$. Thus the set of ω for which the two sides of (8.5) differ by more than ε has measure 0 and hence (8.5) holds almost everywhere.

§9. Deterministic Sequences

9.1. <u>Definition</u>. In the preceding section we restated the problem of prediction for a regular sequence as that of finding a probability measure on the set of all extensions to $\Omega(\xi)$ of a sequence in $\Omega^-(\xi)$. Without assuming any further properties of the prediction measure we cannot expect to determine this measure for a wide class of sequences. However there are cases for which what has been said so far is already sufficient to solve this prediction problem. Namely, it may occur that the given point ξ in $\Omega^-(\xi)$ extends in only one way to $\Omega(\xi)$. It is natural to refer to this situation as the deterministic case.

DEFINITION 9.1. A regular sequence $\xi(n)$ is <u>deterministic</u> if the homomorphism ξ of $A^-(\xi)$ has a unique extension to $A(\xi)$.

In particular if $A^-(\xi) = A(\xi)$ then any homomorphism of $A^-(\xi)$ is already a homomorphism of $A(\xi)$ and $\xi(n)$ must be deterministic. For example in §4.3 it was shown that the left half of an almost periodic sequence satisfies $A^-(\xi) = A(\xi)$; hence such a sequence is deterministic. This sharpens a theorem of Bohr which asserts that two almost periodic sequences (in his case, functions) that agree for negative values are identical. What we have shown is that not only does the past of an almost periodic sequence determine a unique almost periodic sequence, but also a unique L-sequence.

There are examples of deterministic sequences for which $A^-(\xi)$ is a proper subalgebra of $A(\xi)$. For example let

$$\xi(n) = \text{sgn} \ (\cos \ (n\alpha + \beta)) \qquad\qquad n \leq 0$$

where $\text{sgn} \ x = x/|x|$ for $x \neq 0$, α is an irrational multiple of π and β is such that $\cos(n\alpha + \beta)$ does not vanish for any integer n. In this case $A^-(\xi)$ and $A(\xi)$ may be explicitly given, and it may be shown that the point ξ in $\Omega^-(\xi)$ has only one extension in $\Omega(\xi)$ — hence $\xi(n)$ is deterministic — however there are points in $\Omega^-(\xi)$ with two extensions in $\Omega(\xi)$ so that $A^-(\xi) \neq A(\xi)$.

If $\xi(n)$ is a deterministic sequence then theoretically its future may be determined from the past. There is however also an effective procedure for evaluating the future values directly from the sequence. We give the details for the case that Λ is discrete:

THEOREM 9.1. If $\xi(n)$ is a Λ-valued deterministic
sequence and Λ is discrete then there exists an integer N such that $\xi(1)$ will be determined uniquely
by the condition that whenever $(\xi(\ell - N), \xi(\ell - N + 1),$
$\ldots, \xi(\ell)) = (\xi(- N), \xi(- N + 1), \ldots, \xi(0))$ then
$\xi(1) = \xi(\ell + 1)$.

In other words the block of entries $(\xi(- N), \xi(- N + 1),$
$\ldots, \xi(0))$, which occurs infinitely often in $\xi(n)$
by the requirement of regularity, will always be followed
by the same element of Λ and this element will have to
be $\xi(1)$.

PROOF. It is only necessary to prove that for some N the sequence $(\xi(- N), \xi(- N + 1), \ldots, \xi(0)$ is always followed by the same
element, since once this is shown, the condition that the extension of
$\xi(n)$ be an L-sequence guarantees that $\xi(1)$ takes on this value. Assume
then that no such N exists. We will show that $\xi(n)$ is not determin-
istic by exhibiting two distinct points in $\Omega^{-}(\xi)$, ω_1 and ω_2, with the
property that $T\omega_1 = T\omega_2 = \xi$. Then ω_1 and ω_2 have preimages η_1 and
η_2 in $\Omega(\xi)$, i.e., $\beta(\eta_1) = \omega_1$, $\beta(\eta_2) = \omega_2$, and since $\omega_1 \neq \omega_2$, $\eta_1 \neq \eta_2$
and also $T\eta_1 \neq T\eta_2$ since T is 1 - 1 on $\Omega(\xi)$. But $\beta(T\eta_1) = T\beta(\eta_1) =$
$T\omega_1 = \xi$ and $\beta(T\eta_2) = \xi$ so that ξ will have two extensions in $\Omega(\xi)$.

We recall that the topology of Λ_{∞}^{-} is given in terms of pointwise convergence so that for Λ discrete we shall have $\omega^k \longrightarrow \omega$ if
and only if $\omega^k(\ell) = \omega(\ell)$ for each ℓ, for sufficiently large k. Now,
if no N exists with the required property there will be arbitrarily large
blocks $(\xi(- N), \xi(- N + 1), \ldots, \xi(0))$ occurring in the past of $\xi(n)$
followed sometimes by one and sometimes by another element of Λ. Since
anything that occurs once does so infinitely often, we can find two ele-
ments of Λ, say $\underline{a}$ and $\underline{b}$ such that arbitrarily large blocks will occur
followed sometimes by $\underline{a}$ and sometimes by $\underline{b}$. If ω is a Λ-sequence and
$\lambda \in \Lambda$, let (ω, λ) denote the sequence $\ldots, \omega(- 1), \omega(0), \lambda$. What we
have just said implies that there is a sequence of $T^{m_1}\xi$ converging to
$(\xi, \underline{a})$ and a sequence $T^{n_1}\xi$ converging to $(\xi, \underline{b})$. Hence $(\xi, \underline{a})$ and
$(\xi, \underline{b}) \in \Omega^{-}(\xi)$ and these have the properties claimed for ω_1 and ω_2
earlier. This proves the theorem.

9.2. <u>Construction of Deterministic Sequences</u>. The following
theorem shows that the class of deterministic sequences is quite extensive.

THEOREM 9.2. Any regular sequence $\xi(n)$ is a derived
sequence of some deterministic sequence $\xi^{*}(n)$.

PROOF. Suppose that $\xi(n)$ is a Λ-sequence; the space $\Omega(\xi)$
may then be imbedded in Λ_∞. Define Λ^* as the space of all right-infinite Λ-sequences with the property that they extend to doubly-infinite
Λ-sequences in $\Omega(\xi)$. Since the shift operator T is invertible in $\Omega(\xi)$
it follows that Λ^* is closed under the operation S of translating sequences to the left. Now let $\tilde{\xi}(n)$ be any extension of $\xi(n)$ in $\Omega(\xi)$,
i.e., $\tilde{\xi}(n) = \xi(n)$ for $n \leq 0$. From $\tilde{\xi}(n)$ we form the Λ^*-sequence

$$(9.1) \qquad \xi^*(n) = (\tilde{\xi}(n), \tilde{\xi}(n + 1), \ldots) \qquad n \leq 0 \quad .$$

Note that $\xi^*(n + 1) = S(\xi^*(n))$ so that if $\xi^*(n)$ is regular it is
plausible that it will be deterministic. In any case $\xi(n)$ is clearly a
derived sequence of $\xi^*(n)$.

To show that $\xi^*(n)$ is regular we remark that to do so it evidently suffices to show that every numerical derived sequence of $\xi^*(n)$
is regular. Moreover by Theorem 5.2 it suffices to show this for a dense
subset of these derived sequences. Denote by $\lambda^* = (\lambda_1, \lambda_2, \ldots)$ a
typical point of Λ^*. We obtain a dense set of derived sequences by taking
functions φ on Λ_∞^{*-} depending only on finitely many coordinates, and
for each coordinate λ^* depending only on finitely many λ_j, $j = 1, 2, 3,$
$\ldots$. Applying a function φ of this type to $\xi^*(n)$, we shall obtain a
derived sequence which is also a derived sequence of some

$$\xi' = (\ldots, \tilde{\xi}(k - 2), \tilde{\xi}(k - 1), \tilde{\xi}(k)) \quad .$$

It suffices therefore to prove that ξ' is regular. If $k \leq 0$, then
$\xi'(n)$ is a derived sequence of $\xi(n)$, hence is itself regular. If $k > 0$,
$\xi'(n)$ consists of finitely many entries adjoined to $\xi(n)$ so it will at
least be stochastic. Moreover $X(\xi') = X(\xi)$ so that in order to prove
that $\xi'(n)$ is regular it suffices by Theorem 5.1 to show that $\xi' \in \Omega^-(\xi)$.
But $\xi' = \beta(T^{-k}\tilde{\xi})$ and since $\tilde{\xi} \in \Omega(\xi)$ it follows that $\xi' \in \Omega^-(\xi)$ and
hence $\xi'(n)$ and therefore also $\xi^*(n)$ are regular.

To see that $\xi^*(n)$ is deterministic we remark that the fact
that $\xi^*(n + 1) = S(\xi^*(n))$ for all n implies that for all $\omega^* \in \Omega^-(\xi^*)$
and for all n, $\omega^*(n + 1) = S(\omega^*(n))$. Now if $\tilde{\xi}^*(n)$ is an extension
of $\xi^*(n)$ in $\Omega(\xi^*)$ then $\beta(T\tilde{\xi}^*) \in \Omega^-(\xi^*)$ and so $\tilde{\xi}^*(1) = S(\tilde{\xi}^*(0)) =$
$S(\xi^*(0))$ since $\tilde{\xi}^*(n)$ agrees with $\xi^*(n)$ for $n \leq 0$. It follows that
$\tilde{\xi}^*(1)$ is uniquely determined by $\xi^*(n)$ and similarly all the $\tilde{\xi}^*(n)$ are
determined uniquely so that $\xi^*(n)$ is deterministic.

§10.   Prediction Measures and Continuous Predictability

10.1.   <u>Prediction Measures</u>.   To widen the class of "predictable" sequences we shall have to introduce other hypotheses regarding the prediction measure. The assumption made so far, that the doubly-infinite sequences to be considered should be L-sequences, has served to restrict the support of this measure to a subset of $\Omega(\xi)$. In this section we make use of the probability measure induced by $\xi(n)$ on $\Omega(\xi)$ to further restrict the nature of the prediction measure. Essentially our assumption will be that the prediction measure behaves "something like" a conditional probability measure determined by the original probability measure on $\Omega(\xi)$ conditioned by the event that ξ has occurred (or, the conditional probability with respect to the set $\beta^{-1}(\xi) \subset \Omega(\xi)$, β being the canonical map from $\Omega(\xi)$ to $\Omega^-(\xi)$). Since the event $\{\xi\}$ (or the subset $\beta^{-1}(\xi)$) has probability 0 in general we cannot speak strictly of a conditional probability and what we shall do is to require that the prediction measure behave in a manner similar to the ordinary conditional probabilities. Stated precisely, we require that if the conditional expectation of a function on $\Omega(\xi)$ with respect to every set of non-zero measure that is "close to" $\beta^{-1}(\xi)$ is non-negative, then the prediction measure should also assign a non-negative value to the function.

In the above condition we make use only of the E-algebra structure of $A^-(\xi)$ and $A(\xi)$ without reference to the transformation T and the variables x_n of the process $X(\xi)$. For this reason we shall state the formal definition of a prediction measure in terms of an arbitrary pair of E-algebras, A and B, with $A \subset B$. This will be useful because we shall later require a notion of predictability from one algebra to another when the algebras do not necessarily correspond to the past and future of a process.

With the E-algebras A and B, $A \subset B$, we consider the homomorphism spaces Ω_A and Ω_B and the induced map β of Ω_B onto Ω_A . As usual the inclusion $A \subset B$ is intended to imply that the expectation functional on A is the restriction of the one on B, or, in terms of the probability measures $P_A(\Delta) = P_B(\beta^{-1}(\Delta))$, $\Delta \subset \Omega_A$. We therefore denote both P_A and P_B as P. In what follows we shall assume that A and B are separable, so that Ω_A and Ω_B are metrizable and the topologies of Ω_A and Ω_B may be described in terms of convergence of sequences.

The condition given for prediction measures can be stated as follows:

DEFINITION 10.1.   If $\xi \in \Omega_A$ and μ is a probability measure on Ω_B we say that μ is a <u>prediction measure at</u> ξ if whenever $g \in B$ has the property that for some neighborhood $U \subset \Omega_A$ of the point ξ, any non-negative function $f \in A$ and having support in U satisfies $E(fg) \geq 0$, then $\mu(g) \geq 0$.

Two consequences of this definition are given in the following lemma:

> LEMMA 10.1. If μ is a prediction measure at ξ,
> then the support of μ is contained in $\beta^{-1}(\xi)$ and
> for elements $g \in A$, $\mu(g) = g(\xi)$.

PROOF. For the first assertion of the lemma, let g' be a non-negative function in B vanishing on the closed set $\beta^{-1}(\xi)$. We must show that $\mu(g') = 0$. For arbitrarily small $\varepsilon > 0$ there will be a neighborhood U of ξ in Ω_A such that $g = \varepsilon - g'$ is non-negative in $\beta^{-1}(U)$. For otherwise we would have $g'(\omega) > \varepsilon$ for some ω in $\beta^{-1}(U)$ for arbitrarily small U and we could then find an ω with $\beta(\omega) = \xi$ for which $g'(\omega) \geq \varepsilon$ which is impossible since $g'(\omega) = 0$. Since $g \geq 0$ in U the hypotheses of Definition 10.1 are obviously satisfied for g and U so that if μ is a prediction measure, $\mu(g) \geq 0$ or $\mu(g') \leq \varepsilon$. ε being arbitrary this implies $\mu(g') = 0$ as was to be shown.

Now suppose $g \in A$ and consider $g' = g - g(\xi)$. g' vanishes on $\beta^{-1}(\xi)$ since for $\omega \in \beta^{-1}(\xi)$, $g(\omega) = g(\xi)$. By the first part of the lemma $\mu(g') = 0$ so that $\mu(g) = g(\xi)$. This proves the lemma.

Thus we see that the condition of Definition 10.1 requires in particular that the prediction measure have its support in the set $\beta^{-1}(\xi)$ of extensions of ξ from A to B. Corresponding to this condition we may define a form of predictability for sequences $\xi(n)$ that have the property that there is just one prediction measure at ξ satisfying Definition 10.1. This leads to the following definition which is again formulated for abstract E-algebras:

DEFINITION 10.2. The E-algebra A is <u>continuously predictable</u> (c.p.) <u>to</u> B <u>at</u> $\xi \in \Omega_A$ if there is exactly one prediction measure at ξ. A will be said to be continuously predictable to B without reference to a point in Ω_A if it is c.p. at every point of Ω_A. A process X will be called continuously predictable if A_X^- is c.p. to A_X. Finally, a regular sequence $\xi(n)$ is continuously predictable if $A^-(\xi)$ is c.p. to $A(\xi)$ at ξ.

We proceed to investigate the conditions for continuous predictability. For this we shall have to develop certain preliminary notions.

10.2. <u>Preliminary Lemmas</u>

DEFINITION 10.3.   Let A be a separable E-algebra, Ω_A its homomorphism space and let $\varphi \in L(A)$ (Definition 1.2).   The function φ <u>is continuous at</u> $\omega \in \Omega_A$ <u>with value</u> α if for every $\varepsilon > 0$ there is a neighborhood U_ε of ω such that

$$(10.1) \qquad\qquad |E(f\varphi) - \alpha E(f)| < \varepsilon E(f)$$

for all non-negative functions $f \in A$ and having support in U_ε .

We note that if (10.1) holds for $f \in A$ with support in U_ε it also holds for $f \in L(A)$ with support in U_ε .

One easily shows that if $\varphi \in A$, i.e., φ is a continuous function on Ω_A in the ordinary sense, then φ is continuous in the sense of Definition 10.3 at every $\omega \in \Omega_A$ with value $\varphi(\omega)$. Moreover sums and products of functions continuous at ω are continuous at ω . For example if φ and ψ are continuous at ω with values α and β , then $|E(f\varphi\psi) - \alpha\beta E(f)| \leq |E(f\varphi\psi) - \alpha E(f\psi)| + |\alpha||E(f\psi) - \beta E(f)|$ which can be made arbitrarily small if (10.1) applies to φ and ψ , with $f\psi$ replacing f in the first term.

LEMMA 10.2.   If $\varphi \in L(A)$ is continuous at every point $\omega \in \Omega_A$ with value $\alpha(\omega)$ then $\varphi \in A$ and $\varphi(\omega) = \alpha(\omega)$ almost everywhere.

PROOF.   We first show that $\alpha(\omega)$ is a continuous function on Ω_A .   Since A is separable it suffices to prove that for $\omega_n \longrightarrow \omega$, $\alpha(\omega_n) \longrightarrow \alpha(\omega)$.   Choose $\varepsilon_n > 0$ with $\varepsilon_n \longrightarrow 0$ and find neighborhoods U_n of ω_n such that (10.1) is satisfied for $\alpha = \alpha(\omega_n)$, $\varepsilon = \varepsilon_n$, and f having support in U_n .   Also choose U_ε for ω in accordance with Definition 10.3.   Since $\omega_n \longrightarrow \omega$ we may suppose that from some point on, $U_n \subset U_\varepsilon$.   Then if f has support in U_n for large n we have simultaneously $|E(f\varphi) - \alpha(\omega_n)E(f)| < \varepsilon_n E(f)$ and $|E(f\varphi) - \alpha(\omega)E(f)| < \varepsilon E(f)$ whence $|\alpha(\omega_n) - \alpha(\omega)| < \varepsilon + \varepsilon_n$.   It follows that $\alpha(\omega)$ is continuous.

Hence $\alpha \in A$ so that the lemma will be proven if it is shown that $\varphi(\omega) = \alpha(\omega)$ almost everywhere.   For this it suffices to show that $E(f(\varphi - \alpha)) = 0$ for all $f \in A$, or for all non-negative f in A .   For any $\omega \in \Omega_A$ it follows from the continuity of $\alpha(\omega)$ that for a sufficiently small neighborhood U_ε' of ω

$$(10.2) \qquad\qquad |E(f\varphi) - E(f\alpha)| < \varepsilon E(f)$$

if f has its support in U'_ε , since close to ω , $E(f\alpha)$ is close to
$\alpha(\omega)E(f)$. By the compactness of Ω_A we may find a finite number of
neighborhoods U'_ε covering Ω_A and a partition of unity

$$1 = \sum_1^N \psi_n$$

corresponding to this covering, i.e., each ψ_n has its support in some
U'_ε . Then for any non-negative f on Ω_A

$$E(f(\varphi - \alpha)) = E\left(\sum_1^N \psi_n f(\varphi - \alpha)\right) = \sum_1^N E(\psi_n f(\varphi - \alpha))$$

and by (10.2) the latter is dominated by

$$\varepsilon \sum_1^N E(\psi_n f) = \varepsilon E(f)$$

in absolute value. Since ε is arbitrary it follows that $\varphi = \alpha$ a.e.,
as was to be shown.

DEFINITION 10.4. Let A be an E-algebra and ξ a point of Ω_A .
We denote by δ_ξ the measure on Ω_A defined by

$$\delta_\xi(\Delta) = \begin{cases} 1 & \text{if } \xi \in \Delta \\ 0 & \text{if } \xi \notin \Delta \end{cases}$$

or equivalently by $\delta_\xi(f) = f(\xi)$ for $f \in A$.

DEFINITION 10.5. A sequence of non-negative functions $f_n \in A$
is a <u>predicting sequence at</u> ξ if $E(f_n g) \longrightarrow \delta_\xi(f)$ for every $g \in A$.

It will be convenient in what follows to consider the imbedding
of an E-algebra into its dual A^* given by $f \longrightarrow F_f$ where

$$(10.3) \qquad\qquad F_f(g) = E(fg) \quad .$$

We note that this mapping is $1 - 1$ since if $F_f = 0$ then $E(ff^*) = 0$
so that $f = 0$; hence we speak of an imbedding. A thereby inherits the
weak topology of A^* and we may speak of convergence of a sequence in A
in this sense. We refer to such convergence as convergence <u>over</u> A ;
$f_n \in A$ may converge to a functional μ in A^* <u>over</u> A if
$E(f_n g) \longrightarrow \mu(g)$ for every g in A . In this terminology Definition
10.5 says that a predicting sequence converges to δ_ξ over A .

LEMMA 10.3. If A and B are E-algebras with $A \subset B$
and $\xi \in \Omega_A$, then if every predicting sequence at ξ
converges over B to the measure μ on Ω_B (i.e., μ a
functional on B) then for every $g \in B$, the function
$E(g|A) \in L(A)$ is continuous at ξ with value $\mu(g)$.

PROOF. We first observe that if we have a sequence of functions
$f_n \in A$ with supports in neighborhoods U_n of $\xi \in \Omega_A$ and the U_n converge to ξ, then $f_n/E(f_n)$ is a predicting sequence at ξ. Namely,
any $g \in A$ satisfies $|g(\omega) - g(\xi)| < \varepsilon$ for $\omega \in U_n$ if n is
sufficiently large and so $|E(f_n g) - E(f_n g(\xi))| < \varepsilon E(f_n)$ whence
$|E(f_n g)/E(f_n) - g(\xi)| \longrightarrow 0$ as $n \longrightarrow \infty$.

Suppose now that $E(g|A)$ were not continuous at ξ with value
$\mu(g)$. Then for arbitrarily small neighborhoods U_n of ξ there would
be non-negative functions f_n with support in U_n for which (10.1) with
$\alpha = \mu(g)$ is violated. Thus

$$(10.4) \qquad |E(f_n E(g|A)) - E(f_n)\mu(g)| \geq \varepsilon E(f_n)$$

or since $f_n \in A$, $f_n E(g|A) = E(f_n g|A)$ we have

$$(10.5) \qquad |E(f_n g) - E(f_n)\mu(g)| \geq \varepsilon E(f_n) \quad .$$

But letting $n \longrightarrow \infty$, $f_n/E(f_n)$ must tend to μ over B since
$f_n/E(f_n) \longrightarrow \delta_\xi$ and is a predicting sequence. Dividing by $E(f_n)$ in
(10.5) and letting $n \longrightarrow \infty$ we would then get $|\mu(g) - \mu(g)| > \varepsilon$ which
is impossible. This proves the lemma.

LEMMA 10.4. If $A \subset B$, $\xi \in \Omega_A$ and μ is a measure
on Ω_B which is the limit over B of non-negative
functions $f_n \in A$ such that $\mu|A = \delta_\xi$ then μ is
a prediction measure at ξ.

PROOF. Since μ is a limit of positive functions it will be
a positive measure and $\mu|A = \delta_\xi$ implies that it is a probability measure.
To verify the condition of Definition 10.1 suppose that for some $g \in B$,
there is a U with the property that $E(fg) \geq 0$ for all $f \in A$ having
support in U. Now let $h \in A$ be any non-negative function bounded by
1 with support in U satisfying $h(\xi) = 1$. Then $|E(f_n g) - E(f_n hg)| \leq$
$E(f_n(1 - h)|g|) \leq \|g\|_\infty E(f_n(1 - h|)$ since the f_n and h are non-
negative. However $E(f_n(1 - h) \longrightarrow \delta_\xi(1 - h) = 0$ so that

$$\lim_{n \to \infty} E(f_n g) = \lim_{n \to \infty} E(f_n hg) \quad .$$

But since h has support in U so does $f_n h$ so that $E(f_n hg) \geq 0$. It follows that

$$\mu(g) = \lim_{n \to \infty} E(f_n g) \geq 0$$

so that μ is a prediction measure.

> COROLLARY. There always exists a prediction measure
> at each point ξ of Ω_A.

PROOF. We saw in the proof of Lemma 10.3 that a sequence of non-negative functions of A with supports shrinking to ξ determines a predicting sequence at ξ. Now some subsequence of the predicting sequence will converge over B as well since B is separable and the non-negative functions are of bounded L^1-norm. Clearly the restriction of the limit measure to A is δ_ξ and so by the preceding lemma the limit measure is a prediction measure.

10.3. <u>A Criterion for Continuous Predictability</u>. We may now establish the following:

> THEOREM 10.1. A is c.p. to B at $\xi \in \Omega_A$ if and
> only if every predicting sequence at ξ of functions
> in A converges over B.

PROOF. Suppose first that A is c.p. to B at ξ with μ the unique prediction measure at ξ. If f_n is a predicting sequence at ξ which converges over B then by Lemma 10.4 it converges to a prediction measure at ξ, hence to μ. Consequently any subsequence of a predicting sequence that converges over B converges to μ and hence any predicting sequence at ξ converges to μ over B.

Now suppose that every predicting sequence at ξ converges over B; clearly the limit μ will be unique since otherwise a non-convergent predicting sequence could be formed. We shall show that μ is the unique prediction measure at ξ. To do so let μ' be any prediction measure at ξ and let $g \in B$. By Lemma 10.3, $E(g|A)$ is continuous at ξ with value $\mu(g)$. Hence for a sufficiently small neighborhood U of ξ

$$(10.6) \qquad |E(fE(g|A)) - E(f)\mu(g)| < \epsilon E(f)$$

whenever f is non-negative and has support in U. Now (10.6) may be rewritten

$$|E(fg) - E(f)\mu(g)| < \varepsilon E(f),$$

$$|E(f(g - \mu(g))| < \varepsilon E(f),$$

or

$$- \varepsilon E(f) < E(f(g - \mu(g))) < \varepsilon E(f)   .$$

This gives

(10.7)
$$E(f(g - \mu(g) + \varepsilon)) > 0$$
$$E(f(- g + \mu(g) + \varepsilon)) > 0$$

for f with support in U. Applying Definition 10.1 to μ' gives

$$\mu'(g - \mu(g) + \varepsilon) \geq 0, \quad \mu'(- g + \mu(g) + \varepsilon) \geq 0$$

or $|\mu'(g - \mu(g))| \leq \varepsilon$ so that $\mu'(g) = \mu(g)$. This proves the theorem.

> COROLLARY 1. The function $E(g|A)$ is continuous at ξ
> (in the sense of Definition 10.3) for each $g \in B$ if
> and only if A is c.p. to B at ξ.

PROOF. For if $E(g|A)$ is continuous at ξ with value $\alpha(g)$
then the proof of the preceding theorem shows that for any prediction
measure μ at ξ, $\mu(g) = \alpha(g)$. Hence μ is unique. Conversely if A
is c.p. to B at ξ then by the theorem every predicting sequence at ξ
converges over B and by Lemma 12.3, $E(g|A)$ is continuous at ξ.

The foregoing corollary is the reason for the term "continuous"
predictability.

> COROLLARY 2. If A is c.p. to B at ξ, then the
> unique prediction measure at ξ is a limit over B
> of non-negative functions in A.

PROOF. For a predicting sequence at ξ always exists as we
have remarked in the proof of Lemma 10.3. By the theorem this predicting
sequence must converge to the unique prediction measure over B.

10.4. <u>Continuous Predictability at Every Point</u>. In the following theorem we shall find conditions for A to be c.p. to B at every
point, or simply, for A to be c.p. to B. If B is an E-algebra and
M a subset of B we denote by $M^{\perp}$ the set of elements f of B satisfy-
$E(fg) = 0$ for all $g \in M$.

THEOREM 10.2. A is c.p. to B if and only if B
is the direct sum of A and $A^{\perp}$. If A is c.p. to
B then the unique prediction measure μ_{ξ} at ξ depends continuously on $\xi \in \Omega_A$. Moreover $\mu_{\xi}(g) =$
$E(g|A)(\xi)$, $g \in B$, almost everywhere in Ω_A if A
is c.p. to B, and $E(g|A) \in A$ for all $g \in B$
if and only if A is c.p. to B.

PROOF. The last statement is an immediate consequence of
Corollary 1 to Theorem 10.1 and Lemma 10.2, since $E(g|A) \in A$ means
that $E(g|A)$ is continuous **at** every point of Ω_A in the sense of Defi-
nition 10.3. We turn next to the first statement of the theorem. If A
is c.p. to B and $g \in B$ we may write $g = E(g|A) + [g - E(g|A)]$ where
$E(g|A) \in A$ as we have already shown. This gives a decomposition of g
into two functions belonging to A and $A^{\perp}$ respectively. Conversely,
suppose for each g, $g = g_1 + g_2$ with $g_1 \in A$, $g_2 \in A^{\perp}$. If $f \in A$ then
$E(fg_2) = 0$; hence $E(fE(g_2|A)) = 0$ and since f is an arbitrary ele-
ment of A $E(g_2|A) = 0$. It follows that $E(g|A) = E(g_1|A) = g_1 \in A$, or
that $E(g|A)$ belongs to A for all g in B. Again, by what we have
already proven this implies that A is c.p. to B, so the first statement is proven.

Combining Lemma 10.3 with Theorem 10.1 we find that when A is
c.p. to B at ξ then $\mu_{\xi}(g) = E(g|A)(\xi)$ which proves the remaining
statements of the theorem.

10.5. <u>Composition of Prediction Measures</u>. We now consider the
situation arising when three E-algebras A, B, and C are given satisfying $A \subset B \subset C$.

THEOREM 10.3. Let A, B, C be E-algebras, Ω_A, Ω_B, Ω_C
the corresponding homomorphism spaces, and suppose
$A \subset B \subset C$. Denote by β the canonical map of Ω_B onto
Ω_A and let ξ be a given point of Ω_A, $\beta^{-1}(\xi)$ its
preimage in Ω_B. If A is c.p. to C at ξ it is c.p.
to B at ξ. If A is c.p. to B at ξ with prediction measure μ, and if B is c.p. to C at all
points of $\beta^{-1}(\xi)$ with the exception of a set of
μ-measure 0, then A is c.p. to C. In particular
if A is c.p. to B and B is c.p. to C then A
is c.p. to C.

PROOF. For the first part of the theorem we employ the criterion
of Theorem 10.1. Namely, if A is c.p. to C at ξ then any predicting

sequence at ξ converges over C and *a fortiori* over B so that A is c.p. to B.

To prove the second part, let $W \subset \beta^{-1}(\xi)$ be a compact set such that at all points of W, B is c.p. to C, and assume that for a preassigned $\varepsilon > 0$, $\mu(\beta^{-1}(\xi) - W) < \varepsilon$. At each $\omega \in W$ let μ_ω denote the unique prediction measure on Ω_C. Let g be a fixed element of C. For any $\omega \in W$ there will be a neighborhood U_ω of ω such that any non-negative f with support in U_ω satisfies

$$|E(fg) - \mu_\omega(g)E(f)| < \varepsilon E(f) \quad .$$

Since we have a covering of W by open sets U_ω we may select a finite subcovering $\{U_k, \ k = 1, 2, \ldots, N\}$. We also construct a family of non-negative functions $\{\varphi_\ell\}$ of B such that each φ_ℓ has its support inside an open set of $\{U_k\}$ and such that $\Sigma_\ell \varphi_\ell \leq 1$ everywhere and $\Sigma_\ell \varphi_\ell = 1$ in W. The construction of such a set of functions is similar to that of a partition of unity. Now set $\varphi = \Sigma_\ell \varphi_\ell$ and suppose that $\{f_n\}$ is a predicting sequence of A at ξ. We have

$$(10.8) \quad |E(f_n g) - E(f_n \varphi g)| \leq E(f_n |1 - \varphi||g|) \leq \sup |g(\omega)| E(f_n(1 - \varphi)) \quad .$$

Since A is c.p. to B at ξ, the predicting sequence f_n converges over B and so the majorant on the right tends to

$$\sup |g(\omega)| \mu(1 - \varphi)$$

as $n \longrightarrow \infty$. Now since $\varphi = 1$ on W and μ has its support on $\beta^{-1}(\xi)$, $\mu(1 - \varphi) \leq \mu(\beta^{-1}(\xi) - W) < \varepsilon$. Hence by (10.8)

$$(10.9) \qquad \lim_{n \to \infty} \sup |E(f_n g) - E(f_n \varphi g)| < \varepsilon \sup |g(\omega)| \quad .$$

On the other hand, $E(f_n \varphi g) = \Sigma_\ell E(f_n \varphi_\ell g)$, and since $f_n \varphi_\ell$ has its support in one of the U_k about a point ω_k of W,

$$|E(f_n \varphi_\ell g) - E(f_n \varphi_\ell)\mu_{\omega_k}(g)| < \varepsilon E(f_n \varphi_\ell)$$

and therefore

$$|E(f_n \varphi g) - \Sigma_\ell E(f_n \varphi_\ell)\mu_{\omega_k}(g)| < \varepsilon E(f_n \varphi) < \varepsilon E(f_n) \quad .$$

Combining this with (10.9):

$$\limsup_{n \to \infty} |E(f_n g) - \Sigma_\ell E(f_n \varphi_\ell) \mu_{\omega_k}(g)| < \varepsilon \limsup_{n \to \infty} E(f_n) + \varepsilon \sup |g(\omega)| \quad .$$

Since A is c.p. to B at ξ we have

$$E(f_n \varphi_\ell) \longrightarrow \mu(\varphi_\ell), \ E(f_n) = E(f_n \cdot 1) = \mu(1) = 1$$

so that

$$(10.10) \quad \limsup_{n \to \infty} |E(f_n g) - \Sigma_\ell \mu(\varphi_\ell) \mu_{\omega_k}(g)| < \varepsilon \ (1 + \sup |g(\omega)|) \quad .$$

Since $\Sigma_\ell \mu(\varphi_\ell) \mu_{\omega_k}(g)$ does not depend upon the sequence $\{f_n\}$ it follows
that for an arbitrary predicting sequence

$$\limsup_{n \to \infty} (E(f_n g)) - \liminf_{n \to \infty} (E(f_n g)) < 2\varepsilon (1 + \sup|g(\omega)|)$$

and since ε is arbitrary it follows that f_n converges over C. By
Theorem 10.1 it follows that A is c.p. to C at ξ as was to be shown.

> COROLLARY. Suppose $\xi(n)$ and $\eta(n)$ are left-infinite
> sequences with the property that $A^-(\xi) \subset A^-(\eta)$, then
> if $A^-(\xi)$ is c.p. to $A^-(\eta)$ and $A^-(\eta)$ is c.p. to
> $A(\eta)$ (i.e., $X(\eta)$ is c.p.) then $\xi(n)$ is c.p.

PROOF. We use both parts of the preceding theorem. Since $A^-(\xi)$
is c.p. to $A^-(\eta)$ and $A^-(\eta)$ is c.p. to $A(\eta)$ it follows that $A^-(\xi)$
is c.p. to $A(\eta)$ at ξ, and since $A(\xi) \subset A(\eta)$ it follows that $A^-(\xi)$
is c.p. to $A(\xi)$ at ξ.

We remark that in this corollary the notion of continuous predictability for the algebras $A^-(\xi)$ and $A^-(\eta)$ is of significance for
prediction from past to future although these two algebras themselves
both represent the "pasts" of the two processes.

10.6. <u>Realization of Prediction Measures</u>. Before concluding
this chapter we wish to interpret continuous predictability for the case
of regular sequences and discuss how the prediction measure can be obtained in an effective manner if we have a sequence $\xi(n)$ before us.
Suppose, for simplicity, that the range of $\xi(n)$ is a finite set
$\{a_1, a_2, \ldots, a_r\}$. We recall (Theorem 9.1) that in the deterministic case
we could find a block of consecutive terms $\xi(-N), \ldots, \xi(0)$ such that
whenever this block occurs in $\xi(n)$ it is followed by the same element
which is taken as the "determined" value of $\xi(1)$. The same procedure

will give the values of $\xi(2)$, $\xi(3)$, etc.

In the non-deterministic case we look for probabilities $\mu(\xi(1) = a_1)$, $\mu(\xi(1) = a_2)$, ..., $\mu(\xi(1) = a_r)$ (μ representing the prediction measure) in terms of the sequence $\xi(n)$, $n \leq 0$. We shall see that in the continuously predictable case a similar procedure may be used. Namely we take the blocks $\xi(-N)$, $\xi(N+1)$, ..., $\xi(0)$ and consider the positions in $\xi(n)$ where these blocks have occurred before. By the regularity of $\xi(n)$, such blocks occur with positive frequency in the past of $\xi(n)$. It is therefore possible to speak of the "relative frequencies" of a_1, a_2, ..., a_r in the terms following the entries of such a block in the past. Then if $\xi(n)$ is continuously predictable, these relative frequencies converge to limits as $N \longrightarrow \infty$ and these limits are the probabilities $\mu(\xi(1) = a_j)$.

In order to verify this, recall that ξ is a generic point of the process $X(\xi)$ and hence the frequencies observed in $\xi(n)$ are the measures of certain sets in $\Omega^-(\xi)$ and $\Omega(\xi)$. In particular the frequency of $\xi(-N)$, ..., $\xi(0)$ in the past of $\xi(n)$ is the measure of the set in $\Omega^-(\xi)$ defined by $\{x_{-N} = \xi(-N), ..., x_0 = \xi(0)\}$; the frequency of $\xi(-N)$, ..., $\xi(0)$, a_j is the probability of the set in $\Omega(\xi)$ defined bt $\{x_{-N} = \xi(-N), ..., x_0 = \xi(0), x_1 = a_j\}$. Now let $X[\xi(-N), ..., \xi(0)]$ denote the characteristic function of the former of these sets, $X[a_j]$ the characteristic function of the set $\{x_1 = a_j\}$. The relative frequencies obtained by the procedure in question are then

$$(10.11) \qquad \frac{E(X[\xi(-N),...,\xi(0)]X[a_j])}{E(X[\xi(-N),...,\xi(0)])} \; .$$

Since $X[\xi(-N), ..., \xi(0)]/E(X[\xi(-N), ..., \xi(0)])$ is a prediction sequence at ξ, as is easily verified, the relative frequencies in (10.11) converge to the value $\mu(X[a_j]) = \mu(\xi(1) = a_j)$. Hence continuous predictability ensures that the limits of the approximations in (10.11) exist and give the correct value for the prediction.

It should be pointed out that for the continuously predictable case there exist many other procedures for obtaining the prediction measure corresponding to the fact that there exist numerous predicting sequences at a point. In other words we could "close down" upon the point ξ in many more ways. Namely, instead of looking at blocks of consecutive terms of the sequence $\xi(n)$ we could also consider sets of non-consecutive terms and see where these have occurred before. That is to say, for our Nth approximation we consider the set of ℓ for which

$$\xi(\ell - \nu_{N,1}) = \xi(- \nu_{N,1}), \ \xi(\ell - \nu_{N,2})$$

$$= \xi(- \nu_{N,2}), \ \ldots, \ \xi(\ell - \nu_{N,M_N}) = \xi(- \nu_{N,M_N})$$

where the sets of integers $(- \nu_{N,1}, - \nu_{N,2}, \ldots, - \nu_{N,M_N})$ tend to the set of negative integers (in an obvious manner) as $N \longrightarrow \infty$. The probability $\mu(\xi(1) = a_j)$ is then approximated by the relative frequency of those of the foregoing ℓ for which $\xi(\ell + 1) = a_j$. If $A(\xi)$ is c.p. to $A(\xi)$ at ξ then any such method yields the same prediction measure. It is interesting that there are sequences for which the general procedure just described does not work — hence the sequences are not continuously predictable — but the method of consecutive blocks described first does give a meaningful answer. The reason for this is not yet completely clear.

CHAPTER 3. EXAMPLES AND COUNTEREXAMPLES

In this chapter and the two following we investigate various
classes of regular sequences for their predictability properties. In
only a few cases can the terms of the sequences studied be displayed explicitly and most often we shall describe a sequence by the statement that
it is generic for a process of one kind or another. In this chapter we
consider examples of sequences representing the simplest processes and
also the one class of regular sequences capable of explicit construction.

§11. Random and Markoff Sequences

11.1. <u>Preliminaries</u>. The most elementary processes from the
point of view of prediction theory are processes with independent variables — which we call random processes — and Markoff processes. In
fact the former may be described as the processes for which the prediction
of the future is independent of the individual sequence representing the
past, and the latter, as those processes for which the prediction depends
only on the last entry of the sequence. (The precise meaning of this will
become clear as we proceed.) Next in order of complexity are m-Markoff
processes for which the prediction depends upon the last m entries.
More precisely an m-Markoff process is one for which

$$(11.1) \quad P(x_n \in \Delta | x_{n-1}, x_{n-2}, \ldots) = P(x_n \in \Delta | x_{n-1}, x_{n-2}, \ldots, x_{n-m}) \ .$$

DEFINITION 11.1. A left-infinite sequence $\xi(n)$ will be called
a random, Markoff, or m-Markoff sequence if it is regular and $X(\xi)$ is
respectively a random Markoff, or m-Markoff process.

In all these cases the question of predictability is easily
answered after establishing several preliminary lemmas. For the following
lemma we define A_X^+ as the algebra of functions generated by the $\psi(x_n)$
for $n > 0$, $\psi \in C(\Lambda)$ where the x_n are Λ-valued variables for the process
X. Thus for a process X, A_X is the algebra generated by the subalgebras
A_X^- and A_X^+.

75

LEMMA 11.1. For a process X, A_X^- is c.p. to A_X
at $\xi \in \Omega_X^-$ if the functions $E(z|A_X^-)$ in $L(A_X^-)$
are continuous at ξ in the sense of Definition 10.3,
for all $z \in A_X^+$.

PROOF. This differs from Corollary 1 to Theorem 10.1 in that
here we require the continuity of $E(z|A_X^-)$ only for $z \in A_X^+$. However
continuity for all $z \in A_X^+$ really implies continuity for all of A_X.
For the subalgebras A_X^- and A_X^+ generate A_X which means that every
function in A_X is a limit of sums of functions of the form $z = z_1 z_2$
with $z_1 \in A_X^-$ and $z_2 \in A_X$. Clearly it suffices to establish the con-
tinuity of $E(z|A_X^-)$ for z of this form. But $E(z_1 z_2|A_X^-) = z_1 E(z_2|A_X^-)$
and hence it suffices to know continuity for $z \in A_X^+$.

LEMMA 11.2. For a process X, let $T^{-1}A_X^-$ denote the
subalgebra of A_X of z satisfying $Tz \in A_X^-$. Then
A_X^- is c.p. to A_X if and only if A_X^- is c.p. to
$T^{-1}A_X^-$.

PROOF. If A_X^- is c.p. to A_X then by the first part of Theo-
rem 10.3 A_X^- is c.p. to $T^{-1}A_X^- \subset A_X$. Suppose conversely that A_X^- is
c.p. to $T^{-1}A_X^-$. Since T^n is an automorphism of A_X (as an E-algebra)
it follows that for each n, $T^{-n}A_X^-$ is c.p. to $T^{-n-1}A_X^-$. By the second
part of Theorem 10.3 (i.e., transitivity of the c.p. relationship) it
follows that A_X^- is c.p. to $T^{-n}A_X^-$ for all n. Now by Theorem 2.1 the
union of the $T^{-n}A_X^-$ is dense in A_X. By Theorem 10.2 we have
$E(g|A_X^-) \in A_X^-$ for all $g \in T^{-n}A_X^-$ since A_X^- is c.p. to $T^{-n}A_X^-$ and hence
$E(g|A_X^-) \in A_X^-$ for all $g \in A_X$. By Theorem 10.2 again A_X^- is c.p. to
A_X as was to be shown.

COROLLARY. X is c.p. if and only if for every
$\psi \in C(\Lambda)$, Λ the range of the x_n, $E(\psi(x_1)|A_X^-) \in A_X^-$.

PROOF. For the functions of the form $\psi(x_1)$ together with
those of A_X^- generate $T^{-1}A_X^-$.

11.2. <u>Continuous Predictability of Random, Markoff, and
M-Markoff Sequences</u>.

THEOREM 11.1. A random sequence is continuously
predictable.

PROOF. This is an immediate consequence of Lemma 11.1 since for a random process $E(z|A_X^-) = E(z)$ if $z \in A_X^+$.

Just as easily we prove

THEOREM 11.2. Every Markoff and, more generally, every
M-Markoff sequence with discrete state space Λ is
continuously predictable.

PROOF. For if $z \in A^+(\xi)$ where ξ is m-Markoff then $E(z|A^-(\xi))$
is a function of x_0, x_{-1}, ..., x_{-m+1}. Thus it is a function on the
cartesian product Λ^m and since Λ is discrete so is Λ^m and any function on it is continuous.

For the general case of a Markoff sequence we consider the transformation R defined by a Markoff process on the set of continuous functions on the state space of the process. R is defined by

$$(11.1) \qquad R\psi(x_0) = E(\psi(x_1)|A_X^-) .$$

where if ψ is continuous $R\psi$ will be a measurable function on Λ (with
respect to the stationary measure on Λ). When the Markoff process is given
in terms of transition probabilities $p(\lambda, \Delta)$, $\lambda \in \Delta$, $\Delta \subset \Lambda$, we have

$$(11.2) \qquad R(\lambda) = \int_\Lambda \psi(\lambda')p(\lambda, d\lambda') .$$

We now apply the corollary to Lemma 11.2 and obtain

THEOREM 11.3. A Markoff process X is c.p. if and only
if the transformation R of (11.1) or (11.2) transforms
continuous functions on Λ into continuous functions
on Λ.

With regard to Markoff sequences it follows that if R takes
continuous functions into continuous functions (for example, if $p(\lambda, \Delta)$
is continuous as a measure valued function of λ) then any generic sequence for X is continuously predictable. Conversely if $R\psi$ is not
always continuous, say $\lambda_0 \in \Lambda$ is a point of discontinuity, then it is
easy to see that a generic sequence terminating in λ_0 will not be c.p.
When the process is given by transition probabilities $p(\lambda, \Delta)$ and λ_0
is a point of discontinuity of $p(\lambda, \Delta)$ it is not surprising that a
sequence terminating in λ_0 is not c.p. since the prediction measure is
precisely $p(\lambda_0, \Delta)$ which cannot be evaluated from the process if λ_0
is a point of discontinuity of $p(\lambda, \Delta)$.

The case of m-Markoff sequences and processes need not be given

a separate treatment because, as is known ([3], Chapter II), an m-Markoff
process defines an ordinary Markoff process which is in a certain sense
equivalent to it. We make this precise in the following:

> LEMMA 11.3. To any m-Markoff process X there corre-
> sponds a Markoff process Y with $A_X^- = A_Y^-$. To any
> m-Markoff sequence $\xi(n)$ there is a Markoff sequence
> $\eta(n)$ such that $\xi(n)$ is a derived sequence of $\eta(n)$
> and $\eta(n)$ is a derived sequence of $\xi(n)$.

PROOF. Take $y_n = (x_{n-m+1}, x_{n-m+2}, \ldots, x_n) \in \Lambda^m$ and
$\eta(n) = (\xi(n - m + 1), \xi(n - m + 2), \ldots, \xi(n))$. It is clear that $A_X^- = A_Y^-$
and that $\xi(n)$ and $\eta(n)$ are derived sequences of one another. The
fact that Y is a Markoff process is evident from the fact that y_1 is
determined by x_1 and y_0 so that the conditional distribution of y_1
given $y_0, y_{-1}, \ldots$ depends upon y_0 and the conditional distribution
of x_1 given $y_0, y_1, \ldots$ which depends only on $(x_0, x_{-1}, \ldots, x_{-m+1}) =$
y_0.

It follows from this that an m-Markoff process will be c.p. if
and only if the corresponding Markoff process is, and similarly for sequences. For, A_X^- and A_Y^- are identical as are A_X and A_Y. Moreover in the isomorphism of A_X^- and A_Y^- the point $\xi \in A^-(\xi)$ is sent
into the point $\eta \in A^-(\eta)$ so that $A^-(\xi)$ is c.p. to $A(\xi)$ at ξ if
and only if $A^-(\eta)$ is c.p. to $A(\eta)$ at η.

§12.   A Non-Continuously Predictable Sequence

12.1. <u>Construction of the Sequence</u>. In the preceding section
we found that there exist Markoff sequences that are not continuously
predictable. By Theorem 11.2. such a sequence must take its values in an
infinite space. We now give an example of a two-valued regular sequence
that is not continuously predictable. This will be all the more surprising since the sequence in question will be a derived sequence of a
random sequence which is trivially c.p. (Theorem 11.1). It follows that
a derived sequence of a c.p. sequence need not be c.p.

> THEOREM 12.1. Let $\eta(n)$ be a generic sequence for a
> random $\{-1, 1\}$ process Y where $P\{y_n = 1\} = p$,
> $P\{y_n = 1\} = q$, $p + q = 1$ and $p \neq 1/2$. If $\xi(n)$ is
> defined by $\xi(n) = \eta(n)\eta(n - 1)$ then $\xi(n)$ is not
> continuously predictable.

A sequence $\xi(n)$ of this kind might arise in the

following way. We suppose that an infinite sequence of coin-tosses have been made and that instead of recording each result as "heads" or "tails", the experimenter records only whether each toss gives the same or a different result when compared with the preceding one. The available data is then in the form of a two-valued sequence which is equivalent to the $\xi(n)$ of the theorem.

PROOF. We shall first prove that the process $Y = X(\eta)$ is an L-extension (Definition 2.6) of $X(\xi)$, i.e., that $A^-(\eta) \subset L(A^-(\xi))$. By Theorem 4.3 we have that the variables x_n of $X(\xi)$ are related to the variables of Y by

$$x_n = y_n y_{n-1} \; .$$

Consider now the elements of A_Y^- defined by

$$z_n = \frac{1}{n} y_0 (y_{-1} + y_{-2} + \cdots + y_{-n}) \; .$$

Since $y_0 y_{-i} = y_0 y_{-1} \cdot y_{-1} y_{-2} \cdot \cdots \cdot y_{-i+1} y_{-i} = x_0 x_{-1} \cdots x_{-i+1} \in A^-(\xi)$ it follows that $z_n \in A^-(\xi)$. Now by the ergodic theorem

$$\lim_{n \to \infty} z_n$$

exists almost everywhere; hence defines a function in $L(A^-(\xi))$, and since Y is ergodic $\frac{1}{n}(y_{-1} + y_{-2} + \cdots + y_n) \longrightarrow E(y_k)$ so that

$$\lim_{n \to \infty} z_n = y_0 E(y_k) \in L(A^-(\xi)).$$

Since $p \neq 1/2$, $E(y_k) \neq 0$ so it follows that $y_0 \in L(A^-(\xi))$. As before $y_0 y_{-i} \in A^-(\xi)$ so it follows that for all i, $y_{-i} \in L(A^-(\xi))$ and hence $A_Y^- \subset L(A^-(\xi))$.

By Corollary 1 to Theorem 10.1, if $\xi(n)$ is c.p. then for all $z \in A(\xi)$, $E(z|A^-(\xi))$ is continuous at ξ. In particular we would then have $E(x_1|A^-(\xi))$ continuous at ξ. By (1.4) (a) the latter is the same as $E(x_1|A_Y^-) = E(y_1 y_0|A_Y^-) = y_0 E(y_1|A_Y^-) = y_0 E(y_k)$. Hence $\xi(n)$ will be c.p. only if $y_0 \in L(A^-(\xi))$ is continuous at ξ.

Suppose then that y_0 is continuous at ξ with value α (Definition 10.3); let $\varepsilon > 0$ and U_ε be a neighborhood of ξ in $\Omega^-(\xi)$ for which

$$(12.1) \qquad |E(fy_o) - \alpha E(f)| < \varepsilon E(f)$$

for all non-negative $f \in A^-(\xi)$ with support in U . Evidently the same
inequality will hold for all non-negative f in $L(A^-(\xi))$, e.g., in
A_Y^-, with support in U_ε . Now the inclusion $A^-(\xi) \subset A_Y^-$ induces the
canonical map from Ω_Y^- to $\Omega^-(\xi)$ which maps η onto ξ, and also maps
the point η' defined by $\eta'(n) = -\eta(n)$ onto ξ. Since the canonical
map is continuous it follows that there is a neighborhood U_ε' of η
in Ω_Y^- which is mapped into U_ε by this map and hence any function of
A_Y^- with support in U_ε' has, as a function in $L(A^-(\xi))$, support in U_ε .
Similarly there is a neighborhood U_ε'' of η' mapped into U_ε .

     Now let f_n converge to δ_η over A_Y^- where the $f_n \in A_Y^-$ and
with supports shrinking to η . In particular we will eventually reach
f_n with support in U_ε' and for these

$$|E(f_n y_o) - \alpha E(f_n)| \leq \varepsilon E(f_n) \quad .$$

Letting $n \longrightarrow \infty$ we find $|y_o(\eta) - \alpha| < \varepsilon$, or $|\eta(0) - \alpha| \leq \varepsilon$. On
the other hand we may also find a sequence $f_n' \longrightarrow \delta_{\eta'}$ over A_Y^- with
the supports of f_n' eventually interior to U_ε''. This gives
$|\eta'(0) - \alpha| \leq \varepsilon$ and since $\eta'(0) = -\eta(0)$ this will not be possible for
ε sufficiently small. Hence y_o is not continuous at ξ and hence
$\xi(n)$ is not continuously predictable.

     12.2. <u>Application of the Prediction Procedure</u>. It is inter-
esting that if we follow the procedure described in §10.6, we do obtain
a definite prediction whose interpretation will be given in Chapter 6.
For example to find the probability of $x_1 = 1$ given ξ, we take as
an approximation the conditional probability

$$P(x_1 = 1 \,|\, (x_{-n}, \ldots, x_o) = (\xi(-n), \ldots, \xi(0)))$$

$$(12.2)$$

$$= \frac{P((x_{-n}, \ldots, x_o, x_1) = (\xi(-n), \ldots, \xi(0), 1)))}{P((x_{-n}, \ldots, x_o) = (\xi(-n), \ldots, \xi(0)))}$$

and let $n \longrightarrow \infty$. These probabilities are to be computed from the frequencies of the corresponding events in the sequence ξ which may be determined from the probability relations of the process $X(\xi)$. Now
$(\xi(-n), \ldots, \xi(0))$ arises from either of two sequences: $(\eta(-n-1),$
$\ldots, \eta(0))$ and $(-\eta(-n-1), \ldots, -\eta(0))$ and the probability of the
ξ-sequence is the sum of the probabilities of occurrence of the η-sequences.
Let $\nu(n)$ and $\mu(n)$ respectively represent the number of 1's and
-1's in $(\eta(-n-1), \ldots, \eta(0))$. Then

$$P((x_{-n}, \ldots, x_0) = \xi(-n), \ldots, \xi(0)) = \qquad .$$

$$P((y_{-n-1}, \ldots, y_0) = (\eta(-n-1), \ldots, \eta(0)) +$$

$$P((y_{-n-1}, \ldots, y_0) = (-\eta(-n-1), \ldots, \eta(0)) =$$

$$p^{\nu(n)}q^{\mu(n)} + p^{\mu(n)}q^{\nu(n)} \quad .$$

Now suppose that $\eta(0) = 1$. Then $\xi(1) = 1$ if $\eta(n) = 1$ so that $(\xi(-n), \ldots, \xi(0), 1)$ arises from the sequences $(\eta(-n-1), \ldots, \eta(0), 1)$ and $(-\eta(-n-1), \ldots, \eta(0), -1)$, and therefore the denominator of (12.2) is $p^{\nu(n)+1}q^{\mu(n)} + p^{\mu(n)}q^{\nu(n)+1}$. Now by assumption η is generic for Y so that $\nu(n)/n \longrightarrow p$, $\mu(n)/n \longrightarrow q$. Assume that $p > q$. The fraction in (12.2) becomes

$$(12.3) \quad \frac{p^{\nu(n)+1}q^{\mu(n)} + p^{\mu(n)}q^{\nu(n)+1}}{p^{\nu(n)}q^{\mu(n)} + p^{\mu(n)}q^{\nu(n)}} = \frac{p^{\nu(n)-\mu(n)+1} + q^{\nu(n)-\mu(n)+1}}{p^{\nu(n)-\mu(n)} + q^{\nu(n)-\mu(n)}} \quad ,$$

and since $p > q$, $\nu(n) - \mu(n) \longrightarrow \infty$ and so (12.3) tends to $p^{\nu(n)-\mu(n)+1}/p^{\nu(n)-\mu(n)} = p$. Similarly if $\eta(0) = -1$ the conditional probability tends to q. In any case the conditional probabilities converge to a definite limit and by this procedure a prediction measure may be obtained.

We observe that the same probabilities might be obtained in a different way. From the sequence $\xi(n)$ we can construct the two possible sequences that could have given rise to it in the Y process. Notice that one of these sequences η is generic for Y and since $p \neq q$ the remaining sequence η' where 1 and -1 are interchanged cannot be generic. Now suppose that in forming the prediction we take for granted that the generic sequence η is the one that "actually occurred". Suppose first that the last reading of this η is $\eta(0) = 1$. Then $\xi(1)$ will be 1 if and only if $\eta(1)$ is 1. Now this occurs with probability p, the η being a random sequence and thus, supposing that $\eta(n)$ is the sequence that did occur, we may take p as the probability that $\xi(1) = 1$. If $\eta(0) = -1$ then $\xi(1) = 1$ is equivalent to $\eta(1) = -1$ so that in this case the probability is q that $\xi(1) = 1$. Our previous results thus coincide with what we would obtain if we knew that η is the sequence that occurred (and not η'). This general idea will be the basis of statistical predictability.

12.3. <u>Another Form of the Sequence</u>. The example of this section shows that a derived sequence of a random sequence need not be continuously

predictable. It is obvious however that if the derived sequence had the
form

$$(12.4) \qquad\qquad \xi(n) = \psi(\eta(n))$$

where $\eta(n)$ is random, then $\xi(n)$ must be c.p. For if $\eta(n)$ is random,
so is $\xi(n)$ in (12.4) and then Theorem 11.1 applies. The situation is
quite different for Markov sequences for, if $\eta(n)$ is a Markov sequence
and $\xi(n)$ has the form (12.4) then $\xi(n)$ need not be a Markov sequence.
(In terms of processes, if the y_n form a Markov process and $x_n = \psi(y_n)$,
the x_n need not form a Markov process.) Accordingly one may inquire
whether or not all derived sequences of a discrete Markov sequence <u>of the
form</u> (12.4) are c.p. The answer is that they are not and this is also
shown by the example of this section. Namely, form the composite sequence

$$\zeta(n) = (\eta(n), \eta(n-1)) \quad ;$$

$\zeta(n)$ is generic for a process Z with variables $z_n = (y_n, y_{n-1})$. Using
the independence of the y_n it is easily shown that the z_n form a
Markov process. We note for later reference that the transition matrix
of this process is

$$(12.5) \qquad
\begin{array}{c|cccc}
 & (-1,-1) & (-1,1) & (1,-1) & (1,1) \\
\hline
(-1,-1) & q & p & 0 & 0 \\
(-1,1) & 0 & 0 & q & p \\
(1,-1) & q & p & 0 & 0 \\
(1,1) & 0 & 0 & q & p
\end{array}$$

Now define ψ by $\psi(1,1) = \psi(-1,-1) = 1$, $\psi(1,-1) = \psi(-1,1) = -1$.
Then we have $\xi(n) = \psi(\zeta(n)) = \psi(\eta(n), \eta(n-1))$. This shows that a derived sequence of a Markov sequence of the form (12.4) need not be continuously predictable.

§13. A Class of Deterministic Sequences

  13.1 $\mathcal{D}$-<u>Sequences</u>. It was pointed out at the beginning of the
chapter that not many regular sequences are capable of explicit construction. In fact the only regular sequences known whose general term
can be explicitly displayed are (but for minor modifications) special
instances of the following:

  DEFINITION 13.1. We denote by $\mathcal{D}$ the class of left-infinite
sequences that are uniform limits of sequences whose general term is given

by

$$(13.1) \qquad \xi(n) = \sum_{\nu=1}^{k} c_\nu e(p_\nu(n))$$

where $e(t) \equiv e^{2\pi i t}$ and $p_\nu(t)$ is a polynomial with real coefficients. A left-infinite sequence belonging to $\mathscr{D}$ will be referred to as a $\mathscr{D}$-sequence.

The set $\mathscr{D}$ is easily seen to form an algebra closed under the shift T, and containing the almost periodic sequences as a sub-algebra. Not every $\mathscr{D}$-sequence is almost periodic since, for example, $\xi(n) = \cos n^2\alpha$, α/π irrational, is a $\mathscr{D}$-sequence but is not almost periodic. This can be seen by noting that the average of $\xi(n)e^{2\pi i n\lambda}$ for $\xi(n) = \cos n^2\alpha$ vanishes for all real λ (a fact to be proven later) which for almost periodic sequences would imply that $\xi(n) \equiv 0$. With regard to prediction, however, all $\mathscr{D}$-sequences behave as almost periodic sequences, i.e., they are regular and deterministic. This is particularly instructive in view of the fact that a sequence as $\cos n^2\alpha$, for α/π irrational, behaves, on the one hand, like a random sequence from the point of view of linear prediction (since such a sequence has no periodic components) whereas the non-linear theory finds these sequences deterministic.

In investigating $\mathscr{D}$-sequences we shall use the following:

LEMMA 13.1. Let $\xi(n)$ be a $\mathscr{D}$-sequence and $\xi^{(k)}$ a set of translates of $\xi(n)$, $\xi^{(k)} = T^{n_k}\xi$. Then any pointwise limit of the $\xi^{(k)}$ is again a $\mathscr{D}$-sequence.

PROOF. The statement of the lemma is obvious if $\xi(n)$ has the form (13.1) since a pointwise limit of translates of such a sequence again has the same form. Now suppose the sequences $\xi_m(n)$ have the form (13.1) and that $\xi_m(n) \longrightarrow \xi(n)$ uniformly. Also let $\omega(n)$ be a pointwise limit of the translates $\xi^{(k)}(n)$. We wish to show that $\omega \in \mathscr{D}$. Now for some subsequences of the sequence $\{n_k\}$ the translates $T^{n_k}\xi_m$ will converge pointwise for all m so let us suppose that this subsequence has been renumbered as $\{n_k\}$. Also let

$$\lim_{k \to \infty} T^{n_k}\xi_m = \omega_m \quad .$$

We now have for each n

$$|\omega_m(n) - \omega(n)| = \lim_{k \to \infty} |T^{n_k}\xi_m(n) - T^{n_k}\xi(n)| \quad .$$

It follows that $|\omega_m(n) - \omega(n)| \leq \|\xi_m - \xi\|_\infty$ so that $\|\omega_m - \omega\|_\infty \leq \|\xi_m - \xi\|_\infty$. Since the ξ_m converge to ξ uniformly, $\omega_m \longrightarrow \omega$ uniformly. But since the ω_m are $\mathcal{D}$-sequences by our first observation it follows that ω is also a $\mathcal{D}$-sequence.

> LEMMA 13.2. If $p(t)$ is a real polynomial with at
> least one irrational coefficient then the average of
> $e(p(n))$ exists and equals 0.

This lemma is a result of H. Weyl ([16]) to which we shall give an alternative proof in Chapter 7. This lemma is equivalent to the statement that the sequence $e(p(n))$ is equidistributed on the circle $|z| = 1$, if $p(t)$ has an irrational coefficient — or to the statement that in every case, $e(p(n))$ is either periodic or equidistributed. It is by this lemma that one deduces above that all the Fourier coefficients of $\cos n^2\alpha$ vanish if α/π is irrational.

> LEMMA 13.3. If G is a compact abelian group with
> identity e and if $g \in G$ is an arbitrary element,
> then some subsequence of $\{g^n,\ n \geq 1\}$ converges to e.

PROOF. Let the closure of $\{g^n,\ n \geq 1\}$ in G be denoted G_1. G_1 is closed under multiplication by g so that $\{g^n G_1\}$ forms a monotonically decreasing set of closed subsets of G. Since G is compact

$$G_2 = \bigcap_{n \geq 1} g^n G_1$$

is non-empty. Since $g^{-1}(g^n G_1) = g^{n-1} G_1$ it follows that G_2 is closed under multiplication by g^{-1}. Hence G_2 is closed under multiplication by g^{-n} and therefore also by G_1^{-1}. Hence it is closed under multiplication by G_2^{-1} so that G_2 is a group. Hence $e \in G_2 \subset G_1$.

The principal result of this section is

> THEOREM 13.1. Every $\mathcal{D}$-sequence is regular and de-
> terministic.

PROOF. We first observe that the assertion that $\mathcal{D}$-sequences are deterministic will follow from the fact that they are regular. Note first that if all $\mathcal{D}$-sequences are regular then it follows that a $\mathcal{D}$-sequence that vanishes for all $n \leq -k$ must vanish for all n. Hence if two $\mathcal{D}$-sequences agree for all $n \leq -k$, their difference being a $\mathcal{D}$-sequence, it must vanish identically so that the two sequences must be identical. Suppose now that $\xi \in \mathcal{D}$ and that ξ is not deterministic. There then exist two sequences $\tilde{\xi}_1$ and $\tilde{\xi}_2$ in $\Omega(\xi)$ with $\beta(\tilde{\xi}_1) = \beta(\tilde{\xi}_2) = \xi$. If $\tilde{\xi}_1 \neq \tilde{\xi}_2$ there will be a k, $k \geq 0$, for which $\tilde{\xi}_1(k) \neq \tilde{\xi}_2(k)$. This means that the sequences $\beta(T^{-k}\tilde{\xi}_1)$ and $\beta(T^{-k}\tilde{\xi}_2)$ are distinct. Denote these as ξ'_1 and ξ'_2. Since $T^{-k}\tilde{\xi}_1$ and $T^{-k}\tilde{\xi}_2$ are in $\Omega(\xi)$ it follows that ξ'_1 and ξ'_2 belong to $\Omega^-(\xi)$ and hence are (pointwise) limits of translates of $\xi(n)$. By Lemma 13.1, ξ'_1 and ξ'_2 are $\mathcal{D}$-sequences. On the other hand $T^k\xi'_1 = T^k\beta(T^{-k}\tilde{\xi}_1) = \beta T^k(T^{-k}\tilde{\xi}_1) = \beta(\tilde{\xi}_1) = \beta(\tilde{\xi}_2) = T^k\xi'_2$ so that $\xi'_1(n) = \xi'_2(n)$ for $n \leq -k$. But then, as $\mathcal{D}$-sequences, ξ'_1 and ξ'_2 would have to be identical throughout contrary to assumption. Hence $\xi(n)$ must be deterministic.

We turn now to the proof that the sequences of $\mathcal{D}$ are regular. By Theorem 5.2 it suffices to prove that a dense subset of $\mathcal{D}$ is regular and hence we may restrict our attention to sequences of the form (13.1). Suppose than that

$$\xi(n) = \sum_{\nu=1}^{k} c_\nu e(p_\nu(n)) .$$

Let G denote the k-dimensional torus group; that is, the elements of G are k-tuples $(z_1, \ldots, z_k)$ with $|z_j| = 1$, $j = 1, 2, \ldots, k$. Using the exponentials defining $\xi(n)$ we may define a G-sequence $\zeta(n)$ by setting

$$(13.2) \qquad \zeta(n) = (e(p_1(n)), e(p_2(n)), \ldots, e(p_k(n))) .$$

$\xi(n)$ is then a derived sequence of $\zeta(n)$ and to show that $\xi(n)$ is a regular sequence it suffices to show that $\zeta(n)$ is regular. Now it is easily shown that $\zeta(n)$ is a stochastic sequence. Namely, for this we must show that the average of sequences

$$\psi_1(\zeta(n))\psi_2(\zeta(n - 1)) \cdots \psi_r(\zeta(n - r + 1))$$

exist for continuous functions ψ_j on G. Now the continuous functions on G can be approximated by linear combinations of the group characters on G and these have the form $x_{q_1,q_2,\ldots,q_k}(z_1, z_2, \ldots, z_k) = z_1^{q_1} z_2^{q_2} \ldots z_k^{q_k}$ for integers $q_1, q_2, \ldots, q_k$. Thus to prove that $\zeta(n)$ is stochastic it suffices to show that sequences of the form

$$(13.3) \qquad \eta(n) = \prod_{\nu,j} e(p_\nu(n - j))^{q_{\nu,j}}$$

possess averages. But $\eta(n)$ has the form $e(P(n))$ for some polynomial
P, so that $\eta(n)$ is either periodic or equidistributed by Lemma 13.2;
in either case η has an average which shows that $\zeta(n)$ is stochastic.

Hence to show that $\zeta(n)$ is regular we must show that one of
the conditions of Theorem 5.1 is satisfied; we will show that $\zeta \in \Omega^-(\zeta)$,
where $\Omega^-(\zeta)$ is thought of as imbedded in G_∞^-. Suppose that $\omega(n)$ is
a G-sequence and $g \in G$; let us define $g\omega$ as the G-sequence with en-
tries $g \cdot \omega(n)$. We first show that if $\zeta \notin \Omega^-(\zeta)$ then also $g\zeta \notin \Omega^-(\zeta)$
for every $g \in G$. To see this recall that by Theorem 4.2 all the se-
quences of $\Omega^-(\zeta)$ are limits of translates of ζ. If, then, $g\zeta \in \Omega^-(\zeta)$
we would have $T^{n_k}\zeta \longrightarrow g\zeta$ for some sequence $\{n_k\}$. But then $T^{n_k}g\zeta =$
$gT^{n_k}\zeta \longrightarrow g^2\zeta$ so that $g^2\zeta$ and similarly $g^n\zeta$ would belong to $\Omega^-(\zeta)$.
Now by Lemma 13.3 there is a sequence $g^{m_k} \longrightarrow e$, the identity of G so
that if all $g^n\zeta \in \Omega^-(\zeta)$ for $n \geq 1$ it would follow that $\zeta \in \Omega^-(\zeta)$,
contrary to supposition. Thus either $\zeta \in \Omega^-(\zeta)$ or no $g\zeta \in \Omega^-(\zeta)$, $g \in G$.

Consider now the G-sequence $\eta(n)$ defined by

$$(13.4) \qquad\qquad \eta(n) = \zeta(n)\zeta(n-1)^{-1} .$$

By Theorem 4.3, $\eta(n)$ defines a subprocess $X(\eta)$ of $X(\zeta)$ with variables
related to those of $X(\zeta)$ in the same way that $\eta(n)$ is related to $\zeta(n)$:
$y_n = z_n z_{n-1}^{-1}$. This, in turn, implies that for $\omega(n)$ a sequence in $\Omega^-(\zeta)$,
the image sequence in $\Omega^-(\eta)$ has the form $\bar{\omega}(n) = \omega(n)\omega(n-1)^{-1}$. We
also note that since $\Omega^-(\eta)$ is an identification space of $\Omega^-(\zeta)$, the
mapping $\omega \longrightarrow \bar{\omega}$ is onto.

From this we shall deduce that if $\zeta \notin \Omega^-(\zeta)$ then $\eta \notin \Omega^-(\eta)$.
For if $\eta \in \Omega^-(\eta)$ it would be the image of a sequence $\zeta'(n)$ in $\Omega^-(\zeta)$.
That is to say we would have

$$\eta(n) = \zeta'(n)\zeta'(n-1)^{-1} .$$

Comparing this with (13.4) we obtain

$$\zeta'(n)\zeta(n)^{-1} = \zeta'(n-1)\zeta(n-1)^{-1}$$

so that there is a $g \in G$ with $\zeta'(n) = g\zeta(n)$ for all n. But then
$g\zeta \in \Omega^-(\zeta)$ which we showed before was not possible if $\zeta \notin \Omega^-(\zeta)$. Hence
$\eta \notin \Omega^-(\eta)$. Consequently if $\zeta(n)$ is not regular, then neither is $\eta(n)$.

Now from (13.2) we have

$$\eta(n) = (e(p_1(n) - p_1(n - 1)), \ e(p_2(n) - p_2(n - 1)), \ \ldots, \ e(p_k(n) - p_k(n - 1)))$$

$$= (ep_1^*(n), \ ep_2^*(n), \ \ldots, \ ep_k^*(n))$$

where the degree of $p_j^*(t)$ is 1 less than the degree of $p_j(t)$. It
follows that if $\zeta(n)$ were not regular we could obtain by these means
a sequence $(e(p_1(n)), \ \ldots, \ e(p_k(n)))$ that is not regular and where the
$p_j(n)$ all have degree 0. But a constant sequence is obviously regular
so that our original assumption that $\zeta(n)$ is not regular leads to a
contradiction. This completes the proof of the theorem.

CHAPTER 4. SUBPROCESSES OF MARKOFF PROCESSES

The example of §12 of a derived sequence of a Markoff sequence
that is not continuously predictable suggests the problem of investigating
when a derived sequence of a Markoff sequence is continuously predictable.
In this chapter we shall find a sufficient condition for this to be the
case. The derived sequence of the Markoff sequence corresponds to a
certain subprocess of the corresponding Markoff process and the condition
we shall give involves the relationship of these two processes. We shall
deal exclusively with discrete Markoff processes which by Theorem 11.2
are continuously predictable. Using the transitivity of continuous predictability (§10.5) we find that the predictability of the subprocess
depends upon the possibility of prediction from the subprocess to the
Markoff process.

§14. Automorphism Groups

14.1. <u>Definitions</u>. We shall first illustrate the ideas of this
chapter by treating a particularly simple example. Suppose A and B are
two E-algebras, $A \subset B$, and suppose there is an automorphism of B preserving E, say τ, such that τ^2 is the identity and τ leaves A
elementwise fixed. Then, if the <u>only</u> elements left fixed by τ are
those in the subalgebra A it is easy to show that A is c.p. to B.
Namely for $f \in B$ we may write

$$f = \frac{f + \tau f}{2} + \frac{f - \tau f}{2} \quad .$$

Now

$$\tau \frac{f + \tau f}{2} = \frac{\tau f + f}{2}$$

so that

$$\frac{f + \tau f}{2} \in A \quad .$$

Also if $g \in A$

$$E\left(\frac{f - \tau f}{2} \cdot g\right) = \frac{1}{2} E(fg) - \frac{1}{2} E(\tau f \cdot g)$$

$$= \frac{1}{2} E(\tau f \cdot \tau g) - \frac{1}{2} E(\tau f \cdot g)$$

$$= \frac{1}{2} E(\tau f \cdot g) - \frac{1}{2} (\tau f \cdot g) = 0$$

so that

$$\frac{f - \tau f}{2} \in A^{\perp} \ .$$

Hence $B = A + A^{\perp}$ and by Theorem 10.2, A is c.p. to B. The general situation to be expected as a result of this is that if a subalgebra is the set of fixed elements of an appropriate group of transformations of an E-algebra then the subalgebra will be c.p. to the algebra. In this chapter we shall show that this is the case when B is the algebra of a Markoff process and the transformations are restricted in certain ways. This result will then be used to deduce the continuous predictability of a class of sequences derived from a Markoff sequence.

Throughout this chapter X and Y will be two processes with X a subprocess of Y. Eventually Y will be specialized to be a Markoff process with discrete state space. The inclusion $A_X^- \subset A_Y^-$ induces a map Φ of Ω_Y^- onto Ω_X^-. We shall refer to the subsets of Ω_Y^- of the form $\Phi^{-1}(\Delta)$ for $\Delta \subset \Omega_X^-$ as X-<u>sets</u>. Together with an algebraic automorphism of the algebra A_Y^- we shall always have in mind the induced homeomorphism of Ω_Y^- and both will be referred to with the same symbol.

DEFINITION 14.1. The <u>group of</u> Y <u>over</u> X, denoted $G(Y|X)$, is the set of automorphisms τ of A_Y^- satisfying

    (i) τ leaves invariant the remote past, i.e., there
            exists an $n \geq 0$ such that $\tau(T^n x) = T^n x$ for
            every x in A_Y^-.

    (ii) τ is the identity on A_X^-.

From the group $G(Y|X)$ we may work backwards and study the sub-algebra of A_Y^- of elements left fixed by the group. This would lead to a notion of "normal extension" by analogy with the theory of field extensions, where the set of fixed elements is just A_X^-. The following notion is a slight modification of this.

DEFINITION 14.2. Y <u>is normal over</u> X if the only closed sub-sets of Ω_Y^- invariant under the homeomorphisms of $G(Y|X)$ are closed X-sets.

That is to say, if Y is normal over X and $\Delta \subset \Omega_Y^-$ satisfies $\tau\Delta = \Delta$ for all $\tau \in G(Y|X)$ then $\Delta = \Phi^{-1}(\Delta')$, $\Delta' \subset \Omega_X^-$. Clearly sets Δ of this form, i.e., closed X-sets, are left invariant by $G(Y|X)$. For if $g \in A_X^-$ and $\tau \in G(Y|X)$ then $\tau g(\omega) = g(\omega)$ so that $g(\tau\omega) = g(\omega)$ which implies that ω and $\tau\omega$ are identified under the map Φ and if ω belongs to an X-set, so does $\tau\omega$.

To see how this notion of normality is related to the simpler one preceding the definition we remark that if Y is normal over X then the only elements left fixed by $G(Y|X)$ are those of A_X^-. For suppose that $g \in A_Y^-$ and that $\tau g = g$ for all $\tau \in G(Y|X)$. Consider two points ω_1 and ω_2 with $\Phi(\omega_1) = \Phi(\omega_2)$, and consider the closure of the set $\{\omega : \omega = \tau\omega_1, \ \tau \in G(Y|X)\}$. This is a closed set invariant under $G(Y|X)$, hence an X-set and since ω_1 belongs to the set, so does ω_2. It follows that $\tau\omega_1$ comes arbitrarily close to ω_2 and since $g(\tau\omega_1) = g(\omega_1)$ it follows that $g(\omega_2) = g(\omega_1)$. Then whenever $\Phi(\omega_1) = \Phi(\omega_2)$, $g(\omega_1) = g(\omega_2)$ from which it follows that $g = g' \cdot \Phi$ so that $g \in A_X^-$ (§1.1).

However the statement that the only invariant closed sets of Ω_Y^- are X-sets is stronger than the requirement that only elements of A_X^- should be fixed under $G(Y|X)$. That is, a group of homeomorphisms may have a certain subalgebra as fixed set but still have invariant closed sets other than those defined by the subalgebra. Since we shall need the stronger notion we have formulated the definition of normality in terms of homeomorphisms rather than automorphisms.

DEFINITION 14.3. Y <u>is normal over</u> X <u>at a point</u> ξ in Ω_X^- if for any closed set $\Delta \subset \Omega_Y^-$ invariant under the homeomorphisms of $G(Y|X)$ either $\Phi^{-1}(\xi) \cap \Delta = \phi$, the empty set, or $\Phi^{-1}(\xi) \subset \Delta$.

If Y is normal over X at each $\xi \in \Omega_X^-$ then for each ξ, $\Phi^{-1}(\xi)$ is either disjoint from or contained in Δ, if Δ is an invariant set. It follows that Δ must have the form $\Phi^{-1}(\Delta')$ for $\Delta' \subset \Omega_X^-$ and hence normality over X is equivalent to normality at every point of Ω_X^-.

We may now state the principal results of this chapter.

THEOREM 17.1. If X and Y are finitely valued processes, X a subprocess of Y and Y an m-Markoff process, then A_Y^- is c.p. to A_X at a point ξ of Ω_X^- if Y is normal over X at some limit point of translates $T^m\xi$ of the point ξ.

THEOREM 17.2. If X and Y are finitely valued processes, X a subprocess of Y and Y an

m-Markoff process, then X is c.p. if Y is
normal over X.

14.2. <u>An Illustration</u>. Consider the example of §12 where we
formed a subprocess X of a random process that is not c.p.. Y is
here a random $\{-1, 1\}$-process and X is defined by $x_n = y_n y_{n-1}$. Ω_Y^-
and Ω_X^- both consist of left-infinite $\{-1, 1\}$-sequences and there is
a map $\Phi : \Omega_Y^- \longrightarrow \Omega_X^-$ given by $\Phi(\eta)(n) = \eta(n)\eta(n-1)$. The map Φ is
$2 - 1$, i.e., each point of Ω_X^- has exactly two preimages in Ω_Y^-, for
if $\Phi(\eta) = \xi$ then $\Phi(\eta') = \xi$ where $\eta'(n) = -\eta(n)$. Consider now the
group $G(Y|X)$ for this pair of processes. If $\tau \in G(Y|X)$ then since
τ leaves A_X^- invariant the homeomorphism τ must satisfy $\Phi(\eta) = \Phi(\tau\eta)$
for all η in Ω_Y^-. Hence $\tau\eta$ is either η or η'. Suppose that
$\tau\eta = \eta'$ then $\tau y_n(\eta) = y_n(\tau\eta) = y_n(\eta') = -y_n(\eta)$ since $\eta'(n) = -\eta(n)$
for all n. But then condition (i) of Definition 14.1 is violated since
τ would leave no y_n invariant. Hence $\tau\eta = \eta$ and hence $G(Y|X)$ reduces to the identity. This implies that Y is not normal over X
at any point. For if $\xi \in \Omega_X^-$ and $\Phi(\eta) = \xi$ then since any set is in-
variant under $G(Y|X)$, $\{\eta\}$ is an invariant closed set. Since
$\Phi^{-1}(\xi) \cap \{\eta\} \neq \phi$ and also $\Phi^{-1}(\xi) \not\subset \{\eta\}$ the requirements of normality
over ξ is violated. Thus the case considered is one where the Markoff
process is not normal over the subprocess and hence is not covered by
the present theorem. It should be pointed out however that neither does
the present theorem cover the case for which $P\{y_n = 1\} = P\{y_n = -1\} = 1/2$
where the argument of §12 fails and where $x_n = y_n y_{n-1}$ is again random
and hence c.p. Hence the conditions of the theorems stated here are not
necessary. This is not surprising since the conditions are of a topological and algebraic nature and ignore the measure theoretic aspects
of the processes.

§15. Linear Transformations of Cones

15.1. <u>The Projective Metric</u>. In several connections, in particular in the proofs of the above theorems, we shall require certain re-
suls regarding linear transformations of cones. It will be convenient to
collect these results in this section.

A cone V in a linear space D will be a subset closed under
addition and multiplication by positive real numbers. We shall always
suppose that the cones we deal with satisfy

(*) If $x, y \in D$ and $x + \lambda y \in V$ for all real λ, then
$y = 0$. The effect of condition (*) is to exclude cones that contain
complete lines rather than just half-lines. For cones satisfying (*) we

may define a function $\Gamma_V(x, y)$ on pairs of points which acts as a measure of the relative separation of the points x and y in the cone.

DEFINITION 15.1. The <u>projective distance</u> between two points $x, y \in V$, $x, y \neq 0$ is given by $\Gamma_V(x, y) = K_V(x, y)K_V(y, x)$ where

$$(15.1) \qquad K_V(x, y) = \sup \ \{\lambda: x - \lambda y \in V\}$$

Since $x + \lambda y \in V$ for positive λ, by (*) there must be a λ such that $x - \lambda y \notin V$ and hence $K_V(x, y)$ is finite for all x, y, so that Γ_V is well defined.

LEMMA 15.1. $\Gamma_V(x, y)$ satisfies the following:

- (i) $0 \leq \Gamma_V(x, y) \leq 1$
- (ii) $\Gamma_V(\lambda x, \mu y) = \Gamma_V(x, y)$ for all $\lambda, \mu > 0$
- (iii) $\Gamma_V(x, y) = 1$ if and only if $x = \lambda y$ for some $\lambda > 0$.
- (iv) If $L : D \longrightarrow D'$ is a linear transformation of the space D into a space D' and $L(V) \subset W$, a cone in D', then $\Gamma_W(Lx, Ly) \geq \Gamma_V(x, y)$.

PROOF. We first show that $\Gamma_V(x, x) = 1$ for all $x \in V$. Since $K_V(x, x) \geq 1$ we have $\Gamma_V(x, x) \geq 1$. If $K_V(x, x) > 1$ then $- x \in V$, hence $\lambda x \in V$ for all real λ so that by (*) $x = 0$ which is excluded for Γ_V is only defined for $x, y \neq 0$. We next prove the triangle inequality

$$(15.2) \qquad \Gamma_V(x, z) \geq \Gamma_V(x, y)\Gamma_V(y, z) \quad .$$

For this it suffices to prove the analogous inequality for K_V. Now by (15.1), $x - (K_V(x, y) - \varepsilon)y \in V$, $y - (K_V(y, z) - \varepsilon)z \in V$. Multiplying the second by $K_V(x, y) - \varepsilon$ and adding to the first we find that $x - [K_V(x, y) - \varepsilon][K_V(y, x) - \varepsilon]z \in V$ so that $K_V(x, z) \geq [K_V(x, y) - \varepsilon]$ $[K_V(y, x) - \varepsilon]$ for every $\varepsilon > 0$ and from this the result follows.

To prove (i) we observe that $\Gamma_V(x, y) = \Gamma_V(y, x)$ and since $\Gamma_V(x, x) = 1$ then by (15.2), $1 \geq \Gamma_V(x, y)^2$.

(ii) follows from the fact that $K_V(\lambda x, \mu y) = \frac{\lambda}{\mu} K_V(x, y)$.

The "if" portion of (iii) has already been shown, so suppose that we are given $\Gamma_V(x, y) = 1$. Write $K_V(x, y) = \lambda$; then $K_V(x, \lambda y) = 1$, so we may suppose to begin with that $K_V(x, y) = 1$ and therefore $K_V(y, x) = 1$. Then by (15.1)

$$x - (1 - \varepsilon)y \in V, \quad y - (1 - \varepsilon)x \in V \quad ,$$

for arbitrary $\varepsilon > 0$. Thus

$$\frac{(x-y)}{\varepsilon_1} + y \in V, \quad -\frac{(x-y)}{\varepsilon_2} + x \in V$$

for arbitrary $\varepsilon_1, \varepsilon_2 > 0$. Now $1/\varepsilon_1 - 1/\varepsilon_2$ can take on any real value so that adding the preceding elements of V we find

$$x + y + \lambda(x - y) \in V$$

for arbitrary λ and by (*) $x = y$. Thus in general $x = \lambda y$.

Finally (iv) is immediate since if $x - \lambda y \in V$, $Lx - \lambda Ly \in W$ so that by (15.1) $K_W(Lx, Ly) \geq K_V(x, y)$.

Lemma 15.1 shows that if we identify points on the same ray of V, we obtain a topological space with metric given by $-\log \Gamma_V(x, y)$. (iv) shows moreover that the linear transformations of such a space will always be contractions in this metric.

As an example take $D = R^m$, m-dimensional Euclidean space, and V the set of m-tuples with no negative components. V clearly satisfies (*). Now suppose that $x, y \in V$ $x = (x_1, x_2, \ldots, x_m)$, $y = (y_1, y_2, \ldots, y_m)$. If $y_i = 0$ and $x_i \neq 0$ for some i, then clearly $y - \lambda x$ will always have a negative component for $\lambda > 0$ so that $K_V(y, x) = 0$. It follows that unless the two vectors have the same components vanishing $\Gamma_V(x, y) = 0$. Otherwise we have

$$K_V(x, y) = \inf_{x_i \neq 0} y_i/x_i$$

so that

$$(15.3) \qquad \Gamma_V(x, y) = \inf_{x_i, y_j \neq 0} x_j y_i / x_i y_j \quad .$$

15.2. Projectively Bounded Transformations.

DEFINITION 15.2. A linear transformation of a cone $V \subset D$ into a cone $W \subset D'$ is projectively bounded if for all $x, y \in V$ with $Lx, Ly \neq 0$

$$(15.4) \qquad \Gamma_W(Lx, Ly) \geq \delta \quad ,$$

for a fixed $\delta > 0$. The largest δ for which (15.4) holds is the projective bound of L, $\delta = \Gamma(L)$.

In the finite dimensional case, a transformation $L : R^M \longrightarrow R^n$ taking the cone of non-negative vectors of R^m into the corresponding vectors of R^n, may be represented by a matrix with non-negative entries (L_{ij}). The condition that L is projectively bounded may be stated as follows: whenever $L_{i_0 j_0} = 0$ then either all $L_{i_0 j} = 0$ or all $L_{i j_0} = 0$. To see this, suppose first that this condition is fulfilled. Then for all i, j, r, s, either both or neither of $L_{ir} L_{js}$ and $L_{jr} L_{is}$ vanish. Therefore there will be a $\delta > 0$ such that $L_{ir} L_{js} \geq \delta\, L_{jr} L_{is}$ for all i, j, r, s. From this we obtain

$$
(15.5) \qquad \sum_{r,s} L_{ir} L_{js} x_r y_s \geq \delta \sum_{r,s} L_{jr} L_{is} x_r y_s
$$

or setting $u = Lx$, $v = Ly$, $u_i v_j \geq \delta u_j v_i$. It follows now from (15.3) and (15.4) that L is bounded. Conversely if L is bounded (15.5) holds for all choice of non-negative $\{x_r\}$ and $\{y_s\}$ so that in particular $L_{ir} L_{js} \geq \delta L_{jr} L_{is}$ and from this we obtain the above condition. Another form of this condition is that all the entries of (L_{ij}) should vanish but for those for which $i \in I \subset \{1, \dots, m\}$ and $j \in J \subset \{1, \dots, n\}$ for these $L_{ij} > 0$. In particular a matrix with all positive entries is projectively bounded.

The following lemma will be of importance to us.

LEMMA 15.2. If for all x, y ϵ V, (15.4) holds then

$$
(15.6) \qquad \Gamma_W(Lx, \, Ly) \geq \Gamma_V(x, \, y) + \frac{\delta}{1 + \delta\Gamma_V(x,y)}\,(1 - \Gamma_V(x, \, y)^2) \quad .
$$

PROOF. Let $K_\epsilon(x, \, y) = K_V(x, \, y) - \epsilon$ and $\Gamma_\epsilon(x, \, y) = K_\epsilon(x, \, y)\, K_\epsilon(y, \, x)$. Setting $x_\epsilon = x - K_\epsilon(x, \, y)y$, $y_\epsilon = y - K_\epsilon(y, \, x)x$, it follows that x_ϵ, $y_\epsilon \, \epsilon \, V$. By the definition of Γ_W and (15.4) if $\delta' < \delta$ there will be numbers λ , $\mu > 0$ with $\lambda\mu > \delta'$ for which

$$
Lx_\epsilon - \lambda Ly_\epsilon \, \epsilon \, W, \;\; Ly_\epsilon - \mu Lx_\epsilon \, \epsilon \, W \quad .
$$

These two may be rewritten:

$$
Lx - \frac{\lambda + K_\epsilon(x,y)}{1 + \lambda K_\epsilon(y,x)}\, Ly \, \epsilon \, W \quad ,
$$

$$Ly - \frac{\mu + K_\varepsilon(y,x)}{1 + \mu K_\varepsilon(x,y)} \, Lx \ \epsilon \ W \quad .$$

Hence

$$K_W(Lx, \ Ly) \geq \frac{\lambda + K_\varepsilon(x,y)}{1 + \lambda K_\varepsilon(y,x)} \ , \ K_W(Ly, \ Lx) \geq \frac{\mu + K_\varepsilon(y, \ x)}{1 + \lambda K_\varepsilon(x,y)}$$

and

$$(15.7) \qquad \Gamma_W(Lx, \ Ly) \geq \frac{\lambda\mu + \Gamma_\varepsilon(x,y) + \lambda K_\varepsilon(y,x) + \mu K_\varepsilon(x,y)}{1 + \lambda\mu\Gamma_\varepsilon(x,y) + \lambda K_\varepsilon(y,x) + \mu K_\varepsilon(x,y)} \quad .$$

Now for any positive numbers $\alpha, \ \beta, \ \alpha', \ \beta',$ if $\alpha/\beta \geq \alpha'/\beta'$
then

$$\alpha/\beta \geq \frac{\alpha + \alpha'}{\beta + \beta'} \geq \alpha'/\beta' \quad .$$

Since

$$\Gamma_W(Lx, \ Ly) \leq 1 = \frac{\lambda K_\varepsilon(y,x) + \mu K_\varepsilon(x,y)}{\lambda K_\varepsilon(y,x) + \mu K_\varepsilon(x,y)}$$

it follows from (15.7)

$$\Gamma_W(Lx, \ Ly) \geq \frac{\lambda\mu + \Gamma_\varepsilon(x,y)}{1 + \lambda\mu\Gamma_\varepsilon(x,y)} \qquad .$$

Now letting $\varepsilon \longrightarrow 0$ and $\delta' \longrightarrow \delta$ we find

$$(15.8) \qquad\qquad \Gamma_W(Lx, \ Ly) \geq \frac{\delta + \Gamma_V(x,y)}{1 + \delta\Gamma_V(x,y)}$$

and from this the assertion of the lemma follows readily.

   We notice that the expression on the right side of (15.8) is
simultaneously $\geq \delta$ and $\geq \Gamma_V(x, \ y),$ thus yielding two inequalities that
we already knew. Thus whenever (15.4) holds for a transformation L, a
stronger inequality may be inferred.

§16. A Sufficient Condition For Continuous Predictability

   16.1. <u>Automorphisms and Their Adjoints</u>. The process Y will,
in this section and for most of the next be an ergodic Markoff process
with finite state space. Also we take X to be a subprocess of Y

defined by $x_n = \psi(y_n)$ and we shall find conditions for A_X^- to be continuously predictable to A_Y^- at a point $\xi \in \Omega_X^-$. Since Y is c.p. these conditions will imply that ξ itself is a continuously predictable sequence.

Let M be the (finite) state space of Y and Λ that of X so that $\psi : M \longrightarrow \Lambda$. We shall assume that every element of M actually occurs with non-zero probability. The transition matrix of Y will be written $(p(b_i, b_j))$ where b_i and b_j range over M. Ω_Y^- is imbedded in M_∞^- and in terms of the transition matrix of Y it is possible to determine just which sequences occur in Ω_Y^-. Namely $\omega \in \Omega_Y^-$ if and only if $p(\omega(n), \omega(n + 1)) > 0$ for all $n \leq 0$. For if $\omega \in \Omega_Y^-$ then the open set of $\Omega_Y^- : \{\omega' : y_n(\omega') = \omega(n), y_{n+1}(\omega') = \omega(n + 1)\}$ is non-empty and if $p(\omega(n), \omega(n + 1)) = 0$ this set would have measure 0. Conversely if all transitions in ω have positive probability, then the subset of Ω_Y^- given by $\{\omega' : y_n(\omega') = \omega(n), y_{n+1}(\omega') = \omega(n + 1), \ldots, y_0(\omega') = \omega(0)\}$ has positive probability and hence is non-empty. Therefore the intersection of these sets, i.e., the point ω, belongs to Ω_Y^-.

We shall be dealing with both finite and infinite M-sequences of the form $\sigma = (\sigma(q), \sigma(q + 1), \ldots, \sigma(0))$. Here and in the future q will represent an integer ≤ 0. In either case we call σ a <u>possible</u> sequence if $p(\sigma(j), \sigma(j + 1)) > 0$ for j in the domain of definition of σ. Thus the sequences of Ω_Y^- are the possible left-infinite sequences. When ρ is a finite sequence and σ an arbitrary sequence we define $\sigma\rho$ as the juxtaposition of the two sequences

$$\sigma\rho = (\sigma(q), \sigma(q + 1), \ldots, \sigma(0), \rho(r), \rho(r + 1), \ldots, \rho(0)) \ .$$

By Δ_σ^- we denote the subset of M_∞^- of sequences terminating in σ. x_σ^- denotes the characteristic function of Δ_σ^- so that $x_\sigma^-(\omega)$ is 1 or 0 according as ω does or does not terminate in σ. We note the relationship

$$(16.1) \qquad x_\sigma^- = \sum_{b \in M} x_{b\sigma}^- \ .$$

The mapping ψ of M onto Λ induces the map Φ of Ω_Y^- onto Ω_X^-, and we clearly have

$$(16.2) \qquad \Phi(\omega)(n) = \psi(\omega(n)) \ .$$

If $\xi \in \Omega_X^-$ we shall say that an M-sequence σ <u>lies over</u> ξ if

(16.3) $\psi(\sigma(j)) = \xi(j)$

for the j for which σ is defined. The sequences in $\Phi^{-1}(\xi)$ are those left-infinite M-sequences lying over ξ.

A_X^- will be c.p. to A_Y^- at ξ if there is a unique prediction measure at ξ. We shall first determine properties that are common to all prediction measures from A_X^- to A_Y^-. These are derived from the following fundamental lemma:

> LEMMA 16.1. Let A and B be two E-algebras,
> $A \subset B$, and let μ be a prediction measure at some
> point of Ω_A. Suppose τ is a linear transforma-
> tion of B which is the identity on A and $\tau*$
> is its adjoint, in the sense that $E(\tau x \cdot y) =$
> $E(x \cdot \tau*y)$ for all $x, y \in B$. Then $\mu(\tau*x) = \mu(x)$
> for all x in B.

PROOF. For all $y \in A$, $\tau y = y$ so that $E(\tau*x \cdot y) = E(x \cdot \tau y) = E(x \cdot y)$. Hence $E((\tau*x - x)y) = 0$ for all $y \in A$, so that by Definition 10.1 (with $U_\varepsilon = \Omega_A$) $\mu(\tau*x - x) \geq 0$, and also $\mu(x - \tau*x) \geq 0$ so that $\mu(\tau*x) = \mu(x)$.

We return to the processes X and Y and the sequence $\xi \in \Omega_X^-$. Consider two finite M-sequences σ_1 and σ_2 of the same length which begin with the same element and which both lie over ξ.

$$\sigma_1 = (\sigma_1(q), \sigma_1(q + 1), \ldots, \sigma_1(0))$$

$\sigma_1(q) = \sigma_2(q)$, $\psi(\sigma_1(j)) = \psi(\sigma_2(j)) = \xi(j)$, and $p(\sigma_i(j), \sigma_i(j + 1)) > 0$ for $i = 1, 2$ and $j \geq q$. We now define a homeomorphism $\tau_{\sigma_1 \sigma_2}$ of Ω_Y^- as follows:

(16.4)
$$\tau_{\sigma_1 \sigma_2}(\omega)(n) = \begin{cases} \sigma_1(n) & \text{if } n \geq q \text{ and } \omega \in \Delta_{\sigma_2}^- \\ \sigma_2(n) & \text{if } n \geq q \text{ and } \omega \in \Delta_{\sigma_1}^- \\ \omega(n) & \text{otherwise.} \end{cases}$$

Thus $\tau_{\sigma_1 \sigma_2}$ leaves a point fixed unless it terminates in either σ_1 or σ_2 in which case the last $|q|$ terms are to be replaced by those of σ_2 or σ_1 respectively. For $\tau_{\sigma_1 \sigma_2}$ to be well defined it must take possible sequences into possible sequences. However this is clear since σ_1 and σ_2 have the same first term. We also note that if ω lies over a point $\eta \in \Omega_X^-$ then so does $\tau_{\sigma_1 \sigma_2}(\omega)$ since $\tau_{\sigma_1 \sigma_2}$ at worst interchanges the terms of σ_1 and σ_2 and these have the same image under ψ. It follows that $\tau_{\sigma_1\sigma_2}$ gives rise to an automorphism of A_Y^- which leaves A_X^- fixed. To apply Lemma 16.1 we shall have to evaluate the adjoint of $\tau_{\sigma_1\sigma_2}$.

LEMMA 16.2. $\tau_{\sigma_1\sigma_2}$ has an adjoint $\tau^*_{\sigma_1\sigma_2}$ defined by

$$(16.5) \qquad \tau^*_{\sigma_1\sigma_2} f(\omega) = \begin{cases} \tau_{\sigma_1\sigma_2} f(\omega)\, \dfrac{p(\sigma_2(q),\sigma_2(q+1))\ldots p(\sigma_2(-1),\sigma_2(0))}{p(\sigma_1(q),\sigma_1(q+1))\ldots p(\sigma_1(-1),\sigma_1(0))} & \text{if } \omega \in \Delta^-_{\sigma_1} \\[2ex] \tau_{\sigma_1\sigma_2} f(\omega)\, \dfrac{p(\sigma_1(q),\sigma_1(q+1))\ldots p(\sigma_1(-1),\sigma_1(0))}{p(\sigma_2(q),\sigma_2(q+1))\ldots p(\sigma_2(-1),\sigma_2(0))} & \text{if } \omega \in \Delta^-_{\sigma_2} \\[2ex] \tau_{\sigma_1\sigma_2} f(\omega) & \text{otherwise.} \end{cases}$$

PROOF. To establish that

$$(16.6) \qquad E(\tau^*_{\sigma_1\sigma_2} f \cdot g) = E(f \cdot \tau_{\sigma_1\sigma_2} g)$$

for all f and g in A_Y^- it suffices to do so for a set of functions whose linear combinations are dense in A_Y^-. The functions x_σ^-, σ finite, form such a set (an application of the Stone-Weierstrass Theorem) so that it suffices to show

$$(16.7) \qquad E(\tau^*_{\sigma_1\sigma_2} x_\rho^- \cdot x_\sigma^-) = E(x_\rho^- \cdot \tau_{\sigma_1\sigma_2} x_\sigma^-) \ .$$

By (16.1) we see that it suffices to consider σ and ρ with length greater than any preassigned value since the other x_σ^- are linear combinations of those with σ of greater length. In particular we may suppose that the length of σ and ρ is greater than that of σ_1 or σ_2. Moreover increasing lengths again if necessary we may suppose that ρ and σ have the same length. Now both sides of (16.7) reduce to $E(x_\rho^- \cdot x_\sigma^-)$ if neither ρ nor σ terminates in either σ_1 or σ_2, so suppose that $\sigma = \sigma'\sigma_1$. Moreover both sides of (16.7) vanish unless ρ terminates in σ_2 if σ terminates in σ_1 so we take $\rho = \rho'\sigma_2$. Finally both sides of (16.7) will still vanish unless $\rho' = \sigma'$ so we set $\sigma = \sigma'\sigma_1$ and $\rho = \sigma'\sigma_2$.

Define $\theta(\omega)$ as the quotient occurring in the definition of $\tau^*_{\sigma_1\sigma_2}$. The lemma reduces to the equality

$$(16.8) \qquad E(\theta \bar{x}_{\sigma'\sigma_1} \cdot \bar{x}_{\sigma'\sigma_1}) = E(\bar{x}'_{\sigma'\sigma_2} \cdot \bar{x}_{\sigma'\sigma_2})$$

or

$$(16.9) \qquad E(\theta \bar{x}_{\sigma'\sigma_1}) = E(\bar{x}_{\sigma'\sigma_2}) \quad .$$

Since $x_{\sigma'\sigma_1}$ vanishes except for ω terminating in σ_1, we may take for θ in (16.9) the constant value

$$\frac{p(\sigma_2(q),\ \sigma_2(q+1)) \ \ldots \ p(\sigma_2(-1),\ \sigma_2(0))}{p(\sigma_1(q),\ \sigma_1(q+1)) \ \ldots \ p(\sigma_1(-1),\ \sigma_1(0))}$$

and then (16.9) follows immediately by evaluating

$$E(\bar{x}_{\sigma'\sigma_1}) = P(\bar{\Delta}_{\sigma'\sigma_1}), \quad E(\bar{x}_{\sigma'\sigma_2}) = P(\bar{\Delta}_{\sigma'\sigma_2})$$

in terms of the transition probabilities of Y.

We can now prove the following lemma.

LEMMA 16.3. If μ is a prediction measure from $\bar{A}_X$ to $\bar{A}_Y$ at ξ then

$$(16.10) \qquad \mu(\bar{x}_\sigma) = \begin{cases} 0 \ \text{if} \ \sigma \ \text{does not lie over} \ \xi \\[2mm] cp(\sigma(q),\ \sigma(q+1)) \ \ldots \ p(\sigma(-1),\ \sigma(0)) \\ \text{if} \ \sigma \ \text{lies over} \ \xi \ , \end{cases}$$

where c depends on q and $\sigma(q)$ but not upon $\sigma(q+1),\ \sigma(q+2),\ \ldots,\ \sigma(0)$.

PROOF. If σ does not lie over ξ then $\bar{\Delta}_\sigma \cap \Phi^{-1}(\xi)$ is empty and since by Lemma 10.1 a prediction measure at ξ has its support in $\Phi^{-1}(\xi)$, $\mu(\bar{x}_\sigma) = 0$. Now if σ lies over ξ we can assume that σ is a possible sequence for if not then both sides of (16.10) vanish. To complete the proof we have only to show that if σ_1 and σ_2 are two possible sequences lying over ξ, beginning with the same term, then

$$(16.11) \qquad \frac{\mu(\bar{x}_{\sigma_1})}{\mu(\bar{x}_{\sigma_2})} = \frac{p(\sigma_1(q),\sigma_1(q+1))\ldots p(\sigma_1(-1),\sigma_1(0))}{p(\sigma_2(q),\sigma_2(q+1))\ldots p(\sigma_2(-1),\sigma_2(0))} \quad .$$

But by Lemmas 16.1 and 16.2, $\mu(x_{\sigma_1}^-) = \mu(\tau^*_{\sigma_1 \sigma_2} x_{\sigma_1}^-)$ and by (16.5)

$$\tau^*_{\sigma_1 \sigma_2} x_{\sigma_1}^- = x_{\sigma_2}^- \frac{p(\sigma_1(q),\sigma_1(q+1))\ldots p(\sigma_1(-1),\sigma_1(0))}{p(\sigma_2(q),\sigma_2(q+1))\ldots p(\sigma_2(-1),\sigma_2(0))}$$

since for $x_{\sigma_2}^-(\omega) \neq 0$, $\omega \in \Delta_{\sigma_2}^-$. From this the lemma follows directly.

16.2. <u>Construction of the Prediction Measures</u>. Consider now the fixed sequence $\xi \in \Omega_X^-$ and for each $q \leq 0$ let $M_q(\xi)$ represent the subset of elements b of M for which there exists a possible sequence $\omega(n)$ lying over ξ with $\omega(q) = b$. Note that $M_q(\xi) \subset \psi^{-1}(\xi(q))$ but $M_q(\xi)$ is not necessarily the same as $\psi^{-1}(\xi(q))$. (For example, if $b \in \psi^{-1}(\xi(q))$ but no possible successor b' of b satisfies $b' \in \psi^{-1}(\xi(q+1))$, then $b \notin M_q(\xi)$.) Let Π denote the transition matrix of Y, $\Pi(b, b') = p(b, b')$, and let $\Pi_q(\xi)$ be the matrix obtained from Π by replacing all entries by 0 except for the entries $\Pi(b, b')$ with $b \in M_q(\xi)$ and $b' \in M_{q+1}(\xi)$. We shall denote the terms of the matrix $\Pi_q(\xi)$ as $\Pi_q(b, b')$.

Now let μ be a fixed prediction measure at ξ from A_X^- to A_Y^-. Define the functions C_q on M by setting $C_q(b) = 0$ if $b \notin M_q(\xi)$ and taking $C_q(b)$ to be the constant c of Lemma 16.3 if $b \in M_q(\xi)$. That is, if $b \in M_q(\xi)$, then b is the first term of some possible M-sequence lying over ξ and then C is determined by q and b. Finally for $q = 0$ where Lemma 16.3 does not apply, we take $C_0(b) = \mu(x_b^-)$.

LEMMA 16.4. The functions C_q satisfy

(i) $\Sigma_{b \in M} C_0(b) = 1$

(ii) If $\sigma = (\sigma(q), \sigma(q + 1), \ldots, \sigma(0))$ then

$$. \quad \mu(x_\sigma^-) = C_q(\sigma(q))\Pi_q(\sigma(q), \sigma(q + 1)) \ldots \Pi_{-1}(\sigma(-1), \sigma(0))$$

(iii) $\Sigma_{b \in M} C_q(b) \, \Pi_q(b, b') = C_{q+1}(b')$.

PROOF. (i) follows from $\Sigma x_b^- = 1$. For (ii) we suppose first that some $\sigma(j) \notin M_j(\xi)$. Then there is no possible sequence over ξ terminating in σ so that $\mu(x_\sigma^-) = 0$. On the other hand $\Pi_{j-1}(\sigma(j - 1), \sigma(j)) = \Pi_j(\sigma(j), \sigma(j + 1)) = 0$ if $\sigma(j) \notin M_j(\xi)$ so that the right hand side of (ii) also vanishes. We may therefore suppose that all the $\sigma(j) \in M_j(\xi)$ in which case the right hand side of (ii) becomes $c_q(\sigma(q))p(\sigma(q), \sigma(q + 1)) \ldots p(\sigma(-1), \sigma(0))$ and (ii) follows from (16.10).

Finally (iii) follows from (ii) if we use $\Sigma_{b \in M} x^-_{b\sigma} = x^-_\sigma$. Namely by (ii) we have

$$\sum_{b \in M} C_q(b) \Pi_q(b, \; b') \Pi_{q+1}(b', \; \sigma(q+2)) \; \ldots \; \Pi_{-1}(\sigma(-1), \; \sigma(0)) =$$

$$C_{q+1}(b') \Pi_{q+1}(b', \; \sigma(q+2)) \; \ldots \Pi_{-1}(\sigma(-1), \; \sigma(0)) \quad ,$$

for any sequence $\sigma(q + 2)$, ..., $\sigma(0)$. (iii) will follow if there is a sequence for which $\Pi_{q+1}(b', \; \sigma(q + 2)) \ldots \Pi_{-1}(\sigma(- 1), \; \sigma(0)) \neq 0$. But if all these products vanish then for all possible sequences satisfying $\sigma(q + 1) = b'$ we must have $\sigma(j) \not\in M_j(\xi)$ for some j. But this implies that $b' \not\in M_{q+1}(\xi)$ and in this case both sides of (iii) vanish. For $q = -1$ (iii) follows from (ii) and the fact that $\Sigma_{b \in M} \mu(x^-_{bb'}) = \mu(x^-_{b'})$. This completes the proof of the lemma.

The functions C_q on M may be regarded as vectors in R^m if m is the number of elements in M. The $\Pi_q(\xi)$ may furthermore be regarded as transformations of R^m taking the cone of non-negative vectors into itself. We take the $\Pi_q(\xi)$ as operating on row vectors from the right so that for $u \in R^m$, $u\Pi_q(\xi)$ has components $\Sigma_{b \in M} u(b) \Pi_q(b, \; b')$. Equation (iii) of the preceding lemma then reads

$$(16.12) \qquad\qquad C_{q+1} = C_q \Pi_q(\xi) \quad .$$

We can now state the following sufficient condition for continuous predictability:

THEOREM 16.1. If as transformations of the cone V of non-negative m-tuples the matrices $\Pi_q(\xi)$ satisfy

$$(16.13) \qquad\qquad \Gamma(\Pi_q(\xi) \; \Pi_{q+1}(\xi) \; \ldots \; \Pi_\ell(\xi)) \longrightarrow 1$$

as $q \longrightarrow - \infty$ for every $\ell \leq 0$, then A^-_X is continuously predictable to A^-_Y at ξ, and $\xi(n)$ is itself continuously predictable.

PROOF. We must show that the prediction measure μ is unique. Now by Lemma 16.4 (ii), μ is determined by the functions C_q. Theorem 20.1 will be proven if it is shown that the C_q are uniquely determined by (i) and (iii) of Lemma 16.4. Assume that there are two sets of solutions $\{C_q\}$ and $\{C'_q\}$ to (16.12). For every ℓ we have

$$C_\ell = C_q \Pi_q(\xi) \Pi_{q+1}(\xi) \; \ldots \; \Pi_{\ell-1}(\xi); \; C'_\ell = C'_q \Pi_q(\xi) \Pi_{q+1}(\xi) \; \ldots \; \Pi_{\ell-1}(\xi)$$

so that (15.4)

$$\Gamma_V(C_\ell, C'_\ell) \geq \Gamma(\Pi_q(\xi)\Pi_{q+1}(\xi) \ldots \Pi_{\ell-1}(\xi)) \quad .$$

Letting $q \longrightarrow -\infty$ we have by (16. 3) $\Gamma_V(C_\ell, C'_\ell) = 1$. By Lemma 15.1 this implies that $C'_\ell = \lambda C_\ell$ for some λ. This together with Lemma 16.4 (i) implies that $C'_\ell = C_\ell$. This being true for each ℓ, μ is uniquely determined. Hence A_X^- is c.p. to A_Y^- at ξ. By the corollary to Theorem 10.3 the sequence $\xi(n)$ is a c.p. sequence.

§17.   Normality and Continuous Predictability

17.1. <u>Reduction to Markoff Processes</u>. In this section we show how the condition of normality introduced in §14 implies continuous pre-dictability in certain cases. We do this by showing that the theorem of the last section is applicable. Let us suppose then that Y is an m-Markoff process and X a subprocess, that both X and Y are finitely valued and that Y is normal over X at a point ξ of Ω_X^-.

Our first step is to reduce the general case to that of a Markoff process and a subprocess of a special kind, that is to the case where $x_n = \psi(y_n)$ and Y is a Markoff process as in the last section.

> LEMMA 17.1. If Y is a finitely valued m-Markoff
> process and X is a finitely valued subprocess then
> there is a finitely valued Markoff process Z with
> $A_Z^- \cong A_Y^-$ such that X is a subprocess of Z satisfy-
> ing $x_n = \psi(z_n)$. Moreover Y is normal over X at
> a point ξ, if and only if Z is normal over X
> at ξ. Finally A_X^- is c.p. to A_Y^- at ξ if and
> only if A_X^- is c.p. to A_Z^- at ξ.

PROOF. We recall from Lemma 11.3 that if we take $z_n = (y_n, y_{n-1}, \ldots, y_{n-m+1})$ then these variables define a Markoff process Z with $A_Y^- = A_Z^-$, and $A_Y = A_Z$. Clearly the same result is true if we take $z_n = (y_n, y_{n-1}, \ldots, y_{n-k})$ for any $k \geq m - 1$. It is then only necessary to show that k may be chosen so large that $x_n = \psi(z_n)$, for the normality and predictability depend only on the algebras A_X^-, $A_Y^- = A_Z^-$. To write $x_n = \psi(z_n)$ we must establish that each x_n depends only on a finite number of y_{n-j} (the number of y_{n-j} and the dependence relation will be independent of n by stationarity). This however follows directly from the fact that the x_n and y_n are finitely valued. One way of seeing this is to introduce a metric upon $\Omega_Y^- \subset M_\infty^-$:

$d(\omega_1, \omega_2) = \inf \{\frac{1}{k} : \omega_1(0) = \omega_2(0), \ldots, \omega_1(-k) = \omega_2(-k)\}$. This is easily verified to be a metric for the topology of Ω_Y^-. Since x_o is continuous on Ω_Y^- it follows that for $d(\omega_1, \omega_2) < \varepsilon, x_o(\omega_1) = x_o(\omega_2)$. Choosing $k > 1/\varepsilon$ it follows that whenever $\omega_1(0) = \omega_2(0), \ldots, \omega_1(-k) = \omega_2(-k)$ then $x_o(\omega_1) = x_o(\omega_2)$ so that x_o is a function of $y_o, y_1, \ldots, y_{-k}$ and $x_n = \psi(y_n, y_{n-1}, \ldots, y_{n-k})$.

As a result of this lemma it is clear that it suffices to consider ordinary Markoff processes and subprocesses of the special kind: $x_n = \psi(y_n)$.

17.2. <u>Consequences of Normality</u>. Our notation in this section is that of §16.

LEMMA 17.2. If y is normal over X at ξ then $\Phi^{-1}(\xi)$ has the following property: whenever η' and η'' belong to $\Phi^{-1}(\xi)$ then for any $k \geq 0$ there is an $\eta \in \Phi^{-1}(\xi)$ with

$$\eta(j) = \begin{cases} \eta'(j) & \text{if } -k \leq j \leq 0 \\ \eta''(j) & \text{if } -\infty \leq j \leq -N \end{cases}$$

for sufficiently large N.

PROOF. The set $\{\tau\eta'' : \tau \in G(Y|X)\}$ has for its closure a closed set Δ in Ω_Y^- invariant under the group $G(Y|X)$. Since $\eta'' \in \Delta$, $\Phi^{-1}(\xi) \cap \Delta \neq \emptyset$ and by normality at ξ, $\Phi^{-1}(\xi) \subset \Delta$ so that $\eta' \in \Delta$. Hence there is an $\eta = \tau\eta''$ with η arbitrarily close to η' and in particular η can be made to agree with η' between $-k$ and 0. Moreover since $\eta = \tau\eta''$ and τ alters only finitely many y_n we must have $\eta(j) = \eta''(j)$ if $j \leq -N$ and N is sufficiently large. This proves the lemma.

LEMMA 17.3. If Y is normal over X at ξ, then for sufficiently large N, if $q < -N$, the product matrices $\Pi_q(\xi)\Pi_{q+1}(\xi) \cdots \Pi_{-1}(\xi)$ will be projectively bounded as transformations of non-negative m-tuples into non-negative m-tuples (Definition 15.2).

PROOF. Set $\Pi_q^* = \Pi_q(\xi)\Pi_{q+1}(\xi) \cdots \Pi_{-1}(\xi)$. We shall show that Π_q^* is bounded eventually by verifying the condition following Definition 15.2. Namely we shall show that the entries of Π_q^* are eventually 0 but for the terms $\Pi_q^*(b, b')$ with $b \in M_q(\xi)$ and $b' \in M_o(\xi)$ (§16) in which case $\Pi_q^*(b, b') > 0$. Now the fact that $\Pi_q^*(b, b') = 0$ for $b \notin M_q(\xi)$ or

$b' \notin M_O(\xi)$ follows from the definition of the $\Pi_q(\xi)$ so that what has to be shown is that eventually $\Pi_q^*(b, b') > 0$ for $b \in M_q(\xi)$, $b' \in M_O(\xi)$. This we shall do by showing that for $q < - N$, N sufficiently large, there exists a possible sequence σ lying over ξ for every $b \in M_q(\xi)$ and $b' \in M_O(\xi)$ with $\sigma(q) = b$, $\sigma(0) = b'$. To see how this gives the desired result observe that since $b \in M_q(\xi)$, there exists in $\Phi^{-1}(\xi)$ a sequence η with $\eta(q) = b$. Define a new sequence η' by

$$(17.1) \qquad \eta'(n) = \begin{cases} \eta(n) & \text{for } n \leq q \\ \sigma(n) & \text{for } n \geq q \end{cases}.$$

It is clear that η' is a possible sequence if σ is and also lies over ξ. From this we deduce that for each $n \geq q$, $\sigma(n) \in M_n(\xi)$. But then $\Pi_n(\sigma(n), \sigma(n + 1)) = p(\sigma(n), \sigma(n + 1)) > 0$ if σ is possible, and hence $\Pi^*(b, b') \geq p(\sigma(q), \sigma(q + 1)) \ldots p(\sigma(-1), \sigma(0)) > 0$. Thus the lemma reduces to showing that for $- q$ large, and $b \in M_q(\xi)$, $b' \in M_O(\xi)$, there is a possible sequence lying over ξ joining b to b'.

Clearly it suffices to show for a particular b', that eventually each element of $M_q(\xi)$ is joined to it by a possible sequence lying over ξ since if we go far enough this would be satisfied by all $b' \in M_O(\xi)$ simultaneously. So we choose $b' \in M_O(\xi)$ and let Δ_q be the set of $b \in M_q(\xi)$ which cannot be joined to b' by a possible sequence lying over ξ. Also let $\tilde{\Delta}_q$ be the closed set of all $\eta \in \Phi^{-1}(\xi)$ such that $\eta(q) \in \Delta_q$. By the definition of $M_q(\xi)$ it follows that if Δ_q is non-empty so is $\tilde{\Delta}_q$. We observe next that $\tilde{\Delta}_{q-1} \subset \tilde{\Delta}_q$. For suppose $\eta \in \tilde{\Delta}_{q-1}$ so that $\eta(q - 1)$ cannot be joined by a possible sequence over ξ to b'. Clearly $\eta(q)$, at the next stage, cannot be joined to b' by a possible sequence over ξ since if it could then $\eta(q - 1)$ could also have been joined to b'.

If we now suppose that the Δ_q are all non-empty, then the $\tilde{\Delta}_q$ would form a monotonically decreasing sequence of non-empty closed sets so that there would be an $\eta' \in \tilde{\Delta}_q$ for all q. This will lead us to a contradiction. For let η' be the sequence of $\Phi^{-1}(\xi)$ terminating in b'. By the preceding lemma there is an $\eta \in \Phi^{-1}(\xi)$ with $\eta(0) = b'$ and $\eta(n) = \eta''(n)$ for $n \leq - N$ for some N. But then $\eta(n) \in \Delta_n$ for $n \leq - N$ and then passage from $\eta(n)$ to b' over ξ would be impossible. This contradiction shows that the Δ_q are eventually empty and this proves the lemma.

LEMMA 17.4. Let ξ' be another point of $\bar{\Omega}_X$ and consider the corresponding sets $M_q(\xi')$, $q \leq 0$, and matrices $\Pi_q(\xi')$. Suppose for a fixed $q \leq 0$ and

$k \geq 0$, ξ' satisfies

(i) $\xi'(n - k) = \xi(n)$ for $q \leq n \leq 0$

(ii) $M_{q-k}(\xi') \subset M_q(\xi)$.

Then $\Gamma(\Pi_{q-k}(\xi')\Pi_{q-k+1}(\xi') \cdots \Pi_{-1-k}(\xi')) \geq$
$\Gamma(\Pi_q(\xi)\Pi_{q+1}(\xi) \cdots \Pi_{-1}(\xi))$.

PROOF. We observe first of all that (ii) implies

$$(17.2) \qquad\qquad M_{n-k}(\xi') \subset M_n(\xi)$$

for all n with $q \leq n \leq 0$. For suppose that $b \in M_{n-1}(\xi')$ so that there
is a sequence η' in $\Phi^{-1}(\xi')$ with $\eta'(n - k) = b$. Now
$\eta'(q - k) \in M_{q-k}(\xi') \subset M_q(\xi)$ so there is a possible sequence over ξ, say
η, taking on the value $\eta'(q - k)$ at q. The sequence $\sigma = (\eta'(q - k),$
$\eta'(q - k + 1), \ldots, \eta'(- k))$ lies over ξ' from $q - k$ to $- k$ and so
by (i), it lies over ξ from q to 0. Hence combining η with σ as
in (17.1) we obtain a possible sequence η'' over ξ with $\eta''(n) = \eta'(n - k)$
for $q \leq n < 0$. In particular $\eta''(n) = b$ so that $b \in M_n(\xi)$, and so
$M_{n-k}(\xi') \subset M_n(\xi)$.

To prove the lemma we shall show that the elements of the matrix
$\Pi_{q-k}(\xi') \cdots \Pi_{-1-k}(\xi')$ are just those of $\Pi_q^* = \Pi_q(\xi) \cdots \Pi_{-1}(\xi)$ with
the terms $\Pi_q^*(b, b')$ replaced by 0 when $b \notin M_{q-k}(\xi')$ or $b' \notin M_{-k}(\xi')$.
Clearly from the definition of $\Pi_n(\xi')$ the terms in the position (b, b')
must vanish if $b \notin M_{q-k}(\xi')$ or $b' \notin M_{-k}(\xi)$. To show that if
$b \in M_{q-k}(\xi')$ and $b' \in M_{-k}(\xi)$ then the term in question agrees with
$\Pi_q^*(b, b')$ we observe that each term in such a product is the sum of
products $p(b, \sigma(q - k + 1)) \cdots p(\sigma(- k - 1), b')$ taken over all σ beginning with b and terminating in b' satisfying $\sigma(j - k) \in M_{j-k}(\xi')$.
Now there is a $1 - 1$ correspondence between these sequences and those
from q to 0 joining b to b', and satisfying $\sigma(j) \in M_j(\xi)$. Namely,
every one of the former is one of the latter by (17.2). Conversely,
suppose $\sigma(j) \in M_j(\xi)$ for $q \leq j \leq 0$. Since $b \in M_{q-k}(\xi')$ and $b' \in M_{-k}(\xi')$
we can find a possible sequence η' lying over ξ' with $\eta'(q - k) = b$,
and a possible sequence η'' lying over ξ' with $\eta''(- k) = b'$. Now define a sequence η by

$$\eta(n) = \begin{cases} \eta'(n) & n \leq q - k \\ \sigma(n - k) & q - k < n < - k \\ \eta''(n) & - k \leq n \leq 0 \ . \end{cases}$$

$\eta(n)$ is a possible sequence since each transition has positive probability
and also lies over ξ'. It follows that $\sigma(j) \in M_{j-k}(\xi')$. It follows that

every sequence from $q - k$ to $-k$ over ξ' joining b to b' corresponds to a sequence from q to 0 over ξ joining b to b'. This establishes the $1 - 1$ correspondence and shows that the terms of $\Pi_{-q-k}(\xi') \ldots \Pi_{-1-k}(\xi')$ are as asserted.

Now let C be the matrix $(c(b_i, b_j))$ with $c(b_i, b_j) = 1$ if $b_i = b_j \in M_{q-k}(\xi')$ and $c(b_i, b_j) = 0$ otherwise. Let D be the matrix $(d(b_i, b_j))$ with $d(b_i, b_j) = 1$ if $b_i = b_j \in M_{-k}(\xi')$ and $d(b_i, b_j) = 0$ otherwise. From the above we have

$$\Pi_{q-k}(\xi') \ldots \Pi_{-k}(\xi) = C\Pi_q^* D \ .$$

Now by Definition 15.2, and Lemma 15.1

$$\Gamma(C\Pi_q^* D) = \inf_{x,y \in V} \Gamma_V(xC\Pi_q^* D, \; yC\Pi_q^* D)$$

$$\geq \inf_{x,y \in V} \Gamma_V(x\Pi_q^* D, \; y\Pi_q^* D)$$

$$\geq \inf_{x,y \in V} \Gamma_V(x\Pi_q^*, \; y\Pi_q^*) = \Gamma(\Pi_q^*) \ .$$

This completes the proof of the lemma.

COROLLARY. If ξ is a limit point of translates of ξ' (i.e., $T^{n_k}\xi' \longrightarrow \xi$) and Y is normal over X at ξ, then $\Gamma(\Pi_q(\xi')\Pi_{q+1}(\xi') \ldots \Pi_{\ell}(\xi')) \longrightarrow 1$ as $q \longrightarrow -\infty$ for each $\ell \leq 0$.

PROOF. By Lemma 17.3 there is a q such that $\Pi_q(\xi)\Pi_{q+1}(\xi) \ldots \Pi_{-1}(\xi)$ is projectively bounded. Fix this q and let us examine the set $M_q(\xi)$. $M_q(\xi)$ consists of the elements in $\psi(\xi(q)) \subset M$ which occur in sequences of $\Phi^{-1}(\xi)$. An element of $\psi^{-1}(\xi(q)) - M_q(\xi)$ cannot occur as any $\eta(q)$ for $\eta \in \Phi^{-1}(\xi)$. We can therefore write $\psi^{-1}(\xi(q)) - M_q(\xi) = \Delta_1 \cup \Delta_2$ where Δ_1 consists of those elements of $\psi^{-1}(\xi(q))$ that cannot be continued to the right as a possible sequence lying over $\xi(q)$, $\xi(q + 1)$, $\ldots$, $\xi(0)$, and Δ_2 consists of those elements not in Δ_1 which cannot be continued to the left as a possible sequence over $\ldots \xi(q - 2)$, $\xi(q - 1)$, $\xi(q)$. Now there is a bound to how far to the left an element of Δ_2 may be continued (as a possible sequence) for otherwise, as a simple compactness argument shows, there would be elements in Δ_2 that may be continued infinitely far to the left as a possible sequence, and this contradicts the definition of Δ_2. This shows that there is a sequence

$\xi(r)$, $\xi(r + 1)$, ..., $\xi(q)$, ..., $\xi(0)$ such that any sequence η of $\bar{\Omega}_\gamma$ taking values in $\psi^{-1}(\xi(j))$ for $r \le j \le 0$ must satisfy $\eta(q) \in M_q(\xi)$. For if not $\eta(q)$ belongs either to Δ_1 or Δ_2. The former is excluded since $\eta(q)$ is continuable to the right and the latter will be impossible if r is sufficiently negative since $\eta(q)$ can be continued as far as $\eta(r)$.

If ξ is a limit point of translates of ξ' then any neighborhood of ξ contains infinitely many translates $T^n\xi'$. By the definition of the topology in Λ_∞^- this implies that the last $|r| + 1$ terms of infinitely many of the $T^n\xi'$ agree with those of ξ; in other words $(\xi(r), \xi(r + 1), ..., \xi(0))$ occurs infinitely often in ξ'. Let K denote the set of k for which $(\xi'(r - k), \xi'(r - k + 1), ..., \xi'(- k)) = (\xi(r), \xi(r + 1), ..., \xi(0))$. What we have just shown amounts to the statement that $M_{q-k}(\xi') \subset M_q(\xi)$. We see then that for $k \in K$ and for q as before, the hypotheses of Lemma 17.4 are fulfilled. Hence

$$r(\Pi_{q-k}(\xi')\Pi_{q-k+1}(\xi')...\Pi_{-1-k}(\xi')$$

(17.3)

$$\ge r(\Pi_q(\xi)\Pi_{q+1}(\xi)...\Pi_{-1}(\xi))$$

for infinitely many k. Now the right hand side of (17.3) is positive, and equals some $\delta > 0$. We thus find that each $\Pi_j(\xi')\Pi_{j+1}(\xi')...\Pi_\ell(\xi')$ may be written as a product

$$M_j^{(1)}N_j^{(1)}M_j^{(2)}N_j^{(2)}...N_j^{(s)}M_j^{(s)}$$

with $r(N_j^{(1)}) \ge \delta$ and $s \longrightarrow \infty$ as $j \longrightarrow -\infty$. From $r(N_j^{(1)}) \ge \delta$ we deduce that also $r(M_j^{(1)}N_j^{(1)}) \ge \delta$ so that $\Pi_j(\xi')\Pi_{j+1}(\xi')...\Pi_\ell(\xi')$ has the form

$$R_j^{(1)}R_j^{(2)}...R_j^{(s)}R'$$

with $r(R_j^{(1)}) \ge \delta$. The corollary will therefore be proven if we establish that there exists a sequence $\varepsilon_n(\delta) \longrightarrow 0$ such that whenever $r(R^{(1)})$, $r(R^{(2)})$, ..., $r(R^{(n)}) \ge \delta$ then $r(R^{(1)}R^{(2)}...R^{(n)}) > 1 - \varepsilon_n(\delta)$.

The latter assertion is an immediate consequence of Lemma 15.2. Namely if

$$r(R^{(1)}R^{(2)}...R^{(n)}) < \gamma < 1$$

we also have

$$r(R^{(1)}R^{(2)}...R^{(s)}) < \gamma$$

for $s \leq n$. But by Lemma 15.2

$$\Gamma(R^{(1)} \dots R^{(s)}) - \Gamma(R^{(1)} \dots R^{(s-1)}) \geq \frac{\delta}{1+\delta}\,(1 - \gamma^2)$$

which can be true only if

$$n \leq \frac{1+\delta}{\delta}\,\frac{\gamma}{1-\gamma^2}\ .$$

This completes the proof of the corollary.

As a consequence of this corollary, Theorem 16.1, and Lemma 17.1 we have

> THEOREM 17.1. If X and Y are finitely valued processes, X a subprocess of Y and Y an m-Markoff process, then $A_{\bar{X}}$ is c.p. to A_X at a point ξ of $\Omega_{\bar{X}}$ if Y is normal over X at some limit point of translates, $T^m\xi$, of the point ξ.

In particular if ξ is a generic point of X, for example if $X = X(\xi)$, then it suffices that Y be normal over X at <u>any</u> point of $\Omega_{\bar{X}}$ since every such point is a limit of translates of ξ. Also if Y is normal over X (at every point) we have

> THEOREM 17.2. If X and Y are finitely valued processes, X a subprocess of Y and Y an m-Markoff process, then X is continuously predictable if Y is normal over X.

17.3. <u>An Example</u>. As an illustration of the foregoing theorems we let Y be a Markoff process with a transition matrix all of whose entries are <u>strictly positive</u>. Since now every transition is possible it follows that $\Omega_{\bar{Y}} = M_\infty^-$. Let X be a subprocess of Y determined by $x_n = \psi(y_n)$ $\psi : M \longrightarrow \Lambda$. The preceding theorem shows that X is always continuously predictable.

To show that this is so we must verify that Y is normal over X. Since $\Omega_{\bar{Y}} = M_\infty^-$ it follows that for arbitrary sequences σ_1 and σ_2 of the same length and having the same images under ψ and also beginning with the same element, $\tau_{\sigma_1 \sigma_2}$ (§16) is a homeomorphism in $G(Y|X)$. Now let η_1 and η_2 be any infinite M-sequences lying over the same point ξ

of Ω_X^-. Clearly for an appropriate σ_1, σ_2, $\tau_{\sigma_1 \sigma_2} \eta_1$ will come arbitrarily close to η_2. Hence a closed set containing η_1 that is invariant under $G(Y|X)$ will also contain η_2. Hence Y is normal over X at each ξ and so X is continuously predictable.

We note however that this case can never arise if Y is m-Markoff for $m > 1$, or if x_n depends upon more than one of the $y_{n-\ell}$. For in these cases, the Markoff process is obtained by taking a vector process $z_n = (y_n, y_{n-1}, \ldots, y_{n-k})$ and the transition probability from $(\lambda_0, \lambda_{-1}, \ldots, \lambda_{-k})$ to $(\lambda'_0, \lambda'_{-1}, \ldots, \lambda'_{-k})$ is always 0 unless $\lambda_0 = \lambda'_{-1}$, $\lambda_{-1} = \lambda'_{-2}$, $\ldots$, $\lambda_{-k+1} = \lambda'_{-k}$. An example of this was given in §12.3.

CHAPTER 5. STOCHASTIC SEMIGROUPS AND CONTINUOUS PREDICTABILITY

In the last chapter we dealt with the prediction problem for processes derived from certain elementary processes, ordinary and m-Markoff processes. In this chapter we shall develop a procedure for constructing a large variety of processes for which we shall be able to study the prediction problem. Our procedure will be of some interest in itself and we shall show how it is also related to certain problems outside of prediction theory.

<h3 style="text-align:center">§18. Stochastic Semigroups</h3>

18.1. <u>Definitions</u>

DEFINITION 18.1. A <u>stochastic semigroup of order</u> r is a semigroup $\mathcal{S}$ having an identity e, together with a set of r elements a_1, a_2, ..., a_r generating $\mathcal{S}$, and a real-valued function F defined on $\mathcal{S}$ satisfying

    (i) $F(e) = 1$,

    (ii) $F(\sigma) \geq 0$ for each $\sigma \in \mathcal{S}$ and $F(a_i) > 0$, $i = 1, 2, ..., r$,

    (iii) $\Sigma_{i=1}^{r} F(a_i\sigma) = \Sigma_{i=1}^{r} F(\sigma a_i) = F(\sigma)$.

The connection between stationary processes and stochastic semigroups is easily seen. Suppose X is a process with values in a set Λ having r elements a_1, ..., a_r. We may form the (free) semigroup $\mathcal{S}$ of all formal products of the a_i with the empty product taken as the identity e. On $\mathcal{S}$ define F for $\sigma = \sigma(1)\sigma(2)...\sigma(n)$ by

$$(18.1) \qquad F(\sigma) = P(x_1 = \sigma(1),\ x_2 = \sigma(2),\ ...,\ x_n = \sigma(n))$$

and take $F(e) = 1$. To show that $\mathcal{S}$ is a stochastic semigroup only (iii) has to be verified. The equality

$$\sum_{i=1}^{r} F(\sigma a_i) = F(\sigma)$$

follows directly from the fact that P is a probability measure. The remaining half follows from the stationarity of X:

$$\sum_{i=1}^{r} F(a_1\sigma) = \sum_{i=1}^{r} P(x_1 = a_1,\ x_2 = \sigma(1),\ \ldots,\ x_{n+1} = \sigma(n)) =$$

$$P(x_2 = \sigma(1),\ \ldots,\ x_{n+1} = \sigma(n)) = P(x_1 = \sigma(1),\ \ldots,\ x_n = \sigma(n)) = F(\sigma)\ .$$

Conversely suppose $\mathcal{S}$ is a given stochastic semigroup. We then define a process X with values in the set of generators $\{a_1,\ a_2,\ \ldots,\ a_r\}$ by setting

$$P(x_1 = \sigma(1),\ x_2 = \sigma(2),\ \ldots,\ x_n = \sigma(n)) = F(\sigma(1)\sigma(2)\ldots\sigma(n))\ .$$

It is easily verified that this will be a process, the stationarity being a consequence of (iii). We shall not distinguish between isomorphic processes and thus the particular nature of the range space of the x_n is irrelevant. It is however convenient for purposes of notation to identify the range of X here with the generators of $\mathcal{S}$.

DEFINITION 18.2. For an r-valued process X, $\mathcal{S}(X)$ will denote the associated stochastic semigroup of order r. For a stochastic semigroup $\mathcal{S}$, $X(\mathcal{S})$ will denote the corresponding process.

Note that $X(\mathcal{S}(X_0)) = X_0$ whereas $\mathcal{S}(X(\mathcal{S}_0))$ is not necessarily $\mathcal{S}_0$ since, for example, $\mathcal{S}_0$ may not have been a free semigroup.

Together with a stochastic semigroup $\mathcal{S}$ it will be convenient to consider a normed algebra defined in terms of it which we shall denote $A(\mathcal{S})$. First define $A_0(\mathcal{S})$ as the set of all real finite linear combinations $\Sigma\alpha_\sigma\sigma$ with multiplication defined in the natural way: $(\Sigma\alpha_\sigma\sigma)(\Sigma\beta_\rho\rho) = \Sigma\alpha_\sigma\beta_\rho\sigma\rho$. We define a semi-norm $\|\ \|$ on $A_0(\mathcal{S})$ by writing each $u \in A_0(\mathcal{S})$ as a difference $\Sigma\alpha_\sigma\sigma - \Sigma\beta_\rho\rho$ with $\alpha_\sigma,\ \beta_\rho \geq 0$ and the σ and ρ distinct. We then set $\|u\| = \Sigma\alpha_\sigma F(\sigma) + \Sigma\beta_\rho F(\rho)$. If we set $F(u) = \Sigma\alpha_\sigma F(\sigma) - \Sigma\beta_\rho F(\rho)$ we obtain a linear functional F on $A_0(\mathcal{S})$ satisfying $|F(u)| \leq \|u\|$.

By (iii) it follows that if $F(\sigma) = 0$, then $F(a_1\sigma) = 0$ and $F(\sigma a_1) = 0$ for $i = 1, 2, \ldots, r$. Consequently $F(\sigma\rho) = F(\rho\sigma) = 0$ for any $\rho \in \mathcal{S}$ so that the set of elements in $\mathcal{S}$ with vanishing F is closed under multiplication by any element of $\mathcal{S}$. As a result the set of $u \in A_0(\mathcal{S})$ with $\|u\| = 0$ forms an ideal I. We now define $A(\mathcal{S})$ by

$$(18.2) \qquad\qquad\qquad A(\mathcal{S}) = A_0(\mathcal{S})\ /I\ ,$$

with the norm induced by $\| \ \|$. The algebra $A_o(\mathcal{S})$ is generated by the identity and a_1, a_2, ..., a_r and so $A(\mathcal{S})$ is also generated by the identity and the images of a_1, a_2, ..., a_r which we shall again denote by a_1, a_2, ..., a_r. $A(\mathcal{S})$ is a normal algebra and we shall verify that multiplication is continuous. Namely we wish to show that if $u_n \longrightarrow 0$ in $A(\mathcal{S})$ and $u \in A(\mathcal{S})$ then $uu_n \longrightarrow 0$ and $u_nu \longrightarrow 0$. Clearly it suffices to take u to be a generator of $A(\mathcal{S})$, namely $u = a_1$. But since $F(a_1\sigma) \leq F(\sigma)$ we have $\|a_1u_n\| \leq \|u_n\|$ and similarly $\|u_na_1\| \leq \|u_n\|$ which proves our assertion.

18.2. <u>Ergodicity of</u> X <u>and</u> $\mathcal{S}(X)$. To illustrate the relationship between a process and the associated semigroup we prove the following:

THEOREM 18.1. Let X be an r-valued process and $A = A(\mathcal{S})$ the algebra of a stochastic semigroup with $X = X(\mathcal{S})$. (For example, $\mathcal{S}$ could be $\mathcal{S}(X)$.) Let Q be the closure in A of the sum

$$\left(e - \sum_{i=1}^{r} a_i \right) A + A \left(e - \sum_{i=1}^{r} a_i \right) \ .$$

Then X is ergodic if and only if Q is of co-dimension 1 in A.

PROOF. We suppose first that X is ergodic. Since by Definition 18.1 (iii) F vanishes on Q, Q is of codimension ≥ 1. If it is of codimension > 1 then there must exist another continuous linear functional G on A, vanishing on Q and such that G is not a multiple of F. Now if G is continuous we must have $|G(u)| \leq \kappa\|u\|$ for some κ. We observe next that if G is restricted to the image of $\mathcal{S}$ in A we obtain a functional on $\mathcal{S}$ which will satisfy (iii) of Definition 18.1. For in terms of A, (iii) merely states that F vanishes on Q and since G does so, it satisfies (iii). This implies that G defines a measure μ_G on Ω_X. Namely set $\mu_G\{\omega : x_k(\omega) = \sigma(0), x_{k+1}(\omega) = \sigma(1),$ ..., $x_{k+\ell}(\omega) = \sigma(\ell)\} = G(\sigma(0)\sigma(1)...\sigma(\ell))$. That this is a measure is a consequence of (iii). Moreover since $\mu_G\{\omega : x_k(\omega) = \sigma(0),$ ..., $x_{k+\ell}(\omega) = \sigma(\ell)\} = \mu_G\{\omega : x_{k+1}(\omega) = \sigma(0),$ ..., $x_{k+\ell+1}(\omega) = \sigma(\ell)\}$ it follows that for any $\Delta \subset \Omega_X$, $\mu_G(\Delta) = \mu_G(T\Delta)$. Also since $|G(\sigma)| \leq \kappa F(\sigma)$ we have $|\mu_G(\Delta)| < \kappa P(\Delta)$, where P is the probability measure of X. By the Radon-Nikodym Theorem

$$\mu_G(\Delta) = \int_\Delta \varphi(\omega)dP(\omega)$$

where $\varphi(\omega)$ is a bounded measurable function on Ω_X. Then $\mu_G(\Delta) = \mu_G(T\Delta)$
implies that $\varphi(\omega) = \varphi(T\omega)$ a.e., so that φ is invariant under T and
since X is ergodic φ is essentially constant. But then $\mu_G(\Delta) = \gamma P(\Delta)$
for some constant γ so that $G(\sigma) = \gamma F(\sigma)$, contrary to our supposition
that G is not a multiple of F. Therefore if X is ergodic, Q is of
codimension 1.

The proof of the converse proceeds along the same lines. If X
is not ergodic there exists an invariant function φ on Ω_X that is not
a constant. Setting

$$G(\sigma) = \int_{\Delta_\sigma} \varphi(\omega) dP(\omega) \quad,$$

where $\Delta_\sigma = \{\omega : x_0(\omega) = \sigma(0), \ x_1(\omega) = \sigma(1), \ \ldots, \ x_\ell(\omega) = \sigma(\ell)\}$, we obtain a function on $\mathcal{S}$ that satisfies (iii) of Definition 18.1. This G
extends to a functional G on $A(\mathcal{S})$ vanishing on a dense subset of Q.
But since φ is bounded and $|G(\sigma)| \leq \|\varphi\|_\infty P(\Delta_\sigma) = \|\varphi\|_\infty F(\sigma)$ it follows
that G is continuous in the norm of A and so G vanishes on all of
Q. If Q is of codimension 1, G must be a multiple of F which
implies φ is essentially constant contrary to supposition.

18.3. <u>Application to Symmetric Processes</u>. The foregoing may
be applied to symmetric processes ([6]), that is, processes X for which
the order of the variables in $P(x_1 = a_{i_1}, \ x_2 = a_{i_2}, \ \ldots, \ x_k = a_{i_k})$ has
no bearing on the probability in question.

THEOREM 18.2. If X is an r-valued symmetric process,
namely, if $P(x_1 = a_{i_1}, \ x_2 = a_{i_2}, \ \ldots, \ x_k = a_{i_k})$ depends only on the unordered k-tuple $\{a_{i_1}, \ a_{i_2}, \ \ldots,$
$a_{i_k}\}$, then if X is ergodic it must be random, i.e.,
the x_n are independent.

PROOF. We observe that if X is symmetric, then it corresponds
to a commutative stochastic semigroup. For we can take $\mathcal{S}$ as the
commuting set of monomials $a_1^{\nu_1} a_2^{\nu_2} \ldots a_r^{\nu_r}$ and define F on $\mathcal{S}$ in terms
of the probabilities of X, the order of the a_i being irrelevant. Since
$\mathcal{S}$ is commutative, $A(\mathcal{S})$ will also be commutative so that Q in Theorem
18.1 is just the closure of

$$\left(e - \sum_{i=1}^{r} a_i\right) A \quad.$$

If X is ergodic, Q is of codimension 1 and F must be the unique (up to scalar multiple) functional vanishing on

$$\left(e - \sum_{i=1}^{r} a_i \right) A \ .$$

But

$$\left(e - \sum_{i=1}^{r} a_i \right) A$$

is an ideal so that the kernel of F is an ideal. This together with the fact that $F(e) = 1$ implies that F is multiplicative. (For since $u - F(u) \in \ker F$, $uv - F(u)v \in \ker F$ so that $F(uv) = F(u)F(v)$.) However it is easily seen that F is multiplicative if and only if the x_n are independent.

§19.   Realization of Stochastic Semigroups

19.1.   <u>Linear Semigroups</u>.

DEFINITION 19.1. Let D be a linear space, D^* its dual, V a cone in D satisfying (*) of §15. Let $\theta \in V$ and $\theta^* \in D^*$ and suppose that θ^* is non-negative on V. A <u>linear stochastic semigroup</u> $\mathcal{S}$ <u>on</u> (V, θ, θ^*) is one whose elements are linear transformations from V to V satisfying

> (i) $\Sigma a_i \theta = \theta$
> (ii) $\Sigma a_i^* \theta^* = \theta^*$ (where L^* denotes the transformation
>   of D^* adjoint to a transformation L of D)
> (iii) $F(\sigma) = \langle \theta^*, \sigma\theta \rangle$ for $\sigma \in \mathcal{S}$, where $\langle \cdot , \cdot \rangle$ de-
>   notes the dual pairing of D^* and D.
> (iv) $\langle \theta^*, a_i\theta \rangle > 0$ $i = 1, 2, \ldots, r$.

We note that a semigroup $\mathcal{S}$ of transformations satisfying (i) to (iv) does define a stochastic semigroup if $\langle \theta^*, \theta \rangle = 1$. (i) and (ii) of Definition 18.1 are immediate and for (iii) we have

$$\sum_{i=1}^{r} F(a_i\sigma) = \langle \theta^*, \Sigma a_i \sigma\theta \rangle = \langle \Sigma a_i^* \theta^*, \sigma\theta \rangle =$$

$$\langle \theta^*, \sigma\theta \rangle = F(\sigma) \ ,$$

and

$$\sum_{i=1}^{r} F(\sigma a_i) = \langle \theta^*, \ \Sigma \sigma a_i \theta \rangle = \langle \theta^*, \ \sigma \theta \rangle = F(\sigma)$$

using (i) and (ii) of Definition 19.1.

Let us call two stochastic semigroups $\mathcal{S}_1$ and $\mathcal{S}_2$ <u>equivalent</u> if $X(\mathcal{S}_1) = X(\mathcal{S}_2)$. We can then prove a converse to the foregoing remark:

THEOREM 19.1. Every stochastic semigroup is equivalent to a linear stochastic semigroup.

PROOF. Let $\mathcal{S}$ be a stochastic semigroup and $A = A(\mathcal{S})$ the associated algebra. Set

$$\tau = \sum_{i=1}^{r} a_i$$

and consider the quotient module of A modulo the left ideal $A(e - \tau)$:

$$D = A/A(e - \tau) \quad .$$

The elements of A and in particular those of $\mathcal{S}$ operate on D by left multiplication. Let a_i' denote the operator induced by left multiplication by $a_i \in \mathcal{S}$. Take V to be the image in D of the set of elements of A that can be represented as positive linear combinations of elements in $\mathcal{S}$. We denote by $\bar{u}$ the image in D of an element u in A. Set $\theta = \bar{e}$ and let θ^* be the functional induced on D by F on A (F vanishes on $A(e - \tau)$). Then $\Sigma a_i'\theta = \Sigma a_i'\bar{e} = \Sigma \overline{a_i e} = \bar{\tau} = \bar{e} = \theta$, and $\Sigma \langle a_i'^* \theta^*, \ \bar{u} \rangle = \Sigma \langle \theta^* a_i' \bar{u} \rangle = \Sigma \langle \theta^*, \ \overline{a_i u} \rangle = \Sigma F(a_i u) = F(\tau u) = F(u) = \langle \theta^*, \ \bar{u} \rangle$ so that (i) and (ii) of Definition 19.1 are verified. Also $\langle \theta^*, \theta \rangle = F(e) = 1$ and $\langle \theta^*, a_i'\theta \rangle = F(a_i) > 0$ so that (iii) and (iv) are also verified. It follows that the semigroup $\mathcal{S}'$ with elements σ' generated by the a_i' and the function F defined by (iii) is a linear stochastic semigroup. That it is equivalent to $\mathcal{S}$ follows from the fact that $F(\sigma') = \langle \theta^*, \ \sigma'\theta \rangle = F(\sigma)$. This proves the theorem.

19.2. <u>Examples</u>. The theorem just proven shows that every (finitely valued) process can be obtained from a linear stochastic semigroup, if we are willing to admit arbitrary spaces D. However if we wish to investigate the properties of the process it will be necessary to restrict ourselves to spaces of a more special kind.

As a first example take $D = R^m$ the set of m-dimensional column vectors and let V be the cone of non-negative vectors in D. Let θ be the column vector with all entries equal to 1. Let π be a stochastic $m \times m$ matrix in the sense that

$$\sum_{j=1}^{m} \prod_{ij} = 1, \quad i = 1, \ldots, m$$

and take π_i as the matrix derived from π by setting all the elements equal to 0 but for those in the i-th column which are left unaltered: $\pi_i(j, k) = \delta_{ik}\pi(j, k)$. D^* may be represented as the set of all m-dimensional row vectors and θ^* is taken to be a vector left fixed by π, $\Sigma\theta^*(i)\pi(i, j) = \theta^*(j)$. The dual pairing between D^* and D may in this case be denoted by juxtaposition of the row vector and column vector in question representing the matrix product of the two. Our semigroup $\mathcal{S}$ is the set of all matrices generated by the matrices π_i. From $\theta^*\pi = \theta^*$ and $\pi\theta = \theta$ and $\pi = \Sigma\pi_i$ it follows that (i) and (ii) of Definition 19.1 are satisfied. (iv) will be satisfied if no column of π vanishes so that $\mathcal{S}$ will be a stochastic semigroup if we further stipulate that $F(e) = \theta^*\theta = \Sigma\theta^*(j)$.

If we then assume that these conditions are verified $\mathcal{S}$ will be a linear semigroup and it generates a process $X(\mathcal{S})$. For any sequence $\pi_{i_1}, \pi_{i_2}, \ldots, \pi_{i_k}$ we shall have

$$P\left(x_1 = \prod_{i_1}, x_2 = \prod_{i_2}, \ldots, x_k = \prod_{i_k} \right) = \theta^*\prod_{i_1}\prod_{i_2}\cdots\prod_{i_k}\theta =$$

$$\sum_{i, j_1, j_2, \ldots, j_k} \theta^*(i)\prod_{i_1}(i, j_1)\prod_{i_2}(j_1, j_2)\cdots\prod_{i_k}(j_{k-1}, j_k) =$$

$$(19.1) \qquad \sum_{i} \theta^*(i)\prod_{i_1}(i, i_1)\prod_{i_2}(i_1, i_2)\cdots\prod_{i_k}(i_{k-1}, i_k) =$$

$$\sum_{i} \theta^*(i)\prod(i, i_1)\prod(i_1, i_2)\cdots\prod(i_{k-1}, i_k) =$$

$$\theta^*(i_1)\prod(i_1, i_2)\prod(i_2, i_3)\cdots\prod(i_{k-1}, i_k) \ .$$

Now consider the Markoff process with state space $(\pi_1, \ldots, \pi_m)$, with transition probability matrix π and with stationary measure θ^*. It is clear that the probability, in the Markoff process, of the sequence

π_{i_1}, π_{i_2}, ..., π_{i_k} is exactly the product in (19.1). We have shown then that $X(\mathcal{S})$ is a Markoff process with transition probability matrix π. Any finite state Markoff process is then of the form $X(\mathcal{S})$ where $\mathcal{S}$ is a stochastic semigroup of finite dimensional matrices. It is, of course, innessential that the state space of $X(\mathcal{S})$ is a set of matrices since the nature of the elements of the state space is of no significance.

Not only is every finite state Markoff process generated by a stochastic semigroup of matrices, but so also is every finite state subprocess of a Markoff or m-Markoff process. All these cases may be reduced to the case where $x_n = \psi(y_n)$, Y a finite state Markoff process as in §17.1. Let π be the transition matrix of Y and θ^* the stationary measure of Y. Let I_1, I_2, ..., I_r be the sets of indices from 1 to m corresponding to elements in the state space of Y that are "lumped together" under the map ψ.

Now define matrices π_i in terms of π by setting

$$\prod_{i}(j,\ k) = \begin{cases} 0 & \text{if } k \notin I_i \\[2em] \prod(j,\ k) & \text{if } k \in I_i \ . \end{cases}$$

We obtain a linear stochastic semigroup $\mathcal{S}$ by taking as generators the matrices π_i with θ^* the stationary measure and θ the vector with all components equal to 1. Since $\Sigma\pi_i = \pi$ conditions (i) and (ii) of Definition 19.1 will again be verified, and (iv) is verified if, as we obviously may assume, θ^* has no vanishing components. This time the probabilities of the process $X(\mathcal{S})$ are

$$P\left(x_1 = \prod_{i_1},\ x_2 = \prod_{i_2},\ ...,\ x_k = \prod_{i_k}\right) =$$

(19.2)

$$\sum_{i=1}^{m}\ \sum_{i_j \in I_j} \theta^*(i)\prod(i,\ i_1)\prod(i_1,\ i_2)...\prod(i_{k-1},\ i_k)\ .$$

It is not difficult to show that (19.2) also represents the probabilities of X. In fact the sum in (19.2) is just the sum of the probabilities of all the sequences in Y that would give rise to the particular sequence in X under the map $x_n = \psi(y_n)$. (This is the meaning of the restriction $i_j \in I_j$.) Thus we see that any finite state subprocess of a finite state m-Markoff process is obtained as the process corresponding to a stochastic semigroup of finite dimensional matrices.

However using such finite dimensional semigroups we cannot obtain any other processes. Namely:

THEOREM 19.2. If a stochastic semigroup $\mathcal{S}$ is finite dimensional; i.e., isomorphic to a semigroup of non-negative matrices Π_i acting on a finite dimensional space R^m with V the cone of non-negative m-tuples, then $X(\mathcal{S})$ is a subprocess of a finite state Markoff process.

PROOF. Let the generators of the process be Π_1, Π_2, ..., Π_r and set

$$\Pi = \sum_{i=1}^{r} \Pi_i \ .$$

θ^* is an m-dimensional row vector and θ and m-dimensional column vector. We shall first of all show that we may assume that none of the components of θ vanishes.

To begin with suppose that there is an index i such that all the entries in the i-th row and in the i-th column of Π vanish. The same must hold for each Π_j and the i-th components of both θ and θ^* must vanish since $\Pi\theta = \theta$, $\theta^*\Pi = \theta^*$. It follows that the semigroup is equivalent (in the sense of Theorem 19.1) to a semigroup of matrices of 1 dimension less, namely the matrices obtained by deleting the i-th rows and i-th columns from each Π_j.

Now suppose that $\theta(i) = 0$ for some i. We shall show that in this case the semigroup is equivalent to one in which all the i-th rows and i-th columns of all the matrices vanish and by the foregoing argument the dimension could be reduced. To see this recall that the process $X(\mathcal{S})$ is determined by the products $\theta^*\Pi_{j_1} \Pi_{j_2} ... \Pi_{j_\ell} \theta$. Now in such a product it is easy to see that any expression involving the i-th row or i-th column of any Π_{j_k} must vanish. That is, if one of the indices of

$$(19.3) \qquad \theta^*(i_1)\prod_{j_1} (i_1,\ i_2)\prod_{j_2} (i_2,\ i_3)...\prod_{j_\ell} (i_\ell,\ i_{\ell+1})\theta(i_{\ell+1})$$

agrees with i then the product in question must vanish. For

$$\prod_{j_k} (i_k,\ i_{k+1})...\prod_{j_\ell} (i_\ell,\ i_{\ell+1})\theta(i_{\ell+1})$$

is less than the i_k-th component of

$$\prod^{\ell-k+1} \theta = \theta$$

and since $\theta(i) = 0$ if $i_k = i$ this product and hence (19.3) vanishes.
It follows that the product $\theta^* \Pi_{j_1} \Pi_{j_2} \ldots \Pi_{j_\ell} \theta$ would be unaffected if we
replaced each $\Pi_j(i, k)$ and $\Pi_j(k, i)$ by 0. By the preceding argument
we see that if the dimension m is taken minimal (among those giving
semigroups equivalent to $\mathcal{S}$) then we can suppose that $\theta(i) > 0$ for
all i.

Now define the diagonal matrix C by

$$C = \begin{pmatrix} \theta(1) & 0 & 0 & \cdots & 0 \\ 0 & \theta(2) & 0 & \cdots & 0 \\ 0 & 0 & \theta(3) & \cdots & 0 \\ \cdots & \cdots & \cdots & \cdots & \cdots \\ 0 & 0 & 0 & \cdots & \theta(m) \end{pmatrix}$$

and define the $rm \times rm$ matrix $\widetilde{\Pi}$ by

$$\widetilde{\Pi} = \begin{pmatrix} C^{-1}\Pi_1 C & C^{-1}\Pi_2 C & \cdots & C^{-1}\Pi_r C \\ C^{-1}\Pi_1 C & C^{-1}\Pi_2 C & \cdots & C^{-1}\Pi_r C \\ \cdots & \cdots & \cdots & \cdots \\ C^{-1}\Pi_1 C & C^{-1}\Pi_2 C & \cdots & C^{-1}\Pi_r C \end{pmatrix} .$$

First we observe that $\widetilde{\Pi}$ is a stochastic matrix. For the sum
of a row of $\widetilde{\Pi}$ is clearly the same as the sum of a row of $C^{-1}\Pi C$ since
$\Pi = \Sigma\Pi_i$. The (i, j) term of $C^{-1}\Pi C$ is

$$\frac{1}{\theta(i)} \prod(i, j)\theta(j)$$

and since $\Pi\theta = \theta$

$$\sum_j \frac{1}{\theta(i)} \prod(i, j)\theta(j) = 1 \quad .$$

Moreover we note that the rm-dimensional vector $\widetilde{\theta}^* = (\theta^*\Pi_1 C, \theta^*\Pi_2 C, \ldots, \theta^*\Pi_r C)$ is a stationary measure for $\widetilde{\Pi}$:

$$\sum_{i=1}^{r}\left(\theta^*\prod_i C\right)\left(C^{-1}\prod_k C\right) = \sum_{i=1}^{r}\theta^*\prod_i\prod_k C =$$

$$\theta^*\prod\prod_k C = \theta^*\prod_i\prod_k C \quad.$$

The matrix $\tilde{\Pi}$ and the stationary vector $\tilde{\theta}^*$ define a Markoff process and we shall see that the subprocess obtained by "lumping together" the indices $1, \ldots, m;\ m + 1, \ldots, 2m;\ \ldots;\ (r - 1), m + 1, \ldots, rm$, is identical to the process generated by $\mathcal{S}$. Now the subprocess in question corresponds to the semigroup $\tilde{\mathcal{S}}$ generated by the $\tilde{\Pi}_i$ where

$$\tilde{\Pi}_i = \begin{pmatrix} 0 & \cdots & 0 & C^{-1}\pi_i C & 0 & \cdots & 0 \\ 0 & \cdots & 0 & C^{-1}\pi_i C & 0 & \cdots & 0 \\ \cdots & & \cdots & & \cdots & & \cdots \\ 0 & \cdots & 0 & C^{-1}\pi_i C & 0 & \cdots & 0 \end{pmatrix}$$

(0's occurring everywhere but in the columns $(i - 1)m + 1$ to im)) and by $\tilde{\theta}^*$ as given, and $\tilde{\theta}$ an rm-dimensional column vector with all entries $= 1$.

To show all the processes in question are identical we must show that

$$(19.4)\qquad \tilde{\theta}^*\tilde{\Pi}_{i_1}\tilde{\Pi}_{i_2}\cdots\tilde{\Pi}_{i_\ell}\tilde{\theta} = \theta^*\prod_{i_1}\prod_{i_2}\cdots\prod_{i_\ell}\theta \quad.$$

Now we have

$$\tilde{\Pi}_{i_1}\tilde{\Pi}_{i_2}\cdots\tilde{\Pi}_{i_\ell} = \begin{pmatrix} 0 & \cdots & 0 & C^{-1}\pi_{i_1}\pi_{i_2} & \cdots & \pi_{i_\ell}C & 0 & \cdots & 0 \\ 0 & \cdots & 0 & C^{-1}\pi_{i_1}\pi_{i_2} & \cdots & \pi_{i_\ell}C & 0 & \cdots & 0 \\ \cdots & & \cdots & & \cdots & & \cdots & & \cdots \\ 0 & \cdots & 0 & C^{-1}\pi_{i_1}\pi_{i_2} & \cdots & \pi_{i_\ell}C & 0 & \cdots & 0 \end{pmatrix}$$

so that the left side of 19.4 becomes

$$(19.5)\qquad \sum_i \theta^*\prod_i C \cdot C^{-1}\prod_{i_1}\prod_{i_2}\cdots\prod_{i_\ell} Cu$$

where u is the m-dimensional vector with all entries = 1. Clearly
Cu = θ so that (19.5) reduces to

$$\theta^* \textstyle\prod CC^{-1} \prod_{i_1} \prod_{i_2} \cdots \prod_{i_\ell} \theta = \theta^* \prod_{i_1} \prod_{i_2} \cdots \prod_{i_\ell} \theta \quad .$$

This completes the proof of the theorem.

As a result of this theorem we make the following definition:

DEFINITION 19.1. A finitely valued subprocess of a finitely
valued m-Markoff process will be referred to as a __finite dimensional__
process.

19.3. __Further Examples__. By the preceding theorem we see that
finite dimensional stochastic semigroups cannot lead to any new processes.
We need not, however, restrict ourselves to finite dimensional spaces D.
For example we may take D to be the set of all real valued continuous
functions on the interval [0, 1] and V the cone of non-negative functions. Let $\kappa_i(s, t)$, i = 1, 2, ..., r, be a set of continuous non-
negative functions in two variables $s, t \in [0, 1]$, satisfying

$$(19.6) \qquad\qquad \sum_{i=1}^{r} \kappa_i(s, t) = 1 \quad .$$

We then take transformations a_i of D given by

$$a_i(f)(s) = \int_0^1 \kappa_i(s, t) f(t)\, dt \quad .$$

θ is taken as the function identically 1 and θ^* is Lebesgue measure
on [0, 1]. By (19.6) we have

$$\sum a_i \theta(s) = \int_0^1 1\, dt = 1 = \theta(s)$$

so that $\Sigma a_i \theta = \theta$. Moreover $\Sigma a_i^* \theta^*(f) = \theta^*(\Sigma a_i f) = \theta^*(g)$ where g is
the constant function

$$\int_0^1 f(t)\, dt \quad .$$

Then

$$\int_0^1 g(t)dt = \int_0^1 f(t)dt$$

and so $\Sigma a_i^* \theta^* = \theta^*$. Since moreover $\langle \theta^*, \theta \rangle = 1$ we find that as long as no $\kappa_1(s, t)$ vanishes identically, the a_i will generate a stochastic semigroup $\mathcal{S}$.

§20. Continuous Predictability for the Processes
of Stochastic Semigroups

20.1. <u>A Criterion for Continuous Predictability</u>. Let X be a process defined in terms of a stochastic semigroup; the variables of X have as their range the set $\Lambda = \{a_1, \ldots, a_r\}$ of generators of the semigroup $\mathcal{S}$. As in §16 we shall consider finite and infinite Λ-sequences σ defined for the integers $> m$ and ≤ 0 where m can be $- \infty$. If $\sigma = (\sigma(q), \sigma(q + 1), \ldots, \sigma(0))$ there corresponds to σ a particular element of $\mathcal{S}$ denoted $\bar{\sigma}$: $\bar{\sigma} = \sigma(q)\sigma(q+1)\ldots\sigma(0)$.

We will find it convenient to speak of convergence of finite Λ-sequences as well as left-infinite ones. This will have the following meaning. If $\sigma^{(k)}$ are Λ-sequences defined for $q_k \leq n \leq 0$, $\sigma^{(k)} = (\sigma(q_k), \sigma(q_{k+1}), \ldots, \sigma(0)$ and ω is a left-infinite Λ-sequence we say that $\sigma^{(k)} \longrightarrow \omega$ if (i) $q_k \longrightarrow - \infty$ and (ii) for any n, $\sigma^{(k)}(n) \longrightarrow \omega(n)$ as $k \longrightarrow \infty$. (Here $\sigma^{(k)}(n)$ is eventually defined since $q_k \longrightarrow - \infty$.)

The following theorem relates the problem of prediction for $X(\mathcal{S})$ with properties of $\mathcal{S}$.

> THEOREM 20.1. X is continuously predictable at a point $\xi \in \Omega_X^-$ if and only if whenever the $\sigma^{(k)}$ are finite Λ-sequences with $\sigma^{(k)} \longrightarrow \xi$ and $\rho \in \mathcal{S}$ then the sequence

(20.1)
$$F(\overline{\sigma^{(k)}\rho})/F(\overline{\sigma^k})$$

> converges as $k \longrightarrow \infty$. Here F is the function associated with $\mathcal{S}$.

PROOF. For a Λ-sequence $\sigma = (\sigma(q), \sigma(q + 1), \ldots, \sigma(0)$ let $x_{\bar{\sigma}}$ denote the characteristic function of $\{\omega : x_q(\omega) = \sigma(q), x_{q+1}(\omega) = \sigma(q + 1), \ldots, x_0(\omega) = \sigma(0)\} \subset \Omega_X$. Also if τ represents a sequence defined for positive integers $\tau = (\tau(1), \tau(2), \ldots, \tau(\ell))$, x_{τ}^+ will represent the characteristic function of the set $\{\omega : x_1(\omega) = \tau(1),\ x_2(\omega) = \tau(2),\ \ldots,\ x_\ell(\omega) = \tau(\ell)\}$. Also let $\bar{\tau}$ represent the element $\tau(1)\tau(2)\ldots\tau(\ell)$ of $\mathcal{S}$. Now in (20.1) we may assume that $\rho = \bar{\tau}$ for a sequence τ (since the a_i generate $\mathcal{S}$) unless $\rho = e$ in which case (20.1) is 1 throughout. Reference to (18.1) shows that

$$(20.2) \quad F(\bar{\sigma}) = P(x_1 = \sigma(q),\ x_2 = \sigma(q+1),\ \ldots,\ x_{1-q} = \sigma(0)) =$$

$$P(x_q = \sigma(q),\ x_{q+1} = \sigma(q+1),\ \ldots,\ x_0 = \sigma(0)) = E(x_\sigma^-)\ .$$

Also

$$(20.3) \qquad F(\bar{\sigma}\bar{\tau}) = P(x_1 = \sigma(q),\ \ldots,\ x_{1-q} = \sigma(0),\ x_{2-q} =$$

$$\tau(1),\ \ldots,\ x_{\ell-q+1} = \tau(\ell)) =$$

$$P(x_q = \sigma(q),\ \ldots,\ x_0 = \sigma(0),\ x_1 = \tau(1),\ \ldots,\ x_\ell = \tau(\ell)) =$$

$$= E(x_\sigma^- x_\tau^+)\ .$$

The theorem may therefore be restated as follows: X is c.p. at ξ if and only if whenever $\sigma^{(k)} \longrightarrow \xi$

$$(20.4) \qquad\qquad E(x_{\sigma^{(k)}}^- x_\tau^+)/E(x_{\sigma^{(k)}}^-)$$

converges.

Assume now that X is c.p. at ξ, i.e., A_X^- is c.p. to A_X at ξ. By Theorem 10.1 this is equivalent to every predicting sequence of A_X^- at ξ converging over A_X. It is clear, however, that if $\sigma^{(k)} \longrightarrow \xi$, then

$$x_{\sigma^{(k)}}^- / E(x_{\sigma^{(k)}}^-)$$

forms a predicting sequence at ξ, and since $x_\tau^+ \in A_X$ it follows that (20.4) converges.

For the sufficiency of the condition we must show that any predicting sequence at ξ will converge over A_X if the sequence

$$x_{\sigma^{(k)}}^- / E(x_{\sigma^{(k)}}^-)$$

does so for any $\sigma^{(k)} \longrightarrow \xi$. Suppose then that $\{f_m\}$ is a predicting sequence at ξ. Now any function in A_X^- can be approximated uniformly to within an arbitrary ϵ by a function of finitely many x_n, $n \leq 0$. Hence we may assume that each f_m depends only on x_{-s_m}, x_{-s_m+1}, ..., x_0 for some sequence of S_m where the $s_m \longrightarrow \infty$ so rapidly that the f_m have the same limiting properties as the original predicting sequence. We may therefore set

$$f_m = \sum_\sigma \alpha_\sigma^{(m)} x_\sigma^-$$

for sequences σ of length s_m, since any function of x_{-s_m}, x_{-s_m+1}, ..., x_0 has this form.

It is clear from the proof of Lemma 11.1 that a predicting sequence will converge over all of A_X if it converges over the "future" A_X^+ generated by x_1, x_2, x_3, ... Moreover we need only consider functions of finitely many of x_1, x_2, ..., x_n, ... since these are dense in A_X^+. Finally we may restrict our attention to the functions x_τ^+ for $\tau = (\tau(1), \tau(2), ..., \tau(\ell))$ since these span the space of functions of finitely many x_i, $i > 0$. Hence to show that A_X^- is c.p. at ξ we must show that

$$(20.5) \qquad \sum \alpha_\sigma^{(m)} E(x_\sigma^- x_\tau^+)$$

converges as $m \longrightarrow \infty$. By (20.3) the latter becomes

$$(20.6) \qquad \sum \alpha_\sigma^{(m)} F(\bar{\sigma}\bar{\tau}) \quad .$$

Now let $\varepsilon > 0$; by hypothesis there is an S such that whenever the sequence σ agrees with ξ at the last S terms then $F(\bar{\sigma}\bar{\rho})/F(\bar{\sigma})$ differs from a fixed number $\beta(\rho)$ by less than ε. Let us write $\rho = (\xi(-s+1), \xi(-s+2), ..., \xi(0))$ and let Σ_σ' denote summation over those σ terminating in ρ and Σ_σ'' denote summation over the remaining σ. Set $g = 1 - x_\rho^-$. Since $\{f_m\}$ is a predicting sequence at ξ $E(f_m g) \longrightarrow \delta_\xi(g) = g(\xi) = 0$. But clearly $E(f_m g) = \Sigma \alpha_\sigma^{(m)} E(x_\sigma^-(1 - x_\rho^-)) = \Sigma_\sigma'' \alpha_\sigma^{(m)} E(x_\sigma^-) = \Sigma_\sigma'' \alpha_\sigma^{(m)} F(\sigma)$ (where m is large enough that $S_m > S$, assuming as we may that $S_m \longrightarrow \infty$.) Hence as $m \longrightarrow \infty$

$$(20.7) \qquad \sum_\sigma {}'' \alpha_\sigma^{(m)} F(\bar{\sigma}) \longrightarrow 0$$

and <u>a fortiori</u> $\Sigma_\sigma'' \alpha_\sigma^{(m)} F(\bar{\sigma}\bar{\tau}) \longrightarrow 0$ as $m \longrightarrow \infty$. To prove the convergence of (20.6) we are left to consider that of $\Sigma_\sigma' \alpha_\sigma^{(m)} F(\bar{\sigma}\bar{\tau})$. For this we note that since $E(f_m) \longrightarrow 1$, $\Sigma \alpha_\sigma^{(m)} F(\bar{\sigma}) \longrightarrow 1$ and so by (20.7), $\Sigma' \alpha_\sigma^{(m)} F(\bar{\sigma}) \longrightarrow 1$. But

$$\left| \sum_{\sigma}{}' \alpha_\sigma^{(m)} F(\overline{\sigma \tau}) - \beta(\tau) \sum_{\sigma}{}' \alpha_\sigma^{(m)} F(\overline{\sigma}) \right| =$$

(20.8)

$$\left| \sum_{\sigma}{}' \alpha_\sigma^{(m)} F(\overline{\sigma}) \left(\frac{F(\overline{\sigma \tau})}{F(\overline{\sigma})} - \beta(\tau) \right) \right| < \epsilon \sum_{\sigma}{}' \alpha_\sigma^{(m)} F(\overline{\sigma}) \quad .$$

Since ϵ is arbitrary and $\Sigma_\sigma' \alpha_\sigma^{(m)} F(\overline{\sigma}) \longrightarrow 1$ for each ϵ it follows that (20.8) converges to 0 and this completes the proof of the theorem.

20.2. <u>The Linear Case</u>. We now turn to the case that the stochastic semigroup in question is a semigroup of transformations of a cone V in a space D with F given by $F(\rho) = \langle \theta^*, \rho \theta \rangle$. The variables of the process X have as their range a set $\{a_1, a_2, \ldots, a_r\}$ of transformations of V. In particular each sequence $\xi(n)$ of Ω_X^- is a left infinite sequence of transformations of V. We shall see that continuous predictability at a point $\xi \in \Omega_X^-$ now depends upon the behavior of the transformations $\xi(q)\xi(q + 1)\ldots\xi(0)$ as $q \longrightarrow -\infty$.

THOEREM 20.2. If

$$(20.9) \quad \Gamma_V(\xi(q)\xi(q + 1)\ldots\xi(0)u, \; \xi(q)\xi(q + 1)\ldots\xi(0)v) \longrightarrow 1$$

as $q \longrightarrow -\infty$ for any $u, v \in V$, then A_X^- is c.p. to A_X at ξ.

PROOF. We shall show that the criterion of Theorem 20.1 is fulfilled. Suppose therefore that $\sigma^{(k)} \longrightarrow \xi$ and that $\rho \in \mathcal{S}$. We wish to prove that if (20.9) holds then $F(\overline{\sigma^{(k)}}\rho)/F(\overline{\sigma^{(k)}})$ converges. Note that since $\sigma^{(k)} \longrightarrow \xi$ we may write

$$(20.10) \quad \overline{\sigma^{(k)}} = \overline{\sigma_1^{(k)}} \, \xi(q)\xi(q + 1)\ldots\xi(0)$$

for sufficiently large k, if $\sigma_1^{(k)}$ is properly chosen.

Now by the definition of K_V (§15.1) we recall that for any $u_1, u_2 \in V$, $u_1 - (K_V(u_1 u_2) - \epsilon) u_2 \in V$ for any $\epsilon > 0$. In particular for any $\alpha \in \mathcal{S}$, $\alpha u_1 - (K_V(u_1, u_2) - \epsilon) \alpha u_2 \in V$ and since θ^* is non-negative on V.

$$\langle \theta^*, \alpha u_1 \rangle \geq (K_V(u_1, u_2) - \epsilon) \langle \theta^*, \alpha u_2 \rangle \quad .$$

Letting $\epsilon \longrightarrow 0$:

$$(20.11) \qquad \langle \theta^*, \alpha u_1 \rangle \geq K_V(u_1 u_2) \langle \theta^*, \alpha u_2 \rangle \ .$$

Let $u_1 = \xi(q)\xi(q+1)\ldots\xi(0)\rho\theta$, $u_2 = \xi(q)\xi(q+1)\ldots\xi(0)\theta$ and take $\alpha = \overline{\sigma_1^{(k)}}$. Then

$$(20.12) \qquad \begin{aligned} F(\overline{\sigma^{(k)}}\rho) &= \langle \theta^*, \overline{\sigma_1^{(k)}}\xi(q)\xi(q+1)\ldots\xi(0)\rho\theta \rangle \geq \\ &K_V(\xi(q)\xi(q+1)\ldots\xi(0)\rho\theta, \ \xi(q)\xi(q+1)\ldots\xi(0)\theta)F(\overline{\sigma^{(k)}}) \end{aligned}$$

by (20.11). Interchanging u_1 and u_2 and $\alpha = \overline{\sigma_1^{\ell}}$ we find

$$(20.13) \qquad \begin{aligned} F(\overline{\sigma^{(\ell)}}) &\geq \\ &K_V(\xi(q)\xi(q+1)\ldots\xi(0)\theta, \ \xi(q)\xi(q+1)\ldots\xi(0)\rho\theta)F(\sigma^{(\ell)}\rho) \ . \end{aligned}$$

Multiplying these two inequalities we have

$$(20.14) \qquad \begin{aligned} F(\overline{\sigma^{(\ell)}})F(\overline{\sigma^{(k)}}\rho) &\geq \\ &\Gamma_V(\xi(q)\xi(q+1)\ldots\xi(0)\theta, \ \xi(q)\xi(q+1)\ldots\xi(0)\rho\theta)F(\overline{\sigma^{(\ell)}}\rho)F(\overline{\sigma^{(k)}}). \end{aligned}$$

Under the conditions of the theorem then,

$$\frac{F(\overline{\sigma^{(k)}}\rho)/F(\overline{\sigma^{(k)}})}{F(\overline{\sigma^{(\ell)}}\rho)/F(\overline{\sigma^{(\ell)}})}$$

comes arbitrarily close to 1 if k and ℓ are chosen sufficiently large and hence

$$F(\overline{\sigma^{(k)}}\rho)/F(\overline{\sigma^{(k)}})$$

converges as $k \longrightarrow \infty$.

> COROLLARY. If each generator a_i of $\mathscr{S}$ is projectively bounded (Definition 15.2), then the corresponding process $X(\mathscr{S})$ is c.p.

> PROOF. We have already seen in the proof of the corollary to Lemma 17.4 that if the transformations $R^{(1)}, \ldots, R^{(n)}$ are uniformly projectively bounded then if n is sufficiently large $\Gamma(R^{(1)}\ldots R^{(n)})$ will be arbitrarily close to 1. Hence since there is a $\delta > 0$ with

$\Gamma(a_i) > \delta$ for all a_i, $i = 1, \ldots, r$ it follows that $\Gamma(\xi(q)\xi(q + 1)\cdots\xi(0))$ $\longrightarrow 1$ as $q \longrightarrow -1$ and so the conditions of Theorem 20.2 are fulfilled.

20.3. Applications.

EXAMPLE 1. Let $\mathcal{S}$ be the linear semigroup of §19.3, that is, V is the cone of non-negative continuous functions on $[0, 1]$, $\theta(S) \equiv 1$ and θ^* is Lebesgue measure. The a_i are integral transforms with kernels $\kappa_i(s, t)$ satisfying $\Sigma\kappa_i(s, t) = 1$. We now suppose that there are two constants $\alpha, \beta > 0$ that

$$(20.15) \qquad\qquad \alpha \leq \kappa_i(s, t) \leq \beta \quad .$$

The resulting process is then continuously predictable. In fact (20.15) implies that the a_i are projectively bounded and so the preceding corollary applies.

To see this we write

$$\langle \alpha\, \theta^*, f \rangle = \alpha \int_0^1 f(t)dt \leq \int_0^1 \kappa_i(s, t)f(t)dt = a_i(f)(s)$$

$$\leq \beta \int_0^1 f(t)dt = \beta \langle \theta^*, f \rangle$$

for any $f \in V$. Hence

$$a_i(f) - \frac{\alpha}{\beta}\frac{\langle\theta^*,f\rangle}{\langle\theta^*,g\rangle} a_i(g) > \alpha\langle\theta^*, f\rangle - \frac{\alpha}{\beta}\frac{\langle\theta^*,f\rangle}{\langle\theta^*,g\rangle} \cdot \beta\langle\theta^*, g\rangle = 0$$

so that

$$K_V(a_i(f), a_i(g)) \geq \frac{\alpha}{\beta}\frac{\langle\theta^*,f\rangle}{\langle\theta^*,g\rangle} \quad .$$

Similarly:

$$K_V(a_i(g), a_i(f)) \geq \frac{\alpha}{\beta}\frac{\langle\theta^*,g\rangle}{\langle\theta^*,f\rangle}$$

so that

$$\Gamma_V(a_i(f), a_i(g)) \geq \frac{\alpha^2}{\beta^2} > 0 \quad .$$

EXAMPLE 2.   We call an infinite matrix (m_{ij}) with non-negative m_{ij} of _essential rank one_ if there are two non-negative sequences $\{m'_i\}$ and $\{m''_j\}$ and a pair of constants $\alpha, \beta > 0$ such that

$$(20.16) \qquad \alpha m'_i m''_j \le m_{ij} \le \beta m'_i m''_j$$

for all i, j. The reason for this terminology is that if $\alpha = \beta$ then (m_{ij}) would be of rank one in the usual sense.

Now $M = (m_{ij})$ will not in general operate on all non-negative column vectors because the sums of products involved may diverge. We shall show that there is a particular cone V_o of non-negative column vectors such that M takes V_o into V_o and such that on this cone M will define a projectively bounded transformation if M is of essential rank one. Namely we take V_o to be the set of all vectors (x_j) with non-negative entries such that all the sums

$$(20.17) \qquad \sum_{i_1, i_2, \ldots, i_k} m_{ii_1} m_{i_1 i_2} \cdots m_{i_k j} x_j$$

converge for all i and k. We now show that $\Gamma_{V_o}(M) \ge \alpha^2/\beta^2$. Namely if $u, v \in V_o, u = (u_i), v = (v_j)$

$$\sum m_{ij} u_j \ge \alpha m'_i \sum m''_j u_j$$

and

$$\sum m_{ij} v_j < \beta m'_i \sum m''_j v_j$$

so that

$$Mu - \frac{\alpha \Sigma m''_j u_j}{\beta \Sigma m''_j v_j} Mv$$

has positive coefficients (unless the denominator vanishes in which case $Mv = 0$) and since $Mu, Mv \in V_o$ this linear combination also belongs to V_o. It follows that

$$K_{V_o}(Mu, Mv) \ge \frac{\alpha}{\beta} \frac{\Sigma m''_j u_j}{\Sigma m''_j v_j}$$

and similarly

$$K_{V_o}(Mv, Mu) \ge \frac{\alpha}{\beta} \frac{\Sigma m''_j v_j}{\Sigma m''_j u_j}$$

and $\Gamma_{V_0}(Mu, Mv) \geq \alpha^2/\beta^2$.

In a similar manner it may be shown that any finite set of matrices of essential rank 1 are projectively bounded in a common cone. The cone may be trivial, i.e., reduce to {0}, but this will not be the case if the products of the matrices themselves are defined. It follows that any semigroup generated by finitely many matrices of essential rank one leads to a c.p. process.

We may remark that the example given at the end of the last chapter also serves as an illustration of the corollary to Theorem 20.2. There we showed that a subprocess of a Markoff process of the form $x_n = \psi(y_n)$ is c.p. if the y_n have strictly positive transition probabilities. By §19 we know that such a process is associated with the matrix semigroup of transformations generated by a set of matrices formed from the transition probability matrix:

$$
a_\ell = \begin{pmatrix}
0 & \cdots & 0 & p_{1j_\ell} & \cdots & p_{1j_{\ell+1}-1} & 0 & \cdots & 0 \\
0 & \cdots & 0 & p_{2j_\ell} & \cdots & p_{2j_{\ell+1}-1} & 0 & \cdots & 0 \\
\cdot & \cdots & \cdot & \cdots & \cdots & \cdots & \cdot & \cdots & \cdot \\
\cdot & \cdots & \cdot & \cdots & \cdots & \cdots & \cdot & \cdots & \cdot \\
0 & \cdots & 0 & p_{mj_\ell} & \cdots & p_{mj_{\ell+1}-1} & 0 & \cdots & 0
\end{pmatrix} .
$$

By the remarks following Definition 15.2 we recognize that these matrices define projectively bounded transformations. (Alternatively we may deduce this from the fact that the above matrices are of "essential rank one".) The continuous predictability of the process now follows from the above corollary.

It might be pointed out that in the last chapter the continuous predictability of this process was a consequence of Theorem 16.1 which ensures continuous predictability if

$$
\Gamma(\pi_q(\xi)\pi_{q+1}(\xi)\cdots\pi_{-1}(\xi)) \longrightarrow 1 \quad ,
$$

and in this chapter it is a consequence of Theorem 20.2 where the condition

$$
\Gamma_V(\xi(q)\xi(q + 1)\cdots\xi(0)u, \ \xi(q)\xi(q + 1)\cdots\xi(0)v) \longrightarrow 1
$$

is used. It is evident that these conditions are related; however neither theorem includes the other. For in the more restricted case of Theorem 16.1 we obtained continuous predictability from A_X^- to an extension A_Y^- as well, whereas in the more general case considered here we only obtain continous predictability of the process itself.

CHAPTER 6. STATISTICAL PREDICTABILITY

In this chapter we shall broaden the notion of predictability so
as to include certain of the non-continuously predictable sequences. That
a more adequate notion of predictability might be given is suggested by
the example of §12 of a non-continuously predictable sequence for which
our direct procedure for prediction does provide prediction probabilities.
(§12.2). As we shall see, the concept of "statistical predictability"
defined in this chapter, covers this example and it will be our task for
the remainder of this investigation to generalize this fact. Namely, we
shall show, though not till Chapter 9, that any finitely-valued derived
sequence of a Markoff sequence is "statistically predictable".

As in our treatment of continuous predictability we regard the
prediction problem from past to future as a special case of prediction
from one process to another. It will be found that the procedure for
prediction in general may be carried out in two stages. Namely, between
two processes X and Y where X is a subprocess of Y, a third
process $\tilde{X}$ may be interposed, and the prediction is first carried out
from X to $\tilde{X}$ and then from $\tilde{X}$ to Y. The second of these steps will
always be possible since $\tilde{X}$ will be c.p. to Y. The problem is thereby reduced to prediction from X to $\tilde{X}$ and as we shall see, the latter
problem will be simplified as a result of the special relationship $\tilde{X}$
bears to X.

§21. The Continuously Predictable Cover of a Process

21.1. <u>Definitions.</u> Let X and Y be two abstract processes,
either one-sided or two-sided, and let X be a subprocess of Y not
necessarily continuously predictable to Y. We shall find that we can
associate to this pair of processes a process $\tilde{X}$, "slightly larger"
than X such that $\tilde{X}$ will be c.p. to Y. Since $\tilde{X}$ will not necessarily
be a subprocess of Y, $\tilde{X}$ will be c.p. to Y in the following sense.

DEFINITION 21.1. If X and Y are subprocesses of a process
Z then we say that X is c.p. to Y if the algebra A_X is c.p. to the
E-subalgebra (A_X, A_Y) of A_Z generated by X and Y.

We may now state:

THEOREM 21.1. For any pair of processes X, Y with
$A_X \subset A_Y$, there exists a unique process $\tilde{X}$ satisfying

    (i) $\tilde{X}$ is an L-extension of X (Definition 2.6).
    (ii) $\tilde{X}$ is c.p. to Y
   (iii) If X' is any process satisfying (i) and (ii)
        then $\tilde{X}$ is a subprocess of X' ($\tilde{X}$ is the
        minimal process satisfying (i) and (ii)).

PROOF. We shall consider $L(A_X)$ imbedded in $L(A_Y)$ so that
all the algebras under consideration will be subalgebras of $L(A_Y)$. Let
B be the set of all elements of $L(A_X)$ of the form $\bar{z} = E(z|A_X)$ for
$z \in A_Y$. Define $A_{\tilde{X}}$ to be the intersection of all the E-subalgebras of
$L(A_X)$ with the property that they contain A_X and B and are invariant
under T. We note that the algebra $A_{\tilde{X}}$ will be separable since A_X
and A_Y and therefore B have denumerable dense subsets, and these subsets may be used to form a denumerable dense subset of $A_{\tilde{X}}$. It follows
that the E-subalgebra $A_{\tilde{X}}$ defines an abstract process $\tilde{X}$.

The process $\tilde{X}$ clearly satisfies (i). To show that $\tilde{X}$ is
c.p. to Y, by Definition 21.1 we must show that $A_{\tilde{X}}$ is c.p. to
$(A_{\tilde{X}}, A_Y)$. Now by Theorem 10.2 this involves showing that $E(z|A_{\tilde{X}}) \in A_{\tilde{X}}$
for all z in $(A_{\tilde{X}}, A_Y)$. Since every such z is a limit of sums of the
form $z_1 z_2$ with $z_1 \in A_{\tilde{X}}$ and $z_2 \in A_Y$, and $E(z_1 z_2|A_{\tilde{X}}) = z_1 E(z_2|A_{\tilde{X}})$
it suffices to show that $E(z_2|A_{\tilde{X}}) \in A_{\tilde{X}}$ for $z_2 \in A_Y$. But $A_X \subset A_{\tilde{X}} \subset L(A_X)$,
$E(z_2|A_{\tilde{X}}) = E(z_2|A_X)$ by (1.4) (a), and this is in B; this shows that
(ii) is verified.

Now suppose that X' satisfies (i) and (ii) with X' in the
place of $\tilde{X}$. Then $A_{X'}$ contains A_X and for every $z \in A_Y$, since X'
is c.p. to Y, by Theorem 10.2, $E(z|A_{X'}) \in A_{X'}$. But again
$A_X \subset A_{X'} \subset L(A_X)$ so the latter conditional expectation is $E(z|A_X)$.
Hence $B \subset A_{X'}$. But then by the definition of $A_{\tilde{X}}$ we must have $A_{\tilde{X}} \subset A_{X'}$.

DEFINITION 21.2. The process $\tilde{X}$ of Theorem 21.1 is the
<u>continuously predictable cover of</u> X <u>with respect to</u> Y. If X^- and X
are one-sided and two-sided versions of a single process we denote the
c.p. cover of X^- with respect to X by $\tilde{X}^-$. The two-sided version of
this process, $\tilde{X}$, is referred to as the <u>continuously predictable cover</u>
<u>of</u> X.

(For the case of X^- and X, the construction of Theorem 21.1
actually furnishes a one-sided process, but we refer to the entire process
$\tilde{X}$ as the c.p. cover of X.)

We note that $\tilde{X} = X$ if and only if X is c.p. to Y so that
the discrepancy between the processes X and $\tilde{X}$ is a measure of how far

X is from being c.p. to Y.

21.2. <u>An Example</u>. The foregoing definition may be illustrated
by the processes X and Y in the example of §12. Y is a random
$\{-1, 1\}$-process with $P(y_n = 1) = p \neq 1/2$; X is obtained from Y by
$x_n = y_n y_{n-1}$. Y^- and X^- denote the one-sided versions of these processes
and we form $\tilde{X}^-$, the c.p. cover of X^- with respect to X. In this
case $\tilde{X}^-$ is quite easily constructed and in fact $\tilde{X} = Y$. To see this we
show that Y satisfies (i), (ii) and (iii) of Theorem 21.1. Now we already showed in §12 that $A_Y^- \subset L(A_X^-)$ so that Y^- is, in fact, an
L-extension of X^-. Also A_Y^- is c.p. to A_Y since Y is a random
process. Since $A_X \subset A_Y$ it follows that A_Y^- is c.p. to A_X (Theorem
10.3) and hence Y^- satisfies (ii). Finally suppose that X' is another process that is c.p. to X and is an L-extension of X^-. By
Theorem 10.2, $E(x_1 | A_{X'}) \in A_{X'}$ since $x_1 \in A_X$, and since X' and Y^-
are L-extensions of X^- this gives (by (1.4) (a)) $E(x_1 | A_Y^-) \in A_{X'}$. But
$x_1 = y_1 y_0$ so $E(x_1 | A_Y^-) = E(y_1 y_0 | A_Y^-) = y_0 E(y_1)$; since $E(y_1) \neq 0$,
$y_0 \in A_{X'}$. Since $A_{X'}$ is invariant under T it follows that all $y_n \in A_{X'}$
$n \leq 0$ and so $A_Y^- \subset A_{X'}$ which verifies (iii). It follows that Y is
the c.p-cover of the process X.

§22. Statistical Predictability

22.1. <u>Statistical Determination</u>. In the last section we constructed an intermediate process $\tilde{X}$ between X and Y such that prediction may always be carried out from $\tilde{X}$ to Y. In this section we
shall consider the problem of predicting from X to $\tilde{X}$. This will be
dealt with as a special case of the general problem of predicting from a
process X to any L-extension Y of X. Here, when the prediction can
be carried out, it will associate to a point of Ω_X a single point of
Ω_Y rather than a measure over a set in Ω_Y. We first consider the situation for E-algebras in general.

Let A and B be two E-algebras with $A \subset B \subset L(A)$ and let
us consider the question of continuous predictability from A to B at
a point $\omega \in \Omega_A$. It turns out that either A is not c.p. to B at ω
or the prediction is trivial in the sense that ω has a unique extension
to B. This follows from the following remark: if β denotes the
canonical map from Ω_B to Ω_A and η is any point of Ω_B for which
$\beta(\eta) = \omega$, then the point measure δ_η is a prediction measure at ω.
For suppose $g \in B$ is such that for all non-negative f in A with
support in a neighborhood U of ω, $E(fg) \geq 0$. This being true for
f in A it follows for f in L(A) with support in U and since
$B \subset L(A)$ it is true for $f \in B$. It follows that g is non-negative in

$\beta^{-1}(U)$ and in particular $g(\eta) \geq 0$ so that $\delta_\eta(y) \geq 0$; hence δ_η is a prediction measure at ω. It follows then that either there is more than one prediction measure at $\dot\omega$ or ω has a unique extension. In other words, if A is c.p. to B at ω then it is deterministic at ω.

In the general case of E-algebras, when a point ω of Ω_A has more than one extension to an L-extension B of A there will be no way of predicting from A to B at ω. However when A and B are the algebras of processes there is a circumstance under which one may choose a particular extension of ω as the "correct" one. The choice will be made on the grounds that among the possible extensions of ω to B, only one exhibits the "correct statistical behavior".

DEFINITION 22.1.   If X is a subprocess of Y, A_X and A_Y the corresponding algebras, $A_X \subset A_Y \subset L(A_X)$ and $\xi \in \Omega_X$, then X <u>determines</u> Y <u>statistically at</u> ξ if there is a unique point $\eta \in \Omega_Y$ that extends ξ and that is generic for Y. η is then the <u>determined</u> point.

The idea of this definition is as follows. If η is a point in the sample space of a process Y then every element (function) z of A_Y determines a left-infinite sequence: $\eta_z(-n) = T^n z(\eta)$. A point η is generic for Y if each of the sequences η has an average equal to $E(z)$. Thus the generic points of Y are those that exhibit the proper statistical behavior. Now to any point ξ of Ω_X that has occurred there is a subset of points of Ω_Y that are compatible with it, namely the set $\beta^{-1}(\xi)$, with β the canonical map from Ω_Y to Ω_X. The situation envisioned in Definition 22.1 is that only one of these points compatible with ξ has the proper statistical behavior and so it is to be chosen as the one that has occurred.

It will be noted that if X determines Y statistically at $\xi \in \Omega_X$ then ξ itself must be a generic point for X since the restriction of a generic point is again generic. (Theorem 5.6.) On the other hand the assumption that ξ is generic does not guarantee that there exists an extension of ξ to Y that is generic. Thus two requirements are implicit in Definition 22.1; first that a generic extension of ξ should exist and secondly that it should be unique.

We may illustrate Definition 22.1 by returning to the process X of §21.2. We are now concerned with prediction from X^- to $\tilde{X}^- = Y^-$. Suppose first that η is a generic point of the random process Y. Then η', defined by $\eta'(n) = -\eta(n)$, and η both restrict to the same point of $\Omega_{\bar{X}}$: $\xi(n) = \eta(n)\eta(n-1)$. Now if $p = P(y_n = 1) \neq 1/2$, then η' cannot be generic for Y since, for example, the average of η' is the negative of that of η and the latter is not 0. It follows that X^- determines Y^- at ξ, for there is exactly one generic extension η of ξ to Y.

At the same time we should note that not every generic point of X has a generic extension to Y. To see this suppose that ξ is a point of Ω_X^- that does come from a generic point η in Ω_Y^-. We observe, first of all, that (assuming $p \neq 0$, 1) $\Omega_X^- = \Lambda_\infty^-$ where $\Lambda = \{-1, 1\}$, and so any $\{-1, 1\}$-sequence is a point in Ω_X^-. Let us consider then a point $\xi^* \in \Omega_X^-$ obtained from ξ by setting $\xi^*(n) = \xi(n)$ for all n outside a set $\{n_k\}$ to be specified later, and $\xi^*(n_k) = -\xi(n_k)$. Clearly if $\{n_k\}$ has density 0, ξ^* and ξ will have the same averages so that ξ^* as well as ξ will be a generic point of X. However the preimages of ξ^* in Ω_Y^- may differ quite radically from those of ξ. In fact it is possible to arrange that the preimages of ξ^* are not even stochastic.

For this, let η^* be one of the preimages of ξ^* so that $\eta^*(n)\eta^*(n-1) = \xi^*(n)$. Note that unless $n = n_k$, $\eta^*(n)\eta^*(n-1) = \eta(n)\eta(n-1)$ so that for n in an interval $n_{k+1} < n < n_k$ we either have $\eta^*(n) = \eta(n)$ or $\eta^*(n) = -\eta(n)$ holding throughout the interval. Also if one of these is true for the interval (n_{k+1}, n_k) the other must hold in (n_{k+2}, n_{k+1}) since

$$\eta^*(n_{k+1})\eta^*(n_{k+1} - 1) = -\eta(n_{k+1})\eta(n_{k+1} - 1) \ .$$

Now the average of η is $E(y_1) = p - (1 - p) = 2p - 1 \neq 0$. We choose the n_k inductively so that the interval (n_{k+1}, n_k) is so large that the average of η^* from n_{k+1} to 0 is very close alternately to $2p - 1$ and $-(2p - 1)$ according as $\eta^*(n) = \eta(n)$ or $\eta^*(n) = -\eta(n)$ in that interval. It follows that $\eta^*(n)$ cannot possess an average and so it will not be a stochastic sequence. Thus X^- determines Y^- statistically, in this case, at only some of the generic points of Ω_X^-.

22.2. <u>Properties of Statistical Determination</u>. The intuitive notion of a sequence $\xi(n)$ determining that another sequence $\eta(n)$ has occurred would lead one to certain conclusions regarding statistical determination that are not all correct. In particular the following three properties would appear natural:

(a) If $A_X \subset A_Y \subset A_Z$ and X determines Y statistically at ξ and η is the determined point, and Y determines Z statistically at η then X determines Z at ξ.

(b) If X determines Z at ξ and Y is a subprocess of Z then X determines Y at ξ.

(c) If X is a subprocess of Y which in turn is a subprocess of both Z and Z' and X determines both Z and Z' at ξ, then

the restriction to Y of the determined points of Z and Z should be
identical.

Property (c) would follow from (b). However (b) is false and
it appears likely that (c) is false as well. Property (a) is true and
moreover easily demonstrated. For if η is the generic point of Y extending ξ and ζ the generic point of Z extending η then ζ extends ξ. If there were another generic point ζ' of Z extending ξ,
let η' be its restriction to Y. Clearly η' is generic for Y and
extends ξ; hence $\eta' = \eta$. But then both ζ' and ζ extend η and
are generic for Z; hence $\zeta' = \zeta$.

It is possible to obtain a counterexample to (b) by considering
the pair of sequences $\xi(n)$ and $\xi^*(n)$ of the preceding subsection.
Namely they can be so chosen that there will be a subprocess W of X
of which X is an L-extension and such that ξ and ξ^* restrict to the
same sample point of W. Call this point ω. Since both ξ and ξ^* are
generic for X, W does not statistically determine X at ω. On the
other hand neither extension of ξ^* to Y will be generic for Y whereas exactly one extension of ξ will be generic for Y. If, therefore ω
has no other extensions in $\bar{\Omega}_X$ other than ξ and ξ^* it follows that
W determines Y statistically at ω. Since all this may be arranged,
property (b) is false.

It may even be that (c) is false although no counterexample has
been constructed. Thus although X determines information about larger
algebras, A_Z and A_Z , the information about the subalgebra A_Y is
ambiguous. It follows from this that the notion of statistical determination is a relative one, namely, the value of a particular variable z
will be determined at a point ξ only relative to the algebra to which
z is to be thought of as belonging.

One might hope to remedy this by fixing the algebra A_Y so as
to include all the L-subprocesses of X. That is we might attempt to
define an absolute notion of determination at ξ by requiring that A_X
statistically determine all of $L(A_X)$ at ξ and every subalgebra would
then be determined by restriction of the determined homomorphism of $L(A_X)$.
The next theorem shows that such a notion would be vacuous, namely, it
implies that no homomorphism of $L(A_X)$ can be generic.

THEOREM 22.1. If ξ is generic for X, then unless
A_X is finite dimensional in which case $L(A_X) = A_X$,
there exists an L-extension of X such that no extension of ξ is generic for this extension of X.

PROOF. Consider the set of translates $\{T^n\xi\} \subset \Omega_X$. This set, being denumerable, has measure 0 unless $\{T^n\xi\}$ is finite which implies that ξ is periodic in which case X is periodic and A_X finite dimensional. Hence there is in the infinite dimensional case an open set V containing $\{T^n\xi\}$ with $P_X(V) = \gamma < 1$. Let φ be the characteristic function of V and let A_Y be any algebra in $L(A_X)$ containing the measurable function φ. Suppose that η is a generic extension of ξ to A_Y. Now since V is open there is a function in A_X positive everywhere in V and vanishing outside of V, and by properly normalizing this function ψ we shall have $\psi < \varphi$. Hence $\varphi(T^n\eta) > \psi(T^n\eta) = \psi(T^n\xi)$ since η reduces to ξ on functions in A_X. It follows now from $\psi(T^n\xi) > 0$ that $\varphi(T^n\eta) > 0$. But $\varphi^2 = \varphi$ and hence for each homomorphism $\varphi^2(T^n\eta) = \varphi(T^n\eta)$ or $\varphi(T^n\eta) = 0, 1$. However we have already excluded the value $\varphi(T^n\eta) = 0$; hence $\varphi(T^n\eta) = 1$ for all n. But if η is generic this will imply $E(\varphi) = 1$ which is impossible in view of the fact that φ is the characteristic function of a set of measure $\gamma < 1$. Hence ξ has no generic extension to A_Y.

In particular no generic point exists for $L(A_X)$. For such a point would be generic for every subalgebra and this would contradict the theorem.

22.3. <u>Statistical Predictability</u>. We combine the notion of statistical determination for prediction from X to $\tilde{X}$ with that of continuous predictability from $\tilde{X}$ to Y to obtain the notion of "statistical predictability".

DEFINITION 22.2. If X is a subprocess of Y and $\xi \in \Omega_{\bar{X}}$, then X is <u>statistically predictable to</u> Y <u>at</u> ξ if X statistically determines $\tilde{X}$ at ξ, where $\tilde{X}$ is the c.p. cover of X with respect to Y. If $\tilde{\xi}$ is the point of $\tilde{X}$ determined by X at ξ and if $\mu_{\tilde{\xi}}$ is the corresponding prediction measure on $(A_{\tilde{X}}, A_Y)$ then the restriction of $\mu_{\tilde{\xi}}$ to A_Y will be the <u>determined prediction measure</u> at ξ. A regular sequence $\xi(n)$ is <u>statistically predictable</u> if $X^-(\xi)$ is statistically predictable to $X(\xi)$ at the point $\xi \in \Omega^-(\xi)$.

It is interesting that although in general the concept of a statistically determined value is a relative one, when applied to statistical prediction it gives an absolute notion of a determined prediction measure. Stated precisely suppose that X is statistically predictable to Y at ξ and that X' is any L-extension of X that is c.p. to Y and such that X statistically determines X' at ξ, then the prediction measure induced by X' on A_Y agrees with the determined prediction measure induced by $\tilde{X}$.

To see this let $\tilde{\xi}$ be the determined point of $\tilde{X}$ at ξ, ξ'

that of X', and let $\mu_{\tilde{\xi}}$ and $\mu'_{\xi'}$ be the corresponding prediction measures on $(A_{\tilde{X}}, A_Y)$ and $(A_{X'}, A_Y)$ respectively. Now by (iii) of Theorem 21.1, $\tilde{X}$ is a subprocess of X' so that $(A_{\tilde{X}}, A_Y) \subset (A_{X'}, A_Y)$. The restriction of ξ' to $A_{\tilde{X}}$ is a generic extension of ξ to $\tilde{X}$ and hence must be identical to $\tilde{\xi}$. Now consider the restriction μ of $\mu_{\xi'}$ to $(A_{\tilde{X}}, A_Y)$. It suffices to show that $\mu = \mu_{\tilde{\xi}}$ and to do this it suffices to show that μ is a prediction measure at $\tilde{\xi}$ from $A_{\tilde{X}}$ to $(A_{\tilde{X}}, A_Y)$. Suppose therefore that $g \in (A_{\tilde{X}}, A_Y)$ is such that $E(fg) \geq 0$ for all non-negative f in $A_{\tilde{X}}$ with support in a neighborhood U of $\tilde{\xi}$. Since ξ' extends ξ and since X' is an L-extension of $\tilde{X}$ we can find a neighborhood U' of ξ' such that $E(fg) \geq 0$ for f non-negative $f \in A_{X'}$ with support in U'. Hence $\mu_{\xi'}(g) \geq 0$ and since $g \in (A_{\tilde{X}}, A_Y)$, $\mu(g) \geq 0$. It follows that μ is a prediction measure at $\tilde{\xi}$ and hence $\mu = \mu_{\tilde{\xi}}$, the unique prediction measure at $\tilde{\xi}$. This proves our assertion.

We shall presently see an example of a sequence that is statistically predictable but not c.p.. It is however also possible that a sequence be c.p. and not statistically predictable. This is seen by taking $\xi(n)$ to be a Markoff sequence with values in a continuous state space and with transition probabilities which do not transform continuous functions into continuous functions (§11). If $\xi(0)$ is a point of continuity of $p(\lambda, \Delta)$ then $\xi(n)$ will be continuously predictable as we saw in §11. On the other hand it is not difficult to show that if one of the remaining $\xi(n)$, $n \leq -1$, is such a point of discontinuity then there may be more than one generic extension of ξ to the c.p. cover of $X(\xi)$, and ξ will not be statistically predictable.

Turning now to the example of §12 we find that the derived sequence $\xi(n) = \eta(n)\eta(n-1)$ which is not c.p. is, however, statistically predictable. Here $\eta(n)$ is a random $\{-1, 1\}$-sequence with the frequencies of 1 and -1 unequal. The processes $X(\xi)$ and $X(\eta)$ are then the processes X and Y of §21.2. There we found that the c.p. cover of X was Y. Thus ξ will be statistically predictable if X^- determines Y^- statistically at ξ. But in §22.1 we found that this was the case if ξ is the image of a generic point of Y. Since $Y = X(\eta)$ it is clear that η is generic for Y and so $\xi(n)$ is statistically predictable.

It must be kept in mind that we have not shown that <u>every</u> generic sequence of $X = X(\xi)$ is statistically predictable. As a matter of fact we have shown in §22.1 that there exist generic points ξ^* of X at which X^- does not determine Y^- statistically since they have no generic extension to Y^-. Thus, with regard to statistical predictability not all generic points of the same process need exhibit the same behavior.

The central problem concerning us in the remaining chapters will be to show that the case just considred is typical of finitely-valued derived sequences of Markoff sequences. That is, we are given a Markoff sequence $\eta(n)$ and a derived sequence $\xi(n)$; we shall show that if $\xi(n)$ and $\eta(n)$ are finitely-valued then $\xi(n)$ is statistically predictable. In terms of processes the problem considered is this. We suppose that Y is a Markoff process and X a subprocess. If the variables of both X and Y are finitely-valued we show that X^- determines its c.p. cover $\tilde{X}^-$ at those generic points which are images of generic points of Y. This will not be true at every generic point of X since not every generic point of X is the image of a generic point of Y as we have seen.

We point out that the general case differs from the one considered here in one aspect, namely, in the general case the c.p. cover of X will be a third process $\tilde{X}$ whereas in the case just considered, $\tilde{X} = Y$. As a consequence, in proving that X^- statistically determines $\tilde{X}^-$ at ξ it will not be obvious either that ξ has any generic extension to $\tilde{X}^-$ or that it is unique, whereas when $\tilde{X} = Y$, η will already be a generic extension of $\xi(n)$ to $\tilde{X}^-$.

§23. The Continuously Predictable Cover of
a Finitely Valued Process

23.1. <u>An Identity for Prediction Measures</u>. So far, the definition of the continuously predictable cover of a process has been made by construction of the algebra $A_{\tilde{X}}^-$ (see the proof of Theorem 21.1). We now restrict ourselves to the case that prediction is to be from the past to the future of a single process X, and moreover we shall assume that X is finitely-valued. In this case it will be possible to study the sample space of the process $\tilde{X}$ directly, rather than as the homomorphism space of a given E-algebra. By this means we shall be able to determine the existence and uniqueness of generic points of $\tilde{X}$ extending certain points of X and this will lead to the solution of the problem posed in the preceding section.

The variables x_n of X will have their values in a set $\Lambda = \{a_1, a_2, \ldots, a_r\}$. It will be convenient to introduce the notation $[x_k = a]$, where $a \in \Lambda$, for the characteristic function of the set in Ω_X for which $x_k(\omega) = a$. If $k \leq 0$, $[x_k = a]$ will also be used for the characteristic function of the corresponding set in $\Omega_{\tilde{X}}^-$. Let $\tilde{X}$ be the c.p. cover of the process X (i.e., $\tilde{X}^-$ is the c.p. cover of X^- with respect to X) and let $\tilde{\beta}$ denote the canonical map from $\Omega_{\tilde{X}}^-$ to Ω_X^-, $\tilde{X}$ being an extension of X.

LEMMA 23.1. If $\mu_{\widetilde{\omega}}$ denotes the unique prediction measure at a point $\widetilde{\omega} \in \Omega_{\widetilde{X}}^{-}$ to A_X, then for all $f \in A_X$

(23.1)
$$\mu_{T\widetilde{\omega}}([x_1 = a_1])\mu_{\widetilde{\omega}}(f) = \mu_{T\widetilde{\omega}}([x_1 = a_1]T^{-1}f) \ ,$$

at all $\widetilde{\omega}$ satisfying $x_0(\widehat{\beta}(\widetilde{\omega})) = a_1$.

PROOF. We may consider $[x_k = a_1]$ as functions in $A_{\widetilde{X}}^{-}$ for $k \leq 0$ since $A_X^{-} \subset A_{\widetilde{X}}^{-}$. In particular the condition $x_0(\widehat{\beta}(\widetilde{\omega})) = a_1$ may be rewritten as $[x_0 = a_1](\widetilde{\omega}) = 1$. (23.1) now reads

(23.2) $[x_0 = a_1](\widetilde{\omega})\mu_{T\widetilde{\omega}}([x_1 = a_1])\mu_{\widetilde{\omega}}(f) = [x_0 = a_1](\widetilde{\omega})\mu_{T\widetilde{\omega}}([x_1 = a_1]T^{-1}f) \ .$

Since both sides of (23.2) are continuous functions of $\widetilde{\omega}$, it suffices to show that (23.2) holds as an equality almost everywhere of functions in $L(A_{\widetilde{X}}^{-})$. Since $\widetilde{X}$ is an L-extension of X, $L(A_{\widetilde{X}}^{-}) = L(A_X^{-})$ so that both sides of (23.2) correspond to measurable functions on Ω_X^{-} and we are to show that these are equal almost everywhere.

We now recall Theorem 10.2 according to which the value of the prediction measure μ_ξ at a fixed function in the larger algebra, as a function of ξ, is identical to the conditional expectation of the function with respect to the smaller algebra. Hence

(23.3)
$$\mu_{\widetilde{\omega}}(f) = E(f|A_{\widetilde{X}}^{-})(\widetilde{\omega}), \quad \mu_{T\widetilde{\omega}}(g) = TE(g|A_{\widetilde{X}}^{-})(\widetilde{\omega}) \ .$$

We can therefore rewrite (23.2) as

(23.4)
$$[x_0 = a_1] \, T\{E([x_1 = a_1]|A_{\widetilde{X}}^{-})\} \, E(f|A_{\widetilde{X}}^{-}) =$$
$$[x_0 = a_1] \, T\{E([x_1 = a_1]T^{-1}f|A_{\widetilde{X}}^{-})\}$$

both sides of which are functions in $L(A_{\widetilde{X}}^{-}) = L(A_X^{-})$. In (23.4) the conditional expectations with respect to $A_{\widetilde{X}}^{-}$ may, because $L(A_{\widetilde{X}}^{-}) = L(A_X^{-})$, be taken with respect to A_X^{-}. We may also replace an expression of the form $T\{E(f|A_X^{-})\}$ by $E(Tf|TA_X^{-})$. This follows from the definition of $E(\cdot|\cdot)$, and the fact that T acts as an isometry of the Hilbert spaces $L^2(A_X^{-})$ and $L^2(TA_X^{-})$. We thus have

(23.5)
$$[x_0 = a_1] \, E([x_0 = a_1]|TA_X^{-})E(f|A_X^{-}) =$$
$$[x_0 = a_1] \, E([x_0 = a_1]f|TA_X^{-}) \ ,$$

as the equality to be proven.

Suppose first that f has the form $\psi(x_0)$. The left side of (23.5) then becomes

$$[x_0 = a_1]E([x_0 = a_1]|TA_X^-)E(\psi(x_0)|A_X^-) =$$

(23.6)
$$[x_0 = a_1]E([x_0 = a_1]|TA_X^-)\psi(x_0) =$$

$$\psi(a_1)[x_0 = a_1]E([x_0 = a_1]|TA_X^-) \quad ,$$

using the fact that $\psi(x_0) \in A_X^-$, (Equation (1.4)). The right hand side of (23.5) becomes

$$[x_0 = a_1]E([x_0 = a_1]\psi(x_0)|TA_X^-) =$$

$$[x_0 = a_1]E(\psi(a_1)[x_0 = a_1]|TA_X^-) =$$

$$\psi(a_1)[x_0 = a_1]E([x_0 = a_1]|TA_X^-)$$

and so (23.5) holds for $f = \psi(x_0)$.

Now suppose that $f = \psi(x_0)g$ with $g \in TA_X^-$. Since g is in both A_X^- and TA_X^- we find $E(\varphi g|A_X^-) = gE(\varphi|A_X^-)$ and $E(\varphi g|TA_X^-) = gE(\varphi|TA_X^-)$ for any φ. It follows that g simply factors out of both sides of (23.5) so that this case is reduced to the preceding one. Hence (23.5) holds for all f in A_X^-, and, in fact, for all f in $L(A_X^-)$ since these are limits in $L^2(A_X^-)$ of linear combinations of functions with the form considered.

Finally suppose that $f \in A_X$. Then (23.5) holds for $g = E(f|A_X^-)$ which is in $L(A_X^-)$. That is,

$$[x_0 = a_1]E([x_0 = a_1]|TA_X^-)E(E(f|A_X^-)|A_X^-) =$$

$$[x_0 = a_1]E([x_0 = a_1]E(f|A_X^-)|TA_X^-)$$

or

$$[x_0 = a_1]E([x_0 = a_1]|TA_X^-)E(f|A_X^-) =$$

$$[x_0 = a_1]E(E([x_0 = a_1]f|A_X^-)|TA_X^-) =$$

$$[x_0 = a_1]E([x_0 = a_1]f|TA_X^-)$$

which is (23.5). This proves the lemma.

23.2. <u>Construction of</u> $\Omega_X^{\simeq}$. It will be convenient to introduce, at this point, the space Ω_X^+ which bears the same relationship to the future of X as Ω_X^- does to the past. More precisely Ω_X^+ is the identification space of Ω_X obtained by identifying ω_1 with ω_2 in Ω_X if $\omega_1(n) = \omega_2(n)$ for $n \geq 1$. Alternatively Ω_X^+ is the homomorphism space of A_X^+, the algebra generated by the functions $\psi(x_n)$, $n \geq 1$. A_X is then the algebra generated by A_X^- and A_X^+ and Ω_X is a compact subset of $\Omega_X^- \times \Omega_X^+$. The "inverse shift" operators on Ω_X^+ and A_X^+ will be denoted by T^{-1} although the operator T itself is undefined in these spaces.

We shall be interested in certain functionals in A_X^+ induced by the functions of A_X^-:

DEFINITION 23.1. We denote by $\mathcal{M}(X)$ the set of probability measures on Ω_X^+ which as functionals on A_X^+ have the form

$$\mu(g) = \lim_{n \to \infty} E(f_n g)$$

where the f_n are non-negative functions in A_X^-. $\mathcal{M}(X)$ is also provided with the topology of weak convergence of measures on Ω_X^+.

The space $\mathcal{M}(X)$ is used in defining a sequence space $\hat{\Omega}(X)$ which will be essential in studying the sample space of $\tilde{X}$.

DEFINITION 23.2. $\hat{\Omega}(X)$ is the set of pairs of left-infinite sequences (ω, π) where $\omega \in \Omega_X^-$ and $\pi = (\pi_n)$ with $\pi_n \in \mathcal{M}(X)$ and (ω, π) satisfies

$$(23.7) \qquad \pi_n(f)\pi_{n-1}([x_1 = \omega(n)]) = \pi_{n-1}([x_1 = \omega(n)]T^{-1}f)$$

for $n \leq 0$ and $f \in A_X^+$.

As a subset of $\{\Lambda \times \mathcal{M}(X)\}_\infty^-$, the set $\hat{\Omega}(X)$ has defined in it an operator T which, as may be verified, takes $\hat{\Omega}(X)$ into itself. We shall also consider the projection $\hat{\beta}$ of $\hat{\Omega}(X)$ into Ω_X^- given by

$$(23.8) \qquad\qquad \hat{\beta}(\omega, \pi) = \omega$$

Now let $\tilde{\omega} \in \Omega_{\tilde{X}}^-$, the sample space of $\tilde{X}$, and let $\mu_{\tilde{\omega}}$ be the unique prediction measure on A_X at the point $\tilde{\omega}$. Let $\theta(\tilde{\omega})$ be the restriction of $\mu_{\tilde{\omega}}$ to A_X^+ so that $\theta(\tilde{\omega})$ is a probability measure on Ω_X^+. We associate to every $\tilde{\omega}$ in $\Omega_{\tilde{X}}^-$ a pair of left-infinite sequences as follows:

$$(23.9) \qquad \gamma(\tilde{\omega}) = (\hat{\beta}(\tilde{\omega}), \pi) \quad \text{where} \quad \pi_{-n} = \theta(T^n\tilde{\omega}) .$$

THEOREM 23.1. γ imbeds the sample space $\Omega_{\tilde{X}}^-$ into $\hat{\Omega}(X)$ in a manner preserving the operator T and with the projection $\tilde{\beta}$ going over to the map $\hat{\beta}$ of (23.8) (in other words, $\gamma T = T\gamma$ and $\hat{\beta}\gamma = \tilde{\beta}$).

PROOF. We must first show that $\gamma(\tilde{\omega}) \in \hat{\Omega}(X)$ or that $\theta(T^n\tilde{\omega}) \in \mathcal{M}(X)$ and that (23.7) is satisfied. For the former we recall Corollary 2 to Theorem 10.1 according to which the (unique) prediction measure $\mu_{\tilde{\omega}}$ is a limit of functions f_n in $A_{\bar{X}}$. Now a sequence f'_n may be found in $A_{\bar{X}}$ such that $\|f_n - f'_n\|_2 \longrightarrow 0$, $\|\cdot\|_2$ representing the norm of $L^2(A_{\bar{X}})$. Under this condition it is clear that $E(f'_n g)$ converges to the same limit as does $E(f_n g)$, hence to $\mu_{\tilde{\omega}}(g)$. It follows that $\theta(\tilde{\omega}) \in \mathcal{M}(X)$ for any $\tilde{\omega}$ and so too for the translates $T^n\tilde{\omega}$.

In our case (23.7) may be rewritten as

$$
(23.10) \quad
\begin{aligned}
&\mu_{T^n\tilde{\omega}}(f)\mu_{T^{n+1}\tilde{\omega}}([x_1 = \tilde{\beta}(\tilde{\omega})(-n)]) = \\
&\mu_{T^{n+1}\tilde{\omega}}([x_1 = \tilde{\beta}(\tilde{\omega})(-n)]T^{-1}f), \qquad n \geq 0 \ .
\end{aligned}
$$

In this form the equation is a restatement of (23.1) and therefore it will hold if $T^n\tilde{\omega}$ satisfies $x_0(\tilde{\beta}(T^n\tilde{\omega})) = \tilde{\beta}(\tilde{\omega})(-n)$. This however is clear since $\tilde{\beta}(\tilde{\omega})(-n) = x_{-n}(\tilde{\beta}(\tilde{\omega})) = x_0(T^n\tilde{\beta}(\tilde{\omega})) = x_0(\tilde{\beta}(T^n\tilde{\omega}))$.

γ is moreover continuous; to show this we need only consider the coordinate functions $\tilde{\beta}(\tilde{\omega})(-n)$ and $\theta(T^n\tilde{\omega})$. The first is evidently continuous in $\tilde{\omega}$ since $\tilde{\beta}$ is a mapping, and the second is also since $\theta(T^n\tilde{\omega})$ is the restriction of $\mu_{T^n\tilde{\omega}}$ which, by Theorem 10.2, depends continuously on $\tilde{\omega}$. Also the assertion that $\gamma T = T\gamma$ and that $\hat{\beta}\gamma = \tilde{\beta}$ may be directly verified. All that remains then to establish the theorem is to show that γ is actually an imbedding, i.e., that γ is a $1-1$ map.

To show that γ is $1-1$ we shall show that the induced map of $C(\gamma(\Omega_{\bar{X}}))$ into $C(\Omega_{\bar{X}})$ is onto whence by §1.1 it will follow that γ is $1-1$. Let $\hat{A}(X)$ be the algebra of continuous functions in $\gamma(\Omega_{\bar{X}}) \subset \hat{\Omega}(X)$, and let $\Gamma : \hat{A}(X) \longrightarrow A_{\bar{X}}$ denote the map induced by γ. Since γ is onto $\gamma(\Omega_{\bar{X}})$, Γ is $1-1$ and so effects an imbedding of $\hat{A}(X)$ into $A_{\bar{X}}$. We shall prove that $\hat{A}(X)$ is all of $A_{\bar{X}}$ by showing that $\hat{A}(X)$ is the algebra of a process extending X^- which is continuously predictable to X. It will then follow from the definition of $\tilde{X}$ that $\hat{A}(X)$ contains $A_{\bar{X}}$.

The details are as follows. First we note that if $\gamma(\tilde{\omega}_1) = \gamma(\tilde{\omega}_2)$ then $\tilde{\beta}(\tilde{\omega}_1) = \beta(\tilde{\omega}_2)$. Consequently if $\psi \in A_{\bar{X}}$ so that ψ may be considered as a function in $A_{\bar{X}}$ depending only on $\tilde{\beta}(\tilde{\omega})$, then if

$\gamma(\tilde{\omega}_1) = \gamma(\tilde{\omega}_2)$ we will have $\psi(\tilde{\omega}_1) = \psi(\tilde{\omega}_2)$. Hence we may write $\psi = \varphi \cdot \gamma$ (§1.1) or $\psi = \Gamma\varphi$ with φ a function on $\hat{\Omega}(X)$ or $\varphi \in \hat{A}(X)$. We have therefore shown that identifying $\hat{A}(X)$ with its image $\Gamma\hat{A}(X)$ in $A_{\tilde{X}}^-$, we will have

$$(23.11) \qquad\qquad A_{\tilde{X}}^- \subset \hat{A}(X) \ .$$

Now $\hat{A}(X)$ is a subalgebra of $A_{\tilde{X}}^-$ and hence an E-algebra. It is also T-invariant since $\hat{\Omega}(X)$ is, and $\gamma T = T\gamma$. Hence $\hat{A}(X)$ is the algebra of the one-sided version of an abstract process $\hat{X}(A_X^- = \hat{A}(X))$. Since $\tilde{X}$ is an L-extension of X, and $\hat{X}$, which is also an extension of X, is a subprocess of $\tilde{X}$ it follows that $\hat{X}$ is an L-extension of X. If we prove that the process $\hat{X}$ is a c.p. process, i.e., that $A_{\hat{X}}^-$ is c.p. to $A_{\hat{X}}$, then _a fortiori_ $A_{\tilde{X}}^-$ will be c.p. to $A_X \subset A_{\hat{X}}^-$ and so by the definition of $\tilde{X}$, $\tilde{X}$ will be a subprocess of $\hat{X}$, or $\tilde{X} = \hat{X}$.

The abstract process $\hat{X}$ becomes an ordinary process if we introduce variables $\hat{x}_n$ on $\Omega_{\hat{X}}^-$ in accordance with Definition 2.1. Since $\Omega_{\hat{X}}^- = \hat{\Omega}(X)$, the homomorphism space of $\hat{A}(X)$, and the points of $\hat{\Omega}(X)$ are left-infinite $\{\Lambda \times \mathcal{M}(X)\}$-sequences we may define the variables $\hat{x}_n$ as the $\{\Lambda \times \mathcal{M}(X)\}$-valued coordinate functions on $\hat{\Omega}(X)$:

$$(23.12) \qquad \hat{x}_0(\omega, \pi) = (\omega(0), \pi_0), \qquad \hat{x}_{-n}(\omega, \pi) = \hat{x}_0(T^n\omega, T^n\pi) \ .$$

Thus we may write $\hat{x}_n = (x_n, \pi_n)$ where the x_n are the variables of the original process X and are defined on Ω_X^- and thereby on $\Omega_{\hat{X}}^-$ (since X is a subprocess of $\hat{X}$), and π_n are $\mathcal{M}(X)$-valued variables. For each $f \in A_X^+$ we shall let $\pi_n(f)$ denote the numerically valued variables given by

$$\pi_n(f)(\hat{\omega}) = \pi_n(\hat{\omega})(f), \qquad \hat{\omega} \in \Omega_{\hat{X}}^- \ .$$

By the corollary to Lemma 11.2, $\hat{X}$ will be c.p. if $E(\psi(\hat{x}_1)|A_{\hat{X}}^-) \in A_{\hat{X}}^-$ for all ψ in $C(\Lambda \times \mathcal{M}(X))$. It will be more convenient to write this $E(\psi(\hat{x}_0)|TA_{\hat{X}}^-) \in TA_{\hat{X}}^-$. It suffices to prove this for a set of ψ whose linear combinations are dense in $C(\Lambda \times \mathcal{M}(X))$ and so we take $\psi(\hat{x}_0)$ of the form

$$\psi(\hat{x}_0) = \varphi(x_0)\pi_0(f_1)\pi_0(f_2)\dots\pi_0(f_k)$$

where $f_1, f_2, \dots, f_k \in A_X^+$. Since Λ is discrete we may, moreover, take $\varphi(x_0) = [x_0 = a_1]$ for some $a_1 \in \Lambda$ so that

$$\psi(\hat{x}_0) = [x_0 = a_1]\Pi_0(f_1)\Pi_0(f_2)\ldots\Pi_0(f_k) \quad .$$

Now by (23.7) we have

$$\Pi_0(f)\Pi_{-1}([x_1 = a_1]) = \Pi_{-1}([x_1 = a_1]T^{-1}f)$$

on the set of $\hat{\omega}$ for which $x_0(\hat{\omega}) = a_1$. This may be written as

(23.13)
$$[x_0 = a_1]\Pi_0(f)\Pi_{-1}([x_1 = a_1])$$
$$= [x_0 = a_1]\Pi_{-1}([x_1 = a_1]T^{-1}f) \quad .$$

Using $[x_0 = a_1]^k = [x_0 = a_1]$ we have

(23.14)
$$[x_0 = a_1]\Pi_0(f_1)\Pi_0(f_2)\ldots\Pi_0(f_k)\Pi_{-1}([x_1 = a_1])^k$$
$$= [x_0 = a_1]\Pi_{-1}([x_1 = a_1]T^{-1}f_1)\ldots\Pi_{-1}([x_1 = a_1]T^{-1}f_k) \quad .$$

Take the expectations of both sides of (23.14) with respect to $TA_{\bar{X}}^{\approx}$ and note that $\Pi_{-1}(f) = T\Pi_0(f) \in TA_{\bar{X}}^{\approx}$:

(23.15)
$$E([x_0 = a_1]\Pi_0(f_1)\ldots\Pi_0(f_k)|TA_{\bar{X}}^{\approx})\Pi_{-1}([x_1 = a_1])^k =$$
$$E([x_0 = a_1]|TA_{\bar{X}}^{\approx})\Pi_{-1}([x_1 = a_1]T^{-1}f_1)\ldots\Pi_{-1}([x_1 = a_1]T^{-1}f_k) \quad .$$

We may write $E([x_0 = a_1]|TA_{\bar{X}}^{\approx}) = TE([x_1 = a_1]\,A_{\bar{X}}^{\approx})$ and $E([x_1 = a_1]|A_{\bar{X}}^{\approx}) = E([x_1 = a_1]|A_{\hat{X}}^{\approx})$ since $\tilde{X}$ is an L-extension of $\hat{X}$. But $A_{\bar{X}}^{\approx}$ is c.p. to A_X, so by Theorem 10.2, $E([x_1 = a_1]|A_{\bar{X}}^{\approx}) = \mu_\omega([x_1 = a_1])$ and since $[x_1 = a_1] \in A_X^+$ this equals $\theta(\tilde{\omega})([x_1 = a_1])$. By the definition of γ, $\theta(\tilde{\omega})([x_1 = a_1]) = \Pi_0([x_1 = a_1])(\gamma\tilde{\omega})$ so that $E([x_1 = a_1]|A_{\bar{X}}^{\approx})$ as a function on $\Omega_{\bar{X}}^{\approx}$ is in the image of Γ and corresponds to the function $\Pi_0([x_1 = a_1])$ in $A_{\bar{X}}^{\approx}$. Hence $E([x_0 = a_1]|TA_{\bar{X}}^{\approx})$ corresponds to $\Pi_{-1}([x_1 = a_1])$., which is in $TA_{\bar{X}}^{\approx}$. We now have from (23.15)

(23.16)
$$E(\psi(\hat{x}_0)|TA_{\bar{X}}^{\approx})\Pi_{-1}([x_1 = a_1])^k =$$
$$\Pi_{-1}([x_1 = a_1])\Pi_{-1}([x_1 = a_1]T^{-1}f_1)\ldots\Pi_{-1}([x_1 = a_1]T^{-1}f_k) \quad .$$

If we could divide through by $\Pi_{-1}([x_1 = a_1])^k$ in (23.16) we would find that $E(\psi(\hat{x}_0)|TA_{\bar{X}}^{\approx})$ is in $TA_{\bar{X}}^{\approx}$ as was to be shown. However $\Pi_{-1}([x_1 = a_1])^k$ may vanish. To bypass this difficulty we use the following elementary fact:

If g_1 and g_2 are continuous functions on a space Ω such that $|g_1| < \kappa g_2$ for some constant κ, and the zeros of g_2 are also zeros of a third function f on Ω, then the function φ defined by

$$\varphi(\omega) = \begin{cases} fg_1/g_2(\omega) & \text{if } g_2(\omega) \neq 0 \\[2mm] 0 & \text{if } g_2(\omega) = 0 \end{cases}$$

is continuous.

To apply this observe that $|\Pi_{-1}([x_1 = a_1]T^{-1}f_1)| < \|f_1\|\Pi_{-1}([x_1 = a_1])$ so that if

$$g_1 = \Pi_{-1}([x_1 = a_1]T^{-1}f_1)\cdots\Pi_{-1}([x_1 = a_1]T^{-1}f_k)$$

and $g_2 = \Pi_{-1}([x_1 = a_1])^k$ then $|g_1| < \kappa g_2$. Moreover if $\Pi_{-1}([x_1 = a_1]) = 0$ then $E([x_0 = a_1]|T^{-1}A_{\hat{X}}^{\bar{z}}) = 0$ and since $|\psi(x_0)| < \|f_1\|\cdots\|f_k\|[x_0 = a_1]$ it follows that $E(\psi(x_0)|TA_{\hat{X}}^{\bar{z}}) = 0$. Consequently the above remark applies and $E(\psi(\hat{x}_0)|TA_{\hat{X}}^{\bar{z}})$ is in $TA_{\hat{X}}^{\bar{z}}$. This shows that $\hat{X}$ is a c.p. process, hence $\hat{X}$ extends $\tilde{X}$ so that $\hat{X} = \tilde{X}$; this shows that the induced mapping Γ is onto and hence γ is $1 - 1$. This completes the proof of the theorem.

Strictly speaking, this theorem does not give us an alternative construction of the sample space $\Omega_{\tilde{X}}^{\bar{z}}$ of $\tilde{X}$ since we do not know which sequences of $\hat{\Omega}(X)$ are images under γ. It is also possible to give a characterization of these sequences; however we shall not make use of it. What is important for us is that the points of $\Omega_{\tilde{X}}^{\bar{z}}$ may be taken as $[\Lambda \times \mathscr{M}(X)]$-sequences satisfying (23.7).

We point out an interesting corollary to the proof of this theorem. Namely, heretofore, $\tilde{X}$ was characterized as the least L-extension of X for which $A_{\tilde{X}}^{\bar{z}}$ is c.p. to A_X. It is not at all evident from this that $\tilde{X}$ is itself c.p., i.e., that $A_{\tilde{X}}^{\bar{z}}$ is c.p. to the larger algebra A_X. However for the finitely-valued processes taken up in this theorem we saw that $\hat{X}$ is c.p. and therefore so is $\tilde{X}$ which is identical to $\hat{X}$.

§24. Applications to Finite Dimensional Processes

24.1. <u>The Canonical Semigroup</u>. We turn now to the case that X is determined by a semigroup of linear transformations acting on a finite dimensional cone. We have seen in §19 that this is equivalent to the condition that X be a finitely-valued subprocess of an m-Markoff

(or ordinary Markoff) process. For this case the sample space $\Omega_{\bar{X}}$ of
the c.p. cover of X takes on a rather special form. To give our construction for the sample space we shall have to choose from among the
equivalent semigroups giving the same process a particular one that we call
the <u>canonical</u> semigroup.

   We start with a semigroup $\mathcal{S}$ of m × m matrices with non-
negative entries, a row vector θ^* and a column vector θ. The func-
tional F is defined on $\mathcal{S}$ by $F(\sigma) = \theta^*\sigma\theta$ for $\sigma \in \mathcal{S}$. The matrices
of $\mathcal{S}$ may be thought of as operating either on the set of m-dimensional
column vectors or on the set of m-dimensional row vectors. Let V denote the cone of m-dimensional row vectors with non-negative entries, in
particular $\theta^* \in V$. Define an equivalence relation in V by

(24.1) $v_1 \sim v_2$ if $v_1\sigma\theta = v_2\sigma\theta$ for all $\sigma \in \mathcal{S}$.

Clearly $v_1 \sim v_2$ and $v_1' \sim v_2'$ imply that $v_1 + v_1' \sim v_2 + v_2'$; also
$v_1 \sim v_2$ implies that $\lambda v_1 \sim \lambda v_2$ for all $\lambda > 0$. This means that the
quotient space of V modulo this equivalence relation is again a cone.
We shall denote this cone by $\bar{V}$. We remark that V is finitely generated
in the sense that every element of V is a non-negative linear combination
of a fixed finite set of vectors in V. The same property will hold for
$\bar{V}$ and we shall denote the generators of $\bar{V}$ by $\{e_1, e_2, \ldots, e_\ell\}$.

   If $\sigma \in \mathcal{S}$ there is induced an operation $\bar{\sigma}$ on $\bar{V}$ defined by

(24.2) $\bar{u}\bar{\sigma} = \overline{u\sigma}$

where $\bar{u}$ denotes the image in $\bar{V}$ of an element u in V. To see that
this is well defined we note that if $\bar{u}_1 = \bar{u}_2$ or $u_1 \sim u_2$ then
$u_1\sigma\rho\theta = u_2\sigma\rho\theta$ for all ρ in $\mathcal{S}$ and therefore $u_1\sigma \sim u_2\sigma$ or $\overline{u_1\sigma} = \overline{u_2\sigma}$.
Let us denote by $\bar{\mathcal{S}}$ the semigroup of operations on $\bar{V}$ obtained in this
way. In a similar way, the functional θ on V induces a functional $\bar{\theta}$
on $\bar{V}$: $\langle\bar{u}, \bar{\theta}\rangle = u\theta$. Again this is well defined since if $u_1 \sim u_2$ then
$u_1\theta = u_2\theta$. We then have $F(\sigma) = \theta^*\sigma\theta = \langle\bar{\theta}^*\bar{\sigma}, \bar{\theta}\rangle = \bar{F}(\bar{\sigma})$. We conclude
from this that the semigroup $\bar{\mathcal{S}}$ with functional $\bar{F}$ is equivalent to $\mathcal{S}$
with F in the sense that both generate the same process. $\bar{\mathcal{S}}$ will be
referred to as the <u>canonical semigroup</u> of X. It is not difficult to show
that it is uniquely determined by X. However we shall not need this fact
so we shall not consider the proof.

   The cone $\bar{V}$ may be imbedded into a Euclidean space D since
$V \subset R^m$ and the equivalence relation extends to R^m as convergence modulo
a subspace. $\bar{V}$ then appears as a closed subset of a Euclidean space of
dimension $\leq m$, and as such it is endowed with a metric topology. When

speaking of the topology of $\bar{V}$ it is to this that we refer.

We observe that if $\langle \bar{u}, \bar{\theta} \rangle = 0$ for $\bar{u} \in \bar{V}$ then $\bar{u} = 0$. For if $u\theta = 0$ then since

$$\theta = \sum_{i} a_i \theta = \sum_{i,j} a_i a_j \theta = \ldots = \sum_{\sigma \text{ of length } \nu} \sigma\theta$$

and since u is non-negative it follows that $u\sigma\theta = 0$ for all σ and hence $u \sim 0$, $\bar{u} = 0$. Since, in particular, we have $\langle e_i, \bar{\theta} \rangle \neq 0$, we shall suppose that the e_i have been normalized so that $\langle e_i, \bar{\theta} \rangle = 1$.

24.2. <u>The Sample Space</u> $\Omega_{\tilde{X}}^{-}$.

DEFINITION 24.1. For a finite dimensional process X and the corresponding cone $\bar{V}$, $\Omega(X, \bar{V})$ denotes the space of pairs of left-infinite $\Lambda \times \bar{V}$ sequences $\omega' = (\omega, \zeta)$ satisfying

(24.3) $\langle \zeta(n - 1)\omega(n), \bar{\theta} \rangle \cdot \zeta(n) = \zeta(n - 1)\omega(n)$.

Let T denote the shift operator in $\Omega(X, \bar{V})$ (clearly T takes $\Omega(X, \bar{V})$ into $\Omega(X, \bar{V})$) and let β' denote the projection of $\Omega(X, \bar{V})$ into Ω_X^{-} given by $\beta'(\omega') = \omega$.

THEOREM 24.1. If X is finite dimensional, then the sample space of the c.p. cover $\tilde{X}$ of X may be taken as a subset of $\Omega(X, \bar{V})$ with the canonical projection $\tilde{\beta}$ going over into β' and the shift operator T on $\Omega_{\tilde{X}}^{-}$ going into the shift operator of $\Omega(X, \bar{V})$.

PROOF. This theorem will follow from Theorem 23.1 if it is shown that the space $\hat{\Omega}(X)$ may be imbedded into $\Omega(X, \bar{V})$ in a manner preserving T and sending $\hat{\beta}$ into β'. For this let $\hat{\omega} = (\omega, \pi) \in \hat{\Omega}(X)$ where $\omega \in \Omega_X^{-}$ and π is an $\mathcal{M}(X)$-sequence satisfying (23.7). We shall show that the $\mathcal{M}(X)$-sequence π corresponds to a $\bar{V}$-sequence satisfying (24.3) for ω.

We recall that $\mathcal{M}(X)$ consists of measures on Ω_X^{+} induced by functionals of the form $\mu(g) = E(fg)$ with $f \in A_X^{-}$, $E(f) = 1$ and $f \geq 0$, (so that μ is a probability measure), and of weak limits of such measures. Let the variables of X be $\{x_n\}$ and let x_σ^{-} represent the characteristic function of the set $\{\omega : x_q(\omega) = \sigma(q), x_{q+1}(\omega) = \sigma(q + 1), \ldots, x_0(\omega) = \sigma(0)\}$ for a Λ-sequence σ, and Λ, the range of X, i.e., the generators of $\mathcal{S}$. Linear combinations of the x_σ^{-} are dense in A_X^{-} and, in particular, a dense subset of $\mathcal{M}(X)$ may be represented as such linear combinations with non-negative coefficients. We define a map δ of $\mathcal{M}(X)$ into $\bar{V}$ by first defining it on this dense subset of $\mathcal{M}(X)$:

$$(24.4) \qquad \delta(\Sigma\alpha_\sigma x_\sigma^-) = \Sigma\alpha_\sigma\overline{\theta^*\sigma}$$

where $\bar{\sigma}$ is the element of $\bar{\mathcal{S}}$ corresponding to σ: $\bar{\sigma} = \sigma(q)\sigma(q+1)\ldots\sigma(0)$. We shall prove presently that δ extends to all of $\mathcal{M}(X)$: this will imply in particular, that δ is well defined on the dense subset of (24.4). We observe first that if $\Sigma\alpha_\sigma x_\sigma^-$ corresponds to a measure in $\mathcal{M}(X)$, then $E(\Sigma\alpha_\sigma x_\sigma^-) = 1$ and so $\Sigma\alpha_\sigma\bar{F}(\bar{\sigma}) = \Sigma\alpha_\sigma\langle\overline{\theta^*\sigma},\ \bar{\theta}\rangle = 1$. The image of δ is therefore contained in the set of $v \in \bar{V}$ for which $\langle v,\ \bar{\theta}\rangle = 1$. Denote this subset of $\bar{V}$ by $\bar{V}_1$. By the convention adopted at the end of §24.1 the vectors $e_i \in \bar{V}_1$. Since the e_i span $\bar{V}$ it follows that $\bar{V}_1$ is just the set of convex linear combinations of $e_1, e_2, \ldots, e_\ell$. In particular it follows that $\bar{V}_1$ is a compact subset of $\bar{V}$.

We now extend δ to all of $\mathcal{M}(X)$ by setting

$$(24.5) \qquad \delta(\mu) = \lim \delta(f_n)$$

where $f_n \longrightarrow \mu$ is in $\mathcal{M}(X)$ and f_n has the form $\Sigma\alpha_\sigma x_\sigma^-$. Since $\bar{V}_1$ is compact, to prove that the limit in (24.5) exists it suffices to suppose that we have $f_n' \longrightarrow \mu$, $f_n'' \longrightarrow \mu$ with $\delta(f_n') \longrightarrow u'$, $\delta(f_n'') \longrightarrow u''$ and to prove that $u' = u''$. This will also show that δ is unambiguously defined at each μ.

Let ρ be the Λ-sequence $(\rho(1), \rho(2), \ldots, \rho(k))$ and $x_\rho^+ = [x_1 = \rho(1)][x_2 = \rho(2)]\ldots[x_k = \rho(k)] \in A_X^+$ (see §23.1, §23.2). We can form $\bar{\rho} = \rho(1)\rho(2)\ldots\rho(k) \in \bar{\mathcal{S}}$. If f has the form $f = \Sigma\alpha_\sigma x_\sigma^-$ then $E(fx_\rho^+) = \Sigma\alpha_\sigma E(x_\sigma^- x_\rho^+) = \Sigma\alpha_\sigma\bar{F}(\overline{\sigma\rho}) = \Sigma\alpha_\sigma\langle\overline{\theta^*\sigma\rho\theta}\rangle = \langle\overline{\delta f\rho\theta}\rangle$. It follows that if $f_n' \longrightarrow \mu'$ and $\delta(f_n') \longrightarrow u'$ then $\mu(x_\rho^+) = \langle u'\bar{\rho},\ \bar{\theta}\rangle$. Therefore $\langle u'\bar{\rho},\ \bar{\theta}\rangle = \langle u''\bar{\rho},\ \bar{\theta}\rangle$ for all $\bar{\rho} \in \bar{\mathcal{S}}$ and by the definition of the equivalence relation in V, we find $u' = u''$. It follows that δ is everywhere defined and also continuous and that

$$(24.6) \qquad \mu(x_\rho^+) = \langle\delta(\mu)\bar{\rho},\ \bar{\theta}\rangle \quad .$$

Turning to $\hat{\omega} = (\omega,\ \pi)$ we define

$$(24.7) \qquad \Delta(\hat{\omega}) = (\omega,\ \delta(\pi))$$

where $\delta(\pi)$ is the sequence $\{\delta(\pi_n)\}$. We wish to show that if $\hat{\omega} \in \hat{\Omega}(X)$ then $\Delta(\hat{\omega}) \in \Omega(X,\ \bar{V})$. For this we must verify that $\Delta(\hat{\omega})$ satisfies

$$(24.8) \qquad \langle \delta(\pi_{n-1})\omega(n), \bar{\theta} \rangle \; \delta(\pi_n) = \delta(\pi_{n-1})\omega(n) \quad .$$

To do this it suffices to show, denoting the two sides of (24.8) by u_1 and u_2, that $\langle u_1 \bar{\sigma}, \bar{\theta} \rangle = \langle u_2 \bar{\sigma}, \bar{\theta} \rangle$ for all $\bar{\sigma} \in \bar{\mathcal{S}}$; i.e., that

$$(24.9) \qquad \langle \delta(\pi_{n-1})\omega(n), \bar{\theta} \rangle \langle \delta(\pi_n)\bar{\sigma}, \bar{\theta} \rangle \cdot = \langle \delta(\pi_{n-1})\omega(n)\bar{\sigma}, \bar{\theta} \rangle \quad .$$

However by (24.6) this becomes

$$(24.10) \qquad \pi_{n-1}(x^+_{\omega(n)})\pi_n(x^+_\sigma) = \pi_{n-1}(x^+_{\omega(n)\sigma})$$

or

$$
\begin{aligned}
&\pi_{n-1}([x_1 = \omega(n)])\pi_n([x_1 = \sigma(1)]\ldots[x_k = \sigma(k)]) = \\
(24.11) \quad &\pi_{n-1}([x_1 = \omega(n)][x_2 = \sigma(1)]\ldots[x_{k+1} = \sigma(k)]) = \\
&\pi_{n-1}([x_1 = \omega(n)]T^{-1}\{[x_1 = \sigma(1)]\ldots[x_k = \sigma(k)]\}) \quad .
\end{aligned}
$$

This however is a special case of (23.7). Thus Δ maps $\hat{\Omega}(X)$ into $\Omega(X, V)$. It follows by direct verification that $\Delta T = T\Delta$ and that $\beta'\Delta = \hat{\beta}$. To prove the theorem we need only show that Δ is $1-1$. Now Δ will be $1-1$ if δ is $1-1$, and the latter is a consequence of (24.6). For if $\delta(\mu) = \delta(\mu')$ then $\mu(x^+_\rho) = \mu'(x^+_\rho)$ for all ρ. Since the linear combinations of the x^+_ρ are dense in A^+_X it follows that $\mu = \mu'$. This completes the proof of the theorem.

Consider now the process Z defined on the sample space of $\tilde{X}$ as follows. We consider the points of $\Omega_{\tilde{X}}$ identified with the points of $\Omega(X, V)$ as in the preceding theorem so that each $\tilde{\omega} = (\omega, \zeta)$ where $\omega(n)$ is a Λ- sequence and $\zeta(n)$ is a $\bar{V}$-sequence. From the proof of the theorem we know that $\zeta(n)$ is actually a $\bar{V}_1$-sequence (i.e., that $\langle \zeta(n), \bar{\theta} \rangle = 1$) so that $\zeta(n)$ takes its values in a compact space. Then Z is defined by setting

$$(24.12) \qquad z_n(\tilde{\omega}) = \zeta(n), \qquad n \leq 0$$

for $\tilde{\omega} = (\omega, \zeta)$. It is clear that the variables of X together with those of Z separate points in $\Omega_{\tilde{X}}$. We can therefore reformulate the preceding theorem (with some change of notation) as follows:

> THEOREM 24.2. If X is a finite dimensional process
> with values in a set $\Lambda = \{a_1, \ldots, a_r\}$ then Λ can
> be identified with a set of linear transformations of

a cone V into itself where V is a cone with
finitely many generators (or extremal lines); there
exists a linear non-negative functional θ on V
with $\langle u, \theta \rangle = 0$ only for $u = 0$ and a V_1-valued
process Z, where $V_1 = \{u \in V: \langle u, \theta \rangle = 1\}$, such
that the c.p. cover $\tilde{X}$ of X is a composite process
(X, Z) with the variables of X and Z satisfying

(24.13) $\langle z_{n-1}x_n, \theta \rangle z_n = z_{n-1}x_n$

where the $x_n \in \Lambda$ operate on the $z_n \in V$ on the
right.

There are several remarks that should be made as regards the
statement of this theorem. First, it should be pointed out that the set
V_1 as defined here is necessarily compact. This follows from the fact
that V is finitely generated and that $\langle u, \bar{\theta} \rangle = 0$ only for $u = 0$.
Hence V_1 is the convex closure of a finite number of generating vectors,
and so V_1 is compact. Also, we remark that a composite process (X, Z)
is a process with variables $\tilde{x}_n$ with X and Z as subprocesses such
that $\tilde{x}_n = (x_n, z_n)$. We note that one can only speak of "a" composite
process (X, Z) and not "the" composite process since X and Z themselves do not determine the process (X, Z). For example, one can always
take the variables x_n to be independent of the variables z_n although
there will be, in general, other ways of combining the two processes.
Finally we observe that (24.13) is a direct consequence of (24.3) so that
Theorem 24.2 is just a reformulation of the preceding results.

It may we well, at this point, to review the steps that have
been taken. For an arbitrary process X we may define the c.p. cover
$\tilde{X}$ of X. $\tilde{X}$ is defined abstractly by constructing an algebra $A_{\tilde{X}}$ with
certain properties. In case X is finitely valued, and Λ is its range
we found that the sample space of $\tilde{X}$ may be taken as a set of $\{\Lambda \times \mathcal{M}(X)\}$-
valued sequences satisfying the functional equations (23.7). Finally if
X is a finite dimensional process the sample space of $\tilde{X}$ will consist
of $\{\Lambda \times V\}$-valued sequences satisfying the functional equation (24.3).
This may be reinterpreted as in Theorem 24.2, so that $\tilde{X}$ now becomes an
ordinary process with variables $\tilde{x}_n = (x_n, z_n)$, where the x_n and z_n
satisfy the functional equation (24.13).

It is not difficult to explain the significance of these func-
tional equations. Namely, at any sequence $\omega(n)$ in $\Omega_{\tilde{X}}$ we consider a
prediction measure on the future. This measure assigns, in particular,
various probabilities for the different possible values of $\omega(1)$.

Equation (23.7) expresses the requirement that the prediction measure at
the "next" point $\omega' = \{\ldots, \omega(-1), \omega(0), \omega(1)\}$ be the prediction
measure at $\omega = \{\ldots, \omega(-1), \omega(0)\}$ "conditioned" by the event that $\omega(1)$
takes on that value that it does take on. If we require that this be the
case at each $T^k\omega$ then if π_n and π_{n+1} are the prediction measures at
$T^{-n}\omega$ and $T^{-n-1}\omega$ our condition is

$$(24.14) \qquad\qquad \pi_n(f) = \frac{\pi_{n-1}([x_1=\omega(n)]T^{-1}f)}{\pi_{n-1}([x_1=\omega(n)])}$$

and this reduces to (23.7) (which is meaningful even if the denominator
in (24.14) vanishes). Here the function f from the point of view of ω
corresponds to T^{-1} from the point of view of $T\omega$.

Of this, what is essential for our later discussion is that the
c.p. cover of a finite dimensional process X is a composite process with
variables (x_n, z_n) satisfying the functional equation (24.13). On the
basis of this we shall show in Chapter 9 that sequences derived from
Markoff sequences are predictable in the sense described in §22.

CHAPTER 7. INDUCTIVE FUNCTIONS

Before completing our treatment of the prediction problem we
shall find it necessary in this chapter and the next to investigate a
class of problems only indirectly related to prediction theory. The problems revolve about certain stochastic functional equations of the type we
were led to in the preceding chapter in analyzing the c.p. cover of a
process. We refer to Equation (24.13) relating the variables of the
process X and Z that make up the process $\tilde{X}$. The functional equations we shall consider also relate the variables x_n, z_n and z_{n-1} of
two processes; we shall assume however that the equation may be solved for
z_n so that it takes the form

$$z_n = \psi(x_n, \; z_{n-1}) \quad .$$

This is not the case in (24.13) since $< z_{n-1}x_n, \; \bar{\theta} >$ may vanish. Since
this does not happen in (24.13) except with probability 0, as we shall
see, our present analysis will carry over to a large extent to the process
$\tilde{X}$ treated there.

The problems we shall consider have the following form. Suppose
X is a fixed process with variables x_n and $\xi(n)$ is a sequence generic
for X. Consider on the one hand the processes Z whose variables z_n
satisfy the foregoing functional equation, and on the other hand the sequences $\zeta(n)$ satisfying the corresponding equation

$$\zeta(n) = \psi(\xi(n), \; \zeta(n - 1)) \quad .$$

Will every solution sequence $\zeta(n)$ be generic for a process X and conversely does there exist for every process Z a solution sequence $\zeta(n)$
generic for it?

We shall find that, in general, there are two circumstances under
which these questions may be answered affirmatively. In one case the function ψ will have a special form — this is the case we consider in this
chapter. On the other hand it is also possible to answer the questions
affirmatively for a wide class of functions ψ if one assumes that the

original process X is of a special kind. This is the case to be treated
in the next chapter.

In all of these cases it turns out to be a relatively simple
matter to describe the processes Z arising from a functional equation
of the type considered. By the correspondence between processes and
sequences it will become possible to give an exact description of the
sequences $\zeta(n)$ arising through such functional equations. This will,
in particular, give an idea of the behavior of the extensions to a process
(X, Z) of a generic point of X where X and Z are related as before.
We observe that this is just the type of problem that has to be solved to
determine the statistical predictability of a particular sequence $\xi(n)$.
Namely, in order that $\xi(n)$ be statistically predictable there must be
exactly one generic extension of $\xi(n)$ to the sample space of the c.p.
cover of the process $X(\xi)$. Thus if we could describe all such extensions
we would be in a position to ascertain whether $\xi(n)$ is predictable.
Since moreover the simply predictable cover has the form $(X(\xi), Z)$ where
the variables of $X(\xi)$ and Z are related by a functional equation of
the type considered it is clear how our present investigations will be
relevant to prediction. The details of this will be carried out in
Chapter 9.

For the most part we treat the subject matter of these two
chapters as being of independent interest and so we do not restrict ourselves to obtaining just those results needed for Chapter 9. Thus in
this chapter we include results on the equidistribution problem that have
no bearing on prediction theory (except for the application made in Chapter
3 to $\mathcal{D}$-sequences) but where the methods illustrate those used later and
which furthermore lend completeness to our discussion.

$\S25.$ Inductive Functions; An Example

25.1. <u>Preliminaries</u>. We begin by giving the formal definitions
for the situations referred to in our introductory remarks. By the "range"
of the variables of a process, or just "range of a process", is meant the
least closed set in which the variables take ther values with probability 1.

DEFINITION 25.1. A Λ-valued process Z is an <u>inductive func-</u>
tion of a Π-valued process X if there is a process of which both X
and Z are subprocesses and the variables x_n of X and z_n of Z
are related by an equation

$$(25.1) \qquad\qquad z_{n+1} = \psi(x_{n+1}, z_n)$$

where ψ is a continuous function defined of the set of pairs (π, λ)
of $\Pi \times \Lambda$ in the range of (x_{n+1}, z_n) into Λ. When it is necessary to

refer to the mapping ψ connecting X and Z we shall speak of a
ψ-<u>inductive function</u>.

DEFINITION 25.2. A Λ-valued sequence $\zeta(n)$ is an <u>inductive</u>
· <u>function</u> of a Π-valued sequence $\xi(n)$ if there is a mapping ψ on the
subset of $\Pi \times \Lambda$ given by the range of the sequence $(\xi(n + 1), \zeta(n))$
to Λ such that

$$(25.2) \qquad\qquad \zeta(n + 1) = \psi(\xi(n + 1), \zeta(n)) \quad .$$

DEFINITION 25.3. If X is a subprocess of Z and the two are
related by (25.1) we refer to Z as an <u>inductive extension</u> of X.
Similarly if $\xi(n)$ is a derived sequence of $\zeta(n)$ and they are related
by (25.2), $\zeta(n)$ will be an <u>inductive extension</u> of $\xi(n)$.

Whenever Z is an inductive function of the process X then
the composite process $Z^* = (X, Z)$ is an inductive extension of X since

$$(x_{n+1}, z_{n+1}) = (x_{n+1}, \psi(x_{n+1}, z_n)) = \psi^O(x_{n+1}, z_n) =$$

$$\psi^*(x_{n+1}, (x_n, z_n)) \quad .$$

In the case of an inductive extension Z of X it is not necessary to
specify the common extension of X and Z since Z itself is such an
extension. However, in general, if both X and Z are given and Z
is a ψ-inductive function of X, the relationship between the variables
of both processes are not completely determined unless the common extension of X and Z is specified. That is to say, the process
$Z^* = (X, Z)$ must be given. For this reason it is often more convenient
to deal with inductive extensions rather than inductive functions. For
example, it is obvious that an inductive extension of an inductive extension is an inductive extension, but the same statement has no meaning
for inductive functions since the processes in question need have no relationship to one another.

In the course of what follows we shall have to make reference
to certain definitions and lemmas which we collect here.

DEFINITION 25.4. A compact Hausdorff space Ω with a distinguished map $T : \Omega \longrightarrow \Omega$ will be called a T-<u>space</u>.

DEFINITION 25.5. If Ω_1 and Ω_2 are two T-spaces with maps T_1
and T_2 and $\Phi : \Omega_1 \longrightarrow \Omega_2$, then Φ is a T-<u>map</u> if $T_2\Phi = \Phi T_1$.

We shall usually denote distinguished maps by the same symbol
T regardless of the space on which they act.

DEFINITION 25.6. We say a process X <u>exists on</u> a T-space Ω

if there is a 1 - 1 T-map ν from Ω_X^- into Ω.

DEFINITION 25.7. If X and Y are two processes and Φ is a measure preserving T-map of Ω_Y^- onto Ω_X^- we shall write $X = \Phi(Y)$.

Thus whenever X is a subprocess of Y and Φ is the induced map of Ω_Y^- onto Ω_X^- we have $X = \Phi(Y)$. In most cases the T-spaces in question will be represented as left-infinite sequence spaces with T the shift to the right. A T-map is then obtained, say from M_∞^- to Λ_∞^-, by taking any function φ from M_∞^- to Λ and setting

$$\Phi(\omega)(n) = \varphi(\ldots, \omega(n - 2), \omega(n - 1), \omega(n)) \quad .$$

Given a process X and a T-map Φ of a T-space Ω onto Ω_X^- we shall say that an abstract process Y is a <u>solution</u> to the equation $X = \Phi(Y)$ if Y exists on Ω and identifying Ω_Y^- with its image in Ω we have $X = \Phi(Y)$.

> LEMMA 25.1. If X is a given process and Φ a
> T-map of a T-space Ω onto Ω_X^-, then if X is
> ergodic and there exists a unique ergodic solution
> to $X = \Phi(Y)$, then Y is the only solution
> (ergodic or not). Moreover if there exists some
> solution (ergodic or not) there exists an ergodic
> solution.

PROOF. Let $\{Y_\alpha\}$ denote the set of solutions on Ω to the above equation. Each Y_α is uniquely determined by a measure μ_α on Ω and let Σ be the set of all measures on Ω so determined. The μ_α in Σ must be carried into the probability measure P on Ω_X^- if Φ is measure preserving so that $\mu_\alpha(\Phi^{-1}(\Delta)) = P(\Delta)$ for $\Delta \subset \Omega_X^-$. Moreover since each Y_α is a process the mapping T is measure preserving so that $\mu_\alpha(T^{-1}(\Delta)) = \mu_\alpha(\Delta)$ for $\Delta \subset \Omega$. Treating the μ_α as functionals on $C(\Omega)$ we may rewrite these conditions as

$$(25.3) \qquad (i) \quad E(f) = \mu_\alpha(f \circ \Phi) \qquad (ii) \quad \mu_\alpha(f) = \mu_\alpha(Tf) \quad .$$

Conversely if a probability measure μ satisfies these two conditions then $\mu \in \Sigma$ since it defines an abstract process on Ω satisfying $X = \Phi(Y)$. Hence Σ is identical with the set of μ satisfying (25.3) (i) and (25.3) (ii). Now the set of probability measures μ satisfying these two conditions is convex and in the weak topology of measures on Ω it is also compact. It follows that a dense subset of Σ is spanned by the convex linear combinations of the extremals in this convex set (Krein-

Milman theorem). In particular if there is only one extremal the entire
set Σ must reduce to a single point. Furthermore extremals always exist. Our lemma will therefore be proved if it is established that the
extremals of Σ correspond to ergodic processes Y in case X is
ergodic.

Assume therefore that X is ergodic; suppose that Y corresponds to an extremal measure μ in Σ and that Y is not ergodic.
There is then a measurable set Δ in Ω with $\mu(\Delta) \neq 0, 1$ and such
that $T\Delta \subset \Delta$. Let X_Δ be the characteristic function of Δ and let μ'
be a measure defined by

$$(25.4) \qquad\qquad \mu'(f) = \mu(fX_\Delta)/\mu(\Delta) \quad .$$

We shall show that $\mu' \in \Sigma$. First of all since $T\Delta \subset \Delta$ we have $X_\Delta = TX_\Delta$
almost everywhere with respect to μ and so

$$(25.5) \qquad \mu'(Tf) = \mu(Tf \cdot TX_\Delta)/\mu(\Delta) = \mu(fX_\Delta)/\mu(\Delta) = \mu'(f) \quad ,$$

and (25.3) (ii) is satisfied.

As to (i) we note that $\mu'(f \circ \Phi) = \mu(f \circ \Phi \cdot X_\Delta)/\mu(\Delta)$ so that
if $f \geq 0$,

$$\mu'(f \circ \Phi) \leq \mu(f \circ \Phi)/\mu(\Delta) = \frac{1}{\mu(\Delta)} E(f) \quad .$$

It follows that $\mu'(f \circ \Phi) = E'(f)$ where E' is absolutely continuous
with respect to E; hence $\mu'(f \circ \Phi) = E(fg)$ for some $g \in L(A_X^-)$. Now
$(Tf) \circ \Phi(\omega) = f[T(\Phi(\omega))] = f \circ \Phi(T\omega)$ by Definition 25.5 so that

$$\mu'(Tf \circ \Phi) = \mu'(T(f \circ \Phi)) = \mu'(f \circ \Phi) \quad .$$

This implies that $E(f \cdot g) = E(Tf \cdot g)$ for all f in A_X^-. The same will
then be true for $f \in L(A_X^-)$. In particular we have $E(g^2) = E(g \cdot Tg)$
so that

$$E((g - Tg)^2) = E(g^2) - 2E(g \cdot Tg) + E((Tg)^2) = 2E(g^2) - 2E(g \cdot Tg) = 0$$

and $g = Tg$. Since X is ergodic and $g \in L(A_X^-)$ this implies that g
is a constant so that $\mu'(f \circ \Phi) = \alpha E(f)$ for a constant α. If we let
$f = 1$ we find $\alpha = 1$ and this establishes (25.3) (i) for the measure
μ'. Hence $\mu' \in \Sigma$

If we similarly define μ'' by

$$\mu''(f) = \mu(f(1 - X_\Delta))/(1 - \mu(\Delta))$$

the same argument shows that $\mu'' \in \Sigma$. This gives us $\mu(f) = \mu(f\chi_\Delta) +$
$\mu(f(1 - \chi_\Delta)) = \mu(\Delta)\mu'(f) + (1 - \mu(\Delta))\mu''(f)$ and $\mu = \mu(\Delta)\mu' + (1 - \mu(\Delta))\mu''$.
Since $\mu(\Delta) \neq 0, 1$ and μ is extremal, this can be true only if
$\mu' = k\mu$ for a constant k. However taking $\mu'(1) = k\mu(1)$ shows that
$k = 1$ and so $\mu'(\Delta) = \mu(\chi_\Delta^2)/\mu(\Delta) = 1$ implies that $\mu(\Delta) = 1$ contrary
to our supposition. This proves the lemma.

The next lemma is fundamental for all that follows.

LEMMA 25.2. Let X be a process and let $\xi \in \Omega_X^-$
be generic for X (§5.3). If Φ is a T-map of a
T-space Ω onto Ω_X^- there always exists a solution
Y on Ω to $X = \Phi(Y)$. If this solution is unique
and $\eta \in \Omega$ satisfies $\eta \in \Omega_Y^-$ and $\Phi(\eta) = \xi$ then
η is generic for Y. (In other words, if Y is
unique then any point of Ω_Y^- is generic for Y if
it extends a generic point of X.)

PROOF. For the first part of the lemma we let S be a de-
numerable dense subset of $C(\Omega)$ and let $\eta \in \Omega$ be a preimage of ξ
under Φ. For any $f \in C(\Omega)$ we may choose a sequence of integers
$N_k \longrightarrow \infty$ such that

$$(25.6) \qquad \lim_{k \to \infty} \frac{1}{N_k + 1} \sum_{0}^{N_k} f(T^n \eta)$$

exists; hence we may choose a sequence such that the limit in (25.6) exists for any denumerable set of f and hence for all $f \in S$. Since S
is dense it follows that the limit in (25.6) will exist for all of $C(\Omega)$.
Thus (25.6) defines a functional E' on $C(\Omega)$. For $f = g \circ \Phi$ where
$g \in C(\Omega_X^-)$ we then have

$$E'(f) = \lim_{k \to \infty} \frac{1}{N_k + 1} \sum_{0}^{N_k} g \circ \Phi(T^n \eta) =$$

$$\lim_{k \to \infty} \frac{1}{N_k + 1} \sum_{0}^{N_k} g(T^n \xi) = E(g)$$

since $\Phi T^n \eta = T^n \Phi \eta = T^n \xi$ and ξ is generic for X. Hence $E'(g \circ \Phi) =$
$E(g)$. It is evident that $E'(Tf) = E'(f)$ and also that $E'(ff^*) \geq 0$ and
$E'(1) = 1$. It follows from this that E' defines an abstract process Y

on Ω satisfying $X = \Phi(Y)$. (The support of the measure corresponding to E' will be a subset of Ω and for f with support in this subset it is clear that $E(ff^*) = 0$ implies that $f = 0$.) This proves the existence of a solution on Ω to $X = \Phi(Y)$.

In case the solution process Y is unique we note that for any sequence N_k the limit in (25.6) must yield the same functional. This being the case it follows that the ordinary averages

$$\lim_{N \to \infty} \frac{1}{N+1} \sum_{0}^{N} f(T^n \eta)$$

exist for $f \in S$ and have the value $E'(f)$. Since S is dense the same must be true for all $f \in C(\Omega)$ and so η is a generic point for Y. This proves the lemma.

We note that since Φ takes Ω_Y^- onto Ω_X^- (since $\Phi(\Omega_Y^-)$ is a closed set with measure 1) there always exists an η satisfying the conditions of the lemma. Thus in the case of uniqueness every generic point of X extends to a generic point of Y.

25.2. The Equation $z_{n+1} = x_{n+1} + z_n$. Examples

To illustrate the problems referred to at the outset of this chapter we consider the equation

$$(25.7) \qquad\qquad z_{n+1} = x_{n+1} + z_n$$

exhibiting a process Z as an inductive extension of X (since $x_{n+1} = z_{n+1} - z_n$). The variables of X and Z are in this case complex-valued. The corresponding "ordinary" functional equation relating the sample sequences of the processes is

$$(25.8) \qquad\qquad \zeta(n+1) = \xi(n+1) + \zeta(n) \ .$$

We suppose that X is fixed and that $\xi(n)$ is a fixed generic sequence of X. The question is that of the relationship of the numerical solutions to (25.8) to the stochastic solutions of (25.7). Does every $\zeta(n)$ correspond to some Z and does every Z have a corresponding $\zeta(n)$?

Naturally we must assume that (25.8) has some bounded solution, for otherwise no solution $\zeta(n)$ will be bounded and there is no question of $\zeta(n)$ being generic for a process Z. We might point out that it is sufficient that (25.7) have some solution Z. For if (25.7) has a solution with variables z_n, then since the z_n have a compact range $|z_n| < \kappa$

for some κ and so by (25.7) $|x_n + \cdots + x_{n+\ell}|$ will be uniformly bounded for all n and ℓ and since $\xi(n)$ is generic for X this implies that $|\xi(n) + \cdots + \xi(n + \ell)|$ is uniformly bounded. This in turn implies that there is a bounded solution to (25.8).

As a result the questions we have in mind may be reformulated in this way. First of all will a bounded solution to (25.8) necessarily be regular (for if it is then $Z = X(\zeta)$ will clearly satisfy (25.7))? Secondly, do the bounded solutions of (25.8) correspond to all possible stochastic solutions to (25.7)? The sequence $\xi(n)$ is, in every case, assumed to be a regular sequence and so, generic for some process X.

We shall first give examples showing that some hypotheses on the process X, or equivalently, on the sequence $\xi(n)$, must be made in order to answer our questions affirmatively. First of all we point out that if $\xi(n)$ is not assumed regular but only stochastic it is easy to give an example where (25.8) has a bounded solution that is not stochastic. For example, let $\xi(-n) = 0$ for all $n > 0$ but for $n = 2^{2k} - 2$ and $n = 2^{2k} - 1$ where $k \geq 2$ in which case we take $\xi(-n) = 1$, and for $n - 2^{2k+1} - 2$ and $n = 2^{2k+1} - 1, k \geq 1$, for which we take $\xi(-n) = -1$ and we set $\xi(0) = -1$. $\xi(n)$ is clearly stochastic since it is "almost always" 0. Now $\zeta(0) - \zeta(-n) = \xi(0) + \cdots + \xi(-n + 1)$ and $\zeta(-n) = \zeta(0) - [\xi(0) + \cdots + \xi(-n + 1)]$. Hence $\zeta(n)$ will be stochastic only if the sums $\{\xi(0) + \cdots + \xi(-n + 1)\}$ form a stochastic sequence. However the latter takes on the values 1 and -1 over intervals of successively larger length in such a way that the average cannot exist.

We now give an example to show that for a regular sequence $\xi(n)$ there may exist a bounded solution to (25.8) that is not stochastic (and certainly not regular). We shall find it convenient to speak here of right-infinite sequences and although we use terms heretofore defined only for left-infinite sequences there will be no difficulty carrying over the definitions to the situation considered here.

Let $\zeta'(n)$ be a regular right-infinite $\{0, 1\}$-random sequence and set $\zeta''(n) = \zeta'(n) + 1$ so that $\zeta''(n)$ is a $\{1, 2\}$-random sequence. Finally let $\zeta'''(n)$ be a regular right-infinite $\{0, 1, 2\}$-random sequence. Now define

$$(25.9) \qquad \zeta(3n^2 + j) = \begin{cases} \zeta'(n^2 + j) & \text{for } 0 < j \leq 2n + 1 \\ \zeta''(n^2 + j) & \text{for } 2n + 1 < j \leq 4n + 2 \\ \zeta'''(n^2 + j) & \text{for } 4n + 2 < j \leq 6n + 3 \end{cases}$$

so that $\zeta(n)$ is defined for all integers $n > 0$. $\zeta(n)$ is thus a superposition of the three sequences $\zeta'(n)$, $\zeta''(n)$, and $\zeta'''(n)$ and it is not hard to see that $\zeta(n)$ is regular. (Note that this would not have been the case if the range of $\zeta'''(n)$ did not contain the ranges of $\zeta'(n)$ and $\zeta''(n)$.) If we now define $\xi(n)$ by $\xi(n) = \zeta(n) - \zeta(n + 1)$ then $\xi(n)$ is regular and the sequences $\overline{\zeta}(n) = \zeta(-n)$ and $\overline{\xi}(n) = \xi(-n)$ are left-infinite sequences satisfying (25.8).

We shall now modify the sequence $\zeta(n)$ in such a manner that $\xi(n)$ remains regular but $\zeta(n)$ itself does not. Namely in (25.9) for a certain set of n we replace all the $\zeta'(n^2 + j)$ by $\zeta''(n^2 + j)$. Since $\zeta''(n^2 + j) = \zeta'(n^2 + j) + 1$ this will tend to increase the average of $\zeta(n)$ over intervals $1 \leq n \leq N$. Consequently if we alternately make this change in a set of n and leave a large set of terms unchanged it will be possible to have the averages of $\zeta(n)$ from 1 to N oscillate in such a way that these averages do not converge and $\zeta(n)$ becomes non-stochastic. On the other hand the change will affect the sequence $\xi(n)$ only at positions of the form $3n^2$, $3n^2 + 1$, $3n^2 + 2n$, $3n^2 + 2n + 1$, since $\zeta'(n)$ and $\zeta''(n)$ differ by a constant. Since these positions have density 0 this will not affect the regularity of $\xi(n)$. This provides us with an example where $\xi(n)$ is regular but no solution sequence $\zeta(n)$ to (25.8) is regular even though they are all bounded.

The same construction may be used to show that even when the solutions to (25.8) are all regular they need not represent all possible processes Z satisfying (25.7). Namely consider the sequence $\xi(n)$ for the $\zeta(n)$ as originally constructed so that both $\xi(n)$ and $\zeta(n)$ are regular. When $\zeta(n)$ is modified, since $\xi(n)$ is changed only inessentially it will always represent the same process X. However $\zeta(n)$ might be modified in such a way as to represent an entirely different process — for example, all $\zeta'(n)$ terms might be replaced by the corresponding $\zeta''(n)$ terms. Thus the same process X can have two solutions Z to (25.7) and the Z do not just differ by a constant (since in both cases Z is $\{0, 1, 2\}$-valued). But for a fixed $\xi(n)$ all the solutions $\zeta(n)$ can only differ by a constant and thus not all solutions Z can be represented.

25.3. <u>A Condition for Regularity</u>. The sequence $\zeta(n)$ of the foregoing example was a superposition of sequences generic for distinct processes and therefore the derived sequence $\xi(n)$ was also such a superposition. As a result these two sequences are not ergodic — in fact they are not even topologically ergodic (§7.1). We shall see now that it is the latter property that brought about the failure of $\zeta(n)$ to be regular.

We shall need the following lemmas.

LEMMA 25.3. A process X is topologically ergodic
if and only if almost all points ω of Ω_X^- have
the property that the set $\{T^n\omega\}$ is dense in Ω_X^-.

PROOF. Let us call a point ω of Ω_X^- "topologically generic"
if it has the property that $\{T^n\omega\}$ is dense in Ω_X^-. This notion is
analogous to that of a point being generic for it is equivalent to

$$\|f\| = \sup_{n \geq 0} (f(T^n\omega))$$

which corresponds to (5.10) for generic points. Our lemma is then the
analog of Birkhoff's ergodic theorem, but it is much simpler. Suppose
first that X is topologically ergodic. To show that almost every ω
is topologically generic it suffices to show that for a particular open
set U almost all ω have some $T^n\omega$ in U, since Ω_X^- has a denumerable
basis of open sets. Now for a particular U the set of ω with no $T^n\omega$
in U forms a closed set Δ (as the intersection of closed sets) and
Δ satisfies $T\Delta \subset \Delta$. Hence Δ has either measure 0 or measure 1. It
cannot have measure 1 since $U \cap \Delta = \emptyset$ and $P(U) > 0$. Hence Δ has
measure 0 as was to be shown.

Suppose that, conversely, almost every point of Ω_X^- is topologically generic and that $T\Delta \subset \Delta$ where Δ is a closed set Clearly
for $\omega \in \Delta$ all translates of ω remain in Δ so that unless Δ is
identical to Ω_X^- no point in Δ is topologically generic. But then Δ
has measure 0. This completes the proof of the lemma.

LEMMA 25.4. If F is a subset of the complex plane
and D(F) denotes the diameter of F:

$$D(F) = \sup_{z_1, z_2 \in F} |z_1 - z_2| \quad ,$$

and if $F + \alpha = \{z + \alpha : z \in F\}$, then if $\alpha \neq 0$

$$D(F \cup (F + \alpha)) > D(F) \quad .$$

PROOF. Consider the parallelogram with vertices z_1, z_2, $z_2 + \alpha$,
$z_1 + \alpha$ where z_1, $z_2 \in F$. Then

$$2|z_1 - z_2|^2 + 2|\alpha|^2 = |z_2 - (z_1 + \alpha)|^2 + |z_1 - (z_2 + \alpha)|^2 \quad .$$

Letting $|z_1 - z_2| \longrightarrow D(F)$ we find

$$2D(F)^2 + 2|\alpha|^2 \leq 2D(F \cup (F + \alpha))^2 \quad .$$

We now have

THEOREM 25.1. If $\xi(n)$ is topologically ergodic
and $\zeta(n)$ is a bounded sequence satisfying
$\zeta(n + 1) = \xi(n + 1) + \zeta(n)$ then $\zeta(n)$ is regular
and topologically ergodic.

PROOF. Let Λ be the range of $\zeta(n)$. Consider the set Ω of
left-infinite sequences in Λ_∞^- generated by $\zeta(n)$; that is, Ω is the
closure in Λ_∞^- of the translates $T^n\zeta$. Ω is then a T-space. To every
$\omega \in \Omega$ we associate the sequence $\eta(n)$ defined by

$$(25.10) \qquad\qquad \eta(n + 1) = \omega(n + 1) - \omega(n) \quad .$$

Since $\xi(n) = \zeta(n + 1) - \zeta(n)$ and ζ generates Ω it is clear that the
set of sequences $\eta(n)$ formed by (25.10) consists of the limits of
translates of ξ, i.e., it is contained in $\Omega^-(\xi)$. Then (25.10) defines
a T-map Φ of Ω onto $\Omega^-(\xi)$, ("onto" because ξ is in the image). To
prove that $\zeta(n)$ is regular we shall show that the point ζ in Ω is
generic for a process Z on Ω. More precisely we shall show that
Lemma 25.2 applies so that there exists just one solution Z to
$X(\xi) = \Phi(Z)$ and that ζ belongs to Ω_Z^- for this Z.

To do this we establish the following fact: each topologically
generic point of $\Omega^-(\xi)$ has just one preimage in Ω under Φ. To this
end define a function $V(\eta)$ on $\Omega^-(\xi)$ by

$$V(\eta) = \sup_{n,\ell} |\eta(n) + \eta(n + 1) + \ldots + \eta(n + \ell)| \quad .$$

It is easy to verify that $V(\eta)$ so defined is lower semi-continuous, i.e.,
if $\eta^{(k)} \longrightarrow \eta$ then

$$V(\eta) \leq \liminf_{k \to \infty} V(\eta^{(k)}) \quad ,$$

and that $V(T\eta) \leq T(\eta)$. From these two facts it follows that $V(\eta)$ has
the same value at all topologically generic points of $\Omega^-(\xi)$. For if
η_1 and η_2 are both topologically generic we can find sequences n_k and
m_k such that $T^{n_k}\eta_1 \longrightarrow \eta_2$ and $T^{m_k}\eta_2 \longrightarrow \eta_1$. By the first of these we
have $V(\eta_2) \leq V(\eta_1)$ and by the second $V(\eta_1) \leq V(\eta_2)$, and this proves
our assertion. Now suppose $\Phi(\omega) = \eta$ for $\eta \in \Omega^-(\xi)$. Then if $R(\omega)$
denotes the range of the sequence ω in the complex numbers, we find that

the diameter of $R(\omega)$, $D(R(\omega))$, equals the supremum of

$$|\omega(m) - \omega(n)| = |\eta(m) + \eta(m - 1) + \ldots + \eta(n + 1)| \qquad (\text{for } n \leq m)$$

and so $D(R(\omega)) = V(\eta)$. It follows that if $\Phi(\omega)$ is topologically generic then $D(R(\omega) = V(\eta) = V(\xi) = D(R(\zeta))$.

Now suppose that $\Phi(\omega_1) = \Phi(\omega_2) = \eta$ and that η is topologically generic. If $\omega_1 \neq \omega_2$ we must have by (25.10) $\omega_2(n) = \omega_1(n) + \alpha$ for some $\alpha \neq 0$. Then $R(\omega_2) = R(\omega_1) + \alpha$. Now since ω_1 is a limit of translates of ζ it follows that $R(\omega_1) \subset R(\zeta)$. Similarly $R(\omega_2) \subset R(\zeta)$. But $R(\omega_2) = R(\omega_1) + \alpha$ and so $R(\omega_1) \cup (R(\omega_1) + \alpha) \subset R(\zeta)$ which by Lemma 25.4 implies that $D(R(\omega_1)) < D(R(\zeta))$ contrary to the above. Hence ω_1 and ω_2 must be equal, and a topologically generic point of $\Omega^-(\xi)$ has only one preimage in Ω.

We proceed with the proof of the theorem. Suppose Z is a process on Ω satisfying $\Phi(Z) = X(\xi)$. Z is determined by a measure μ on Ω or alternatively by a functional $\mu(f)$ for $f \in C(\Omega)$. We shall prove that the functional μ is uniquely determined by the requirement that $\mu(g \circ \Phi) = E(g)$ for $g \in A^-(\xi)$. Since the topologically generic points of $\Omega^-(\xi)$ have measure 1 we can find compact sets W_ε in $\Omega^-(\xi)$ with $P(W_\varepsilon) > 1 - \varepsilon$ such that all points of W_ε are topologically generic. Now let $F \in C(\Omega)$ and consider f restricted to $\Phi^{-1}(W_\varepsilon)$. This set is homeomorphic to W_ε since Φ is $1 - 1$ here so there exists a continuous function on W_ε, say f'_ε, with $f = f'_\varepsilon \circ \Phi$ on $\Phi^{-1}(W_\varepsilon)$. f'_ε may be extended to a function f_ε on all of $\Omega^-(\xi)$ without changing its supremum norm. Thus f agrees with $f_\varepsilon \circ \Phi$ on $\Phi^{-1}(W_\varepsilon)$. Now $\mu(\Phi^{-1}(W_\varepsilon)) = P(W_\varepsilon) > 1 - \varepsilon$ and hence

$$\mu(f) = \lim_{\varepsilon \to 0} \mu(f_\varepsilon \circ \Phi) = \lim_{\varepsilon \to 0} E(f_\varepsilon) \ .$$

It follows that μ is determined by E, the expectation functional of $X(\xi)$. Hence μ is unique and so is the solution Z to $\Phi(Z) = X(\xi)$.

To complete the proof of the regularity of $\zeta(n)$ we have only to show that ζ lies in Ω^-_Z for the unique solution Z on Ω. But $\Phi(\Omega^-_Z)$ must be all of Ω^-_X since it is a compact set with measure 1. Hence Φ is onto so that ξ has a preimage in Ω^-_Z. But Φ is $1 - 1$ over topologically generic points of $\Omega^-(\xi)$ and so ξ has only one preimage ζ in Ω, not to mention in Ω^-_Z. Hence $\zeta \in \Omega^-_Z$ and ζ is generic for Z.

We have now only to prove that $\zeta(n)$ is topologically ergodic, or that the process Z determined above is topologically ergodic. This however follows immediately from the fact that Φ is $1 - 1$ over almost all points of $\Omega^-(\xi)$. For from this we conclude that every point lying over such a point in $\Omega^-(\xi)$ is itself topologically ergodic in Ω_Z^-, or that almost all points of Ω_Z^- are topologically generic. But then by Lemma 25.3, Z is topologically ergodic. This completes the proof.

> COROLLARY. If all points of $\Omega^-(\xi)$ are topologically
> generic then the inductive extension $\zeta(n)$ of $\xi(n)$
> is a derived sequence of $\xi(n)$.

PROOF. For then $\Omega^-(\xi) \cong \Omega_Z^-$ and the algebra $A_Z^- = A^-(\xi)$ and since ζ is determined by the sequences in A_Z^- it follows that $\zeta(n)$ is a derived sequence of $\xi(n)$.

For example, if ξ is almost periodic then it is easily shown that all points of $\Omega^-(\xi)$ are topologically generic (since each sequence in $\Omega^-(\xi)$ is a <u>uniform</u> limit of translates of ξ it follows that for $\omega \in \Omega^-(\xi)$ $A^-(\omega) \cong A^-(\xi)$ as E-algebras and so ω is generic). It follows that if $\zeta(n)$ is bounded and satisfies $\zeta(n + 1) - \zeta(n) = \xi(n)$ where $\xi(n)$ is almost periodic then $\zeta(n)$ is a derived function of $\xi(n)$; in particular $\zeta(n)$ is almost periodic. This is the discrete analogue of the theorem that the indefinite integral of an almost periodic function is almost periodic if it is bounded ([4]).

25.4. <u>Existence of Representative Sequences</u>. We still consider the equation $z_{n+1} = x_{n+1} + z_n$ and we ask whether for a given generic sequence $\xi(n)$ of a process X there exist generic sequences satisfying $\zeta(n + 1) = \xi(n + 1) + \zeta(n)$ for every solution z_n to the above equation. The answer is given in

> THEOREM 25.2. If X is a topologically ergodic
> process and $\xi(n)$ is generic for X then for every
> topologically ergodic Z satisfying $z_{n+1} = x_{n+1} + z_n$
> there exists a sequence $\zeta(n)$ generic for Z
> satisfying $\zeta(n + 1) = \xi(n + 1) + \zeta(n)$.

PROOF. We remark that by the theorem just proven, if $\xi(n)$ is topologically ergodic then the solution sequence will be so too and so Z must be assumed topologically ergodic for the conclusion to hold. Now assume that Z is topologically ergodic and satisfies $z_{n+1} = x_{n+1} + z_n$ and let Φ be the map from Ω_Z^- to Ω_X^-. Since the z_n are bounded we conclude as before that there exists a bounded solution $\zeta'(n)$, to $\zeta'(n + 1) = \xi(n + 1) + \zeta'(n)$. By the preceding theorem $\zeta'(n)$ is regular and topologically ergodic; hence $\zeta'(n)$ is generic for a topologically ergodic process Z'. Let Φ' denote the induced map from $\Omega_{Z'}^-$ to Ω_X^-. Now since almost every point of Ω_Z^- is topologically generic in Ω_Z^- and similarly for $\Omega_{Z'}^-$, it follows that there is a pair of sequences $\bar{\zeta} \in \Omega_Z^-$ and $\bar{\zeta}' \in \Omega_{Z'}^-$ with $\Phi(\bar{\zeta}) = \Phi'(\bar{\zeta}')$. This implies that $\bar{\zeta}(n) = \bar{\zeta}'(n) + \alpha$ for some constant α since $\bar{\zeta}(n + 1) - \bar{\zeta}(n) = \bar{\zeta}'(n + 1) - \bar{\zeta}'(n)$. Since $\bar{\zeta}'$ is topologically generic in $\Omega_{Z'}^-$ and $\bar{\zeta}$ in Ω_Z^- it follows that the map $\omega' \longrightarrow \omega$ where $\omega(n) = \omega'(n) + \alpha$ sends $\Omega_{Z'}^-$ onto Ω_Z^-. Now take $\zeta(n) = \zeta'(n) + \alpha$; then ζ is topologically generic for Ω_Z^- since ζ' is topologically generic for $\Omega_{Z'}^-$. However by the preceding theorem ζ is generic for a process Z^* satisfying $X = \Phi(Z^*)$. Moreover if Ω is the closure of the set of translates of ζ then Z is the only solution to $X = \Phi(Z^*)$ on Ω. But $\Omega \subset \Omega_Z^-$ since $\zeta \in \Omega_Z^-$. It follows that $Z^* = Z$ so that $\zeta(n)$ is generic for Z. This completes the proof of the theorem.

§26. Group-Valued Inductive Functions

26.1. <u>The Equation</u> $z_{n+1} = x_{n+1} z_n$. In this section we consider the analogue of the functional equation $z_{n+1} = x_{n+1} + z_n$ for compact abelian groups. Here the problem changes essentially, the reason being that the processes X and Z are no longer "almost" identical as they were in the foregoing section. Thus the ranges of two solutions $\zeta(n)$ and $\zeta'(n)$ to the equation $\zeta(n + 1) = \xi(n + 1) + \zeta(n)$ had to be distinct since $\zeta'(n) = \zeta(n) + \alpha$ for some $\alpha \neq 0$. In the case of G-valued sequences, where G is a compact abelian group, two solutions $\zeta(n)$ and $\zeta'(n)$ to $\zeta(n + 1) = \xi(n + 1)\zeta(n)$ would be related by $\zeta'(n) = \alpha\zeta(n)$ for some $\alpha \in G$ and the two solutions might still have the same range. As a result of this more stringent conditions have to be imposed on the sequence $\xi(n)$ in order that a solution $\zeta(n)$ be regular. To formulate these conditions we shall require some preliminary notions.

To begin with for the remainder of this chapter G will denote a compact abelian group. The elements of G will be denoted ρ, σ, τ, $\ldots$ If $\omega \in G_\infty^-$, i.e., ω is a G-sequence, and $\sigma \in G$, then $\sigma\omega$ will denote the G-sequence with entries $\sigma\omega(n) = \sigma \cdot \omega(n)$.

DEFINITION 26.1. If X is a G-valued process and $\sigma \in G$, then σX denotes the G-valued process with variables x_n' given by $x_n' = \sigma x_n$.

DEFINITION 26.2. A G-valued process X is <u>equidistributed</u> in G if for each $\sigma \in G$, $\sigma X \cong X$.

By the isomorphism of two processes $X' \cong X$ we mean that there is a measure preserving homeomorphism of $\Omega_{X'}^-$ with Ω_X^- (or $\Omega_{X'}$ with Ω_X) such that the functions x_n' are carried into the x_n. In other words the joint distributions of the variables x_n' are identical to those of the corresponding x_n.

For example if G is the group of reals mod 1 and X is a random process with each x_n having a uniform distribution on G then X is equidistributed. Or if we take for G the cyclic group of two elements $\{-1, 1\}$ and let X represent a random $\{-1, 1\}$-sequence then X is equidistributed if and only if $P(1) = P(-1) = 1/2$.

In order to prove the main result in this section we shall need the following lemma.

LEMMA 26.1. If X is a subprocess of Y and Φ is the induced map of $\Omega_{\bar{Y}}$ onto $\Omega_{\bar{X}}$, then if there is a subset S of $\Omega_{\bar{Y}}$ with $P(S) = 1$ such that whenever $\omega_1, \omega_2 \in S$ then $\Phi(\omega_1) \neq \Phi(\omega_2)$ then Y is an L-extension of X.

PROOF. We think of both $A_{\bar{X}}$ and $A_{\bar{Y}}$ as algebras of functions on $\Omega_{\bar{Y}}$. It suffices to show that for every $g \in A_{\bar{Y}}$ there is a sequence $\{g_n\}$ of functions in $A_{\bar{X}}$ with $g_n \longrightarrow g$ in measure. For then a subsequence of the g_n converges almost everywhere in $\Omega_{\bar{X}}$ to a function in $L(A_{\bar{X}})$ and this will give an imbedding of A_Y into $L(A_{\bar{X}})$. We proceed by choosing a sequence of compact sets $W_\epsilon \subset S$ with $P(W_\epsilon) > 1 - \epsilon$. Since Φ is $1-1$ on W_ϵ it maps W_ϵ homeomorphically onto $\Phi(W_\epsilon)$. Consequently the restriction of a function g in $A_{\bar{Y}}$ to W_ϵ has the form $g'_\epsilon \circ \Phi$ where g'_ϵ is a continuous function on $\Phi(W_\epsilon)$. We extend g'_ϵ to a continuous function g''_ϵ on all of $\Omega_{\bar{X}}$. We then construct the functions $g_\epsilon = g''_\epsilon \circ \Phi$; g_ϵ will be identical to g on W_ϵ. Hence $P(|g - g_\epsilon| > \epsilon) < \epsilon$ so that $g_\epsilon \longrightarrow g$ in measure. Since $g_\epsilon = g''_\epsilon \circ \Phi$, the $g_\epsilon \in A_{\bar{X}}$ and this proves the lemma.

We may now prove

THEOREM 26.1. Let X be a G-valued process which is ergodic and λ-ergodic for every rational λ. If X is equidistributed in G then there is a unique process Z whose variables satisfy $z_{n+1} = x_{n+1} z_n$ and if $\xi(n)$ is generic for X then any solution $\zeta(n)$ to $\zeta(n+1) = \xi(n+1)\zeta(n)$ is generic for this Z. In case G is the circle group it suffices that X be ergodic. The resulting process Z will in both cases again be ergodic and equidistributed in G.

PROOF. Let Ω be the set of all left-infinite G-sequences $\omega(n)$ in G_∞^- for which $\omega(n)\omega(n-1)^{-1}$ is a sequence in $\Omega_{\bar{X}}$. Also let Φ be

the T-map of Ω onto $\Omega_{\overline{X}}$ defined by $(\Phi\omega)(n) = \omega(n)\omega(n - 1)^{-1}$. Then
any solution Z to the functional equation in question must exist on Ω
and must satisfy $X = \Phi(Z)$. We shall show that under the hypotheses of
the theorem there is a unique ergodic solution Z to $X = \Phi(Z)$ on Ω
and that this Z is equidistributed itself. By Lemma 25.1 it will
follow that there is a unique solution Z to $X = \Phi(Z)$ (ergodic or not)
and so we may apply Lemma 25.1 which asserts that if $\Phi(\zeta') = \xi$ and
$\zeta' \in \Omega_{\overline{Z}}$ for this unique Z then ζ' is generic for Z. But if it has
been shown that Z is equidistributed then any $\sigma\zeta'$, for $\sigma \in G$, is
generic for $\sigma Z \cong Z$. Hence since any solution $\zeta(n)$ to $\zeta(n + 1) =$
$\xi(n + 1)\zeta(n)$ has the form $\zeta(n) = \sigma\zeta'(n)$ where ζ' is a particular
solution, it follows that $\zeta(n)$ is itself generic for Z. Thus it is
only necessary to prove that there is a unique ergodic Z satisfying
$\Phi(Z) = X$ and that this Z is equidistributed. Moreover these two
statements are both equivalent. For suppose we know that Z is the
unique solution to $\Phi(Z) = X$. Then since each σX for $\sigma \in G$ is another solution it follows that $\sigma X \cong X$. Conversely suppose Z is an
equidistributed ergodic solution to $\Phi(Z) = X$, and suppose Z' were
some other ergodic solution. Since both Z and Z' are ergodic and
almost every sample point of an ergodic process is generic for it there
will be a pair of points $\overline{\zeta}$ and $\overline{\zeta}'$ in Ω the first generic for Z
the second for Z' with $\Phi(\overline{\zeta}) = \Phi(\overline{\zeta}')$. But then $\overline{\zeta}' = \sigma\overline{\zeta}$ and since $\overline{\zeta}$
is generic for Z, $\overline{\zeta}'$ is generic for $\sigma Z \cong Z$. Thus $Z' \cong Z$ and the
solution to $\Phi(Z) = X$ is unique. We may therefore suppose that Z is
some ergodic solution to $\Phi(Z) = X$; to prove the theorem we have only to
prove that Z is equidistributed.

  By the definition of Ω it is clear that if $\omega \in \Omega$ and $\sigma \in G$
then $\sigma\omega \in \Omega$ also; hence we may think of G as operating on Ω and
on $C(\Omega)$. We shall denote the latter operation by $\overline{\sigma}$ for $\sigma \in G$:
$\overline{\sigma}f(\omega) = f(\sigma\omega)$. The mapping $g \longrightarrow g \circ \Phi$ for $g \in A_{\overline{X}}$ imbeds the algebra $A_{\overline{X}}$ into $C(\Omega)$ and we shall denote this imbedding by $j : A_{\overline{X}} \longrightarrow C(\Omega)$
so that $j(g) = g \circ \Phi$. If f has the form $f = j(g)$ then $\overline{\sigma}f(\omega) =$
$f(\sigma\omega) = g(\Phi(\sigma\omega))$ and $\Phi(\sigma(\omega) = \Phi(\omega)$, so that $\overline{\sigma}f(\omega) = f(\omega)$. Conversely
if $\overline{\sigma}f = f$ for all $\sigma \in G$ we may conclude that f is in the image of
j. For if $\Phi(\omega_1) = \Phi(\omega_2)$ then $\omega_1 = \sigma\omega_2$ for some $\sigma \in G$ so that if
$\overline{\sigma}f = f$ then $f(\omega_1) = f(\omega_2)$. Then since $\Phi(\omega_1) = \Phi(\omega_2)$ implies that
$f(\omega_1) = f(\omega_2)$ it follows that $f = g \circ \Phi = j(g)$ as claimed.

  Let us return to the process Z satisfying $\Phi(Z) = X$ and
suppose that Z is ergodic. As a process on Ω, Z determines a functional-
al E_Z on $C(\Omega)$. The process Z will be equidistributed if and only if
$E_Z(\overline{\sigma}f) = E_Z(f)$ for all $\sigma \in G$ and $f \in C(\Omega)$. For the process σZ with
variables σz_n will have the same distributions as the process Z if and

only if the functions $f(z_{i_1}, \ldots, z_{i_r})$ and $f(\sigma z_{i_1}, \ldots, \sigma z_{i_r}) = \bar{\sigma}f(z_{i_1}, \ldots, z_{i_r})$ have the same expectations and this means that $E_Z(f) = E_Z(\bar{\sigma}f)$ for all f in $C(\Omega)$.

We may therefore assume that the closed subgroup H of G defined by

$$H = \{\sigma \in G : E_Z(\bar{\sigma}f) = E_Z(f) \text{ for all } f \in C(\Omega)\}$$

does not coincide with G. From this we shall derive a contradiction. First we note that if $\sigma \in H$ then $\bar{\sigma}$ may be thought of as acting on A_Z^-. For A_Z^- is the quotient algebra of $C(\Omega)$ by the ideal of functions f for which $E_Z(ff^*) = 0$ (i.e., the functions with support outside of the support of the measure P_Z corresponding to E_Z). But if $E_Z(ff^*) = 0$ then $E_Z(\bar{\sigma}f \cdot \bar{\sigma}f^*) = 0$ if $\sigma \in H$ so that the ideal of functions is carried into itself by H and therefore so is A_Z^-.

This being the case we may consider the subset of A_Z^- of functions left fixed by H:

$$A_Z^H = \{f \in A_Z^- : \bar{\sigma}f = f \text{ for all } \sigma \in H\} \ .$$

We thus have $j(A_X^-) \subset A_Z^H \subset A_Z^-$. We prove next that $A_Z^H \subset L(j(A_X^-))$ or that A_Z^H is an L-extension of $j(A_X^-)$. For this suppose that ω_1 and ω_2 are two points of Ω generic for Z. If $\Phi(\omega_1) = \Phi(\omega_2)$ then $\omega_2 = \sigma(\omega_1)$ for some $\sigma \in G$. From this we may conclude that $\sigma \in H$. For since ω_1 and ω_2 are generic for Z we have

$$E_Z(\bar{\sigma}f) = \lim_{N \to \infty} \frac{1}{N+1} \sum_0^N \bar{\sigma}f(T^n\omega_1) = \lim_{N \to \infty} \frac{1}{N+1} \sum_0^N f(T^n\omega_2)$$

$$= E_Z(f)$$

since $\sigma T^n = T^n \sigma$. Consequently $f(\omega_1) = f(\omega_2)$ for all f in A_Z^H. Thus if two generic points of Ω agree on all functions of $j(A_X^-)$ they agree for all functions in A_Z^H. It follows that two homomorphisms of A_Z^H that are restrictions of generic points of Z and are distinct, must remain distinct when restricted to $j(A_X^-)$. Taking S to be the set of points of the homomorphism space of A_Z^H that are restrictions of generic points of Z we find that S has measure 1 since Z is generic and on S the canonical map to the homomorphism space of the subalgebra $j(A_X^-)$ is $1 - 1$. Hence we may apply Lemma 26.1 to infer that $A_Z^H \subset L(j(A_X^-))$.

Since G is an abelian group and $H \neq G$ there exists a character χ on G which is identically 1 on H but not identically 1 throughout G. We consider the numerical variables g_n defined by $g_n = \chi \circ z_n$, the z_n being G-valued. If $\sigma \in H$ then

$$\bar{\sigma}g_n(\omega) = \chi(z_n(\sigma\omega)) = \chi(\sigma z_n(\omega)) = \chi(\sigma)\chi(z_n(\omega)) = \chi(\sigma)g_n(\omega)$$

and since $\chi(\sigma) = 1$ for $\sigma \in H$, $\bar{\sigma}g_n = g_n$. Hence $g_n \in A_Z^H$ and so also $g_n \in L(j(A_X^-))$.

We now recall that the process X is equidistributed so that if $\omega \in \Omega_X^-$ and $\sigma \in G$ then $\sigma\omega \in \Omega_X^-$ and if $\hat{\sigma}$ denotes the induced operation on A_X^- then $\hat{\sigma}$ preserves the expectation functional. Since A_X^- is isomorphic to $j(A_X^-)$ we may suppose the operations σ are defined on $j(A_X^-)$. The operations $\hat{\sigma}$ and $\bar{\sigma}$ are not to be confused; the former arise from multiplication of group elements with sequences in Ω_X^- and the latter arise from multiplication with sequences in Ω. In particular each $\bar{\sigma}$ leaves $j(A_X^-)$ fixed whereas $\hat{\sigma}$ in general does not.

Now the fact that $\hat{\sigma}$ preserves expectation implies that it can be extended to $L(j(A_X^-))$; in fact, $\hat{\sigma}$ gives an isometry of a dense subspace of the Hilbert space $L^2(j(A_X^-))$ and so extends to all of $L^2(j(A_X^-))$ as an isometry and in particular it is defined on $L(j(A_X^-))$. As we have shown $g_n \in A_Z^H \subset L(j(A_X^-))$ and so it follows that $\hat{\sigma}$ is defined on the functions g_n. Since the character χ is not identically 1 there is an element $\rho \in G$ with $\chi(\rho) \neq 1$. Recalling that $z_n z_{n-1}^{-1} = j(x_n)$ by definition of j we will have

$$\hat{\rho}(g_n g_{n-1}^{-1}) = \hat{\rho}((\chi \circ z_n)(\chi \circ z_{n-1}^{-1})) = \hat{\rho}(\chi \circ (j(x_n))) \quad .$$

Clearly the operation $\hat{\rho}$ on A_X^- satisfies $\hat{\rho}(f(x_n)) = f(\rho x_n)$ and so $\hat{\rho}(\chi \circ (j(x_n))) = \chi(\rho \cdot j(x_n)) = \chi(\rho)\chi(j(x_n))$. On the other hand $g_n g_{n-1}^{-1} = \chi(j(x_n))$ so that we have

$$\hat{\rho}(g_n g_{n-1}^{-1}) = \chi(\rho)g_n g_{n-1}^{-1} \quad ,$$

and this may be rewritten as

$$(26.1) \qquad \hat{\rho}(g_n)g_n^{-1} = \chi(\rho)\hat{\rho}(g_{n-1})g_{n-1}^{-1} = \chi(\rho)T\{\hat{\rho}(g_n)g_n^{-1}\} \quad .$$

But g_n and $\hat{\rho}(g_n)$ are both in $L(j(A_X^-))$ and $|g_n| = 1$ so that $\hat{\rho}(g_n)g_n^{-1} \neq 0$. (26.1) therefore shows that X is not λ-ergodic for λ satisfying $\chi(\rho)^{-1} = e^{2\pi i\lambda}$. To complete the proof of the theorem for the case of an arbitrary compact abelian group G we show that ρ may be chosen so that λ is rational. Now equation (26.1) holds for any $\rho \in G$. Hence if $x(\rho)$ is not already a root of unity the powers $x(\rho^n) = x(\rho)^n$ may be made to approximate any value on the unit circle and in particular a root of unity. So if $x(\rho)^{n_k} \longrightarrow \zeta$ where ζ is a root of unity then some subsequence of ρ^{n_k} will converge in G to an element $\rho_0 \in G$ and $x(\rho_0) = \zeta$. Hence in every case there will be in $L(A_X^-) \cong L(j(A_X^-))$ a solution to $Tf = e^{2\pi i \lambda}f$ with λ rational.

In case G is the circle group, $G = \{z : |z| = 1\}$, the only proper closed subgroups H of G are those consisting of the qth roots of unity for some q. We may then take for x the character $x(\sigma) = \sigma^q$ and as ρ ranges over G, $x(\rho)$ will range over the unit circle. Thus unless $H = G$ (26.1) will show that X is not λ-ergodic for any λ.

However it is easy to see that the set of λ for which X is not λ-ergodic is at most denumerable. For, the Hilbert space $L^2(A_X^-)$ is separable and hence a set of mutually orthogonal elements in $L^2(A_X^-)$ can be at most denumerably infinite. But if $Tf_\alpha = e^{2\pi i \alpha}f_\alpha$ for a set $\{f_\alpha\}$ then for f_α and f_β

$$(26.2) \qquad E(Tf_\alpha \cdot \overline{Tf_\beta}) = e^{2\pi i (\lambda_\alpha - \lambda_\beta)} E(f_\alpha \overline{f_\beta})$$

and also

$$(26.3) \qquad E(Tf_\alpha \cdot \overline{Tf_\beta}) = E(f_\alpha \cdot \overline{f_\beta})$$

since T is measure preserving. Hence for $\lambda_\alpha \neq \lambda_\beta$, $E(f_\alpha \cdot \overline{f_\beta}) = 0$ and so the f_α for distinct λ_α are mutually orthogonal and hence can be at most denumerably infinite in number. This completes the proof of the theorem.

In case G is not the circle group we remark that the condition of ergodicity of X is, in general, not sufficient for the conclusion of the theorem. For example if we take $\zeta(n)$ to be the periodic sequence

$$\zeta : \ldots 1, 1, 1, -1, 1, 1, 1, -1, 1, 1, 1, -1$$

then $Z = X(\zeta)$ is not an equidistributed process in the group $G = \{1, -1\}$. However the sequence $\xi(n) = \zeta(n)\zeta(n-1)^{-1}$ is given by

$$\xi : \ldots 1, 1, -1, -1, 1, 1, -1, -1, 1, 1, -1, -1$$

which is equidistributed. Hence $X = X(\xi)$ is ergodic and equidistributed but $Z = X(\zeta)$ is not equidistributed although X and Z satisfy $z_{n+1} = x_{n+1}z_n$. In this case the conclusion of the theorem does not hold because $X(\xi)$ is not $1/4$-ergodic as may easily be verified. Thus the assumption of λ-ergodicity is essential in general.

26.2. <u>Applications to Equidistribution</u>. From the theorem proven in the preceding subsection we can deduce a result of H. Weyl regarding equidistribution ([13]). Namely we can show that if $p(t)$ is a polynomial with real coefficients of which at least one is irrational and $\xi(n) = e^{2\pi i p(n)}$ then $\xi(n)$ is equidistributed on the unit circle. By "equidistributed" we mean that for any continuous function $f(z)$ on the circle $|z| = 1$ we have

$$(26.4) \qquad \lim_{N \to \infty} \frac{1}{N+1} \sum_{n=0}^{N} f(\xi(n)) = \frac{1}{2\pi} \int_{0}^{2\pi} f(e^{i\theta})d\theta$$

Suppose that $p(t) = a_0 t^m + a_1 t^{m-1} + \ldots + a_m$ and that a_r is the first irrational coefficient. The coefficients $a_0, a_1, \ldots, a_{r-1}$ being rational, we may choose an integer k such that $ka_0, ka_1, \ldots, ka_{r-1}$ are all integers. It follows that $p(kn + \ell) \equiv a_0 \ell^m + \ldots + a_{r-1}\ell^{m-r+1} + a_r(kn + \ell)^{m-r} + \ldots + a_m$ modulo 1 and hence for ℓ fixed

$$(26.5) \qquad p(kn + \ell) \equiv b_r n^{m-r} + \ldots + b_m \; (\text{mod } 1)$$

where the leading coefficient b_r is irrational. Now set $\xi_\ell(n) = \xi(kn + \ell)$. Clearly if (26.4) holds for each of the sequences $\xi_\ell(n)$, $\ell = 0, 1, \ldots, k - 1$ then it will hold for $\xi(n)$. Thus it suffices to prove that each $\xi_\ell(n)$ is equidistributed on the unit circle. We now have, however, that $\xi_\ell(n)$ is given by a polynomial $p_\ell(n)$ whose leading coefficient in irrational ((26.5)). The theorem of H. Weyl will therefore follow from the following theorem.

THEOREM 26.2. If $p(t)$ is a polynomial whose coefficients are real and whose leading coefficient is irrational then $\xi(n) = e^{2\pi i p(n)}$ is "strongly equidistributed" in the sense that $\xi(n)$ is regular and generic for a process X that is equidistributed in the circle group.

PROOF. By Theorem 26.1 we know that $\xi(n)$ will be generic for a process that is equidistributed and ergodic if this holds for the derived sequence $\xi'(n) = \xi(n)\xi(n - 1)^{-1}$. However $\xi'(n)$ corresponds to the polynomial $p'(t) = p(t) - p(t - 1)$ in the same way as $\xi(n)$

corresponds to $p(t)$. Moreover $p'(t)$ is a polynomial with the same leading coefficient as $p(t)$ but of degree one less than that of $p(t)$. It follows that if the theorem is proven for polynomials of degree 1 it will follow for all polynomials (of degree > 0). However for $p(t) = \alpha t + \beta$ and α irrational, the theorem is easily proven. Namely consider the process X with variables x_n having values in $G = \{z : |z| = 1\}$ defined by: $x_{n+1} = e^{2\pi i \alpha} x_n$ and x_0 is uniformly distributed in G, i.e., $P(x_n \in \Delta) = m(\Delta)$ where m is the normalized Lebesgue measure on the circle. X is obviously equidistributed and it is ergodic since Ω_X may be identified with G and T consists of an irrational rotation of G ([7]). Clearly $\xi(n) = e^{2\pi i p(n)}$ is a sample sequence of X, so to prove the theorem it suffices to show that $\xi(n)$ is generic for X. But since X is ergodic some points of Ω_X^- must be generic for X and since X is equidistributed if $\omega(n)$ is generic for X so is $\sigma\omega(n)$ for $\sigma \in G$ and hence every point of Ω_X^- is generic since any point of Ω_X^- can be obtained from any other point by multiplication by a group element.

Before closing this section we state another theorem similar to Theorem 26.1 but a good deal easier to prove. The details of the proof will be evident from the proof of Theorem 26.1 and will therefore be omitted.

> THEOREM 26.3. Let X and Z be G-valued processes and suppose that $\xi(n)$ is generic for X. If Z satisfies $z_{n+1} = x_{n+1} z_n$ and Z is ergodic and equidistributed in G then Z is the only solution to the functional equation and any G-sequence $\zeta(n)$ satisfying $\zeta(n + 1) = \xi(n + 1)\zeta(n)$ is generic for Z.

§27. Periodic Subsequences of Regular Sequences

27.1. <u>Connection with Inductive Functions</u>. By a periodic subsequence of a sequence $\xi(n)$ we mean a sequence $\zeta'(n)$ given by

$$(27.1) \qquad\qquad \xi'(n) = \xi(kn + \ell)$$

for fixed k and ℓ. Suppose that $\xi(n)$ is a regular sequence; it is natural to inquire about the behavior of a periodic subsequence $\xi'(n)$. In particular will $\xi'(n)$ again be regular and if so what is the relationship between $X(\xi')$ and $X(\xi)$? In this section we shall show how the problem may be dealt with by the methods of the preceding sections. We shall first see how this problem is related to inductive functions of processes and sequences.

Suppose that X is a process with variables x_n and that Z is a composite process with variables (x_n, y_n) where the y_n are G-valued for a certain compact abelian group G and the y_n satisfy a functional relationship

$$(27.2) \qquad\qquad\qquad y_{n+1} = \sigma y_n$$

where σ is a fixed element of G. The process Z is then an inductive extension of X since

$$(27.3) \quad z_{n+1} = (x_{n+1}, y_{n+1}) = (x_{n+1}, \sigma y_n) = \psi(x_{n+1}, (x_n, y_n)) = \psi(x_{n+1}, z_n).$$

DEFINITION 27.1. An inductive extension Z of X with variables (x_n, y_n) satisfying (27.2) for a given group G and $\sigma \in G$ will be called a σ-<u>process over</u> X.

The sample sequences of Z will have the form $(\xi(n), \eta(n))$ where $\eta(n)$ is a G-sequence satisfying

$$(27.4) \qquad\qquad\qquad \eta(n + 1) = \sigma \eta(n) \quad .$$

We shall refer to a σ-sequence of this kind as a σ-<u>sequence over</u> $\xi(n)$; every such sequence has the form

$$\zeta : \ldots (\xi(-2),\ \sigma^{-2}\tau),\ (\xi(-1),\ \sigma^{-1}\tau),\ (\xi(0),\ \tau)$$

for some $\tau \in G$. Clearly there will be a σ-sequence over $\xi(n)$ for every element of G.

Consider now the case that G is the group of kth roots of unity and σ is a generator of G. Suppose we have a regular sequence $\xi(n)$ such that each of the k σ-sequences over $\xi(n)$ is regular and suppose moreover that they are all generic for the same σ-process over $X(\xi)$. Let $\zeta(n) = (\xi(n), \eta(n))$ be a particular σ-sequence over $\xi(n)$. Since $\zeta(n)$ is regular the averages of the sequences

$$(27.5) \quad \theta_\nu(n) = (\eta(n)\eta(0)^{-1}\sigma^{-\ell})^\nu \varphi(\xi(n),\ \xi(n-k),\ \ldots,\ \xi(n-rk))$$

exist for $\nu = 0, 1, \ldots, k-1$, where φ is a continuous function on Λ^{r+1}, Λ being the range of X. Moreover these averages are the same for $\zeta(n)$ as for any other of the σ-sequences over $\xi(n)$, and so if we replace $\eta(n)$ (but not $\eta(0)!$) by $\sigma\eta(n)$ the resulting sequences will have the same averages as $\theta_\nu(n)$. But for $\nu = 1, \ldots, k - 1$, the resulting sequences will be $\sigma^\nu \theta_\nu(n)$ which cannot have the same averages as $\theta_\nu(n)$ unless each of these $\theta_\nu(n)$ has average 0. It follows from this that the sequence $\theta(n) = \theta_0(n) + \theta_1(n) + \ldots + \theta_{k-1}(n)$ has the same average as does the sequence $\theta_0(n)$. But the general term of $\theta(n)$ will be 0 unless

$$\sum_{\nu=0}^{k-1} (\eta(n)\eta(0)^{-1}\sigma^{-\ell})^\nu \neq 0$$

i.e., unless $\eta(n)\eta(0)^{-1}\sigma^{-\ell} = 1$ or $\eta(n) = \eta(0)\sigma^\ell$ and $n = \ell \pmod{k}$ in which case $\theta(n) = k\varphi(\xi(n), \xi(n - k), \ldots, \xi(n - rk))$. If we now define $\xi'(n)$ as the periodic subsequence of $\xi(n)$ given in (27.1) we find that the sequence $\theta(n)$ consists of the entries $k\varphi(\xi'(n), \xi'(n - 1), \ldots, \xi'(n - r))$ with $k - 1$ zeros between successive terms. What we have then shown is that if $\xi(n)$ has the properties assumed then the average of

$$\varphi(\xi'(n), \xi'(n - 1), \ldots, \xi'(n - r))$$

is identical to that of

$$\varphi(\xi(n), \xi(n - k), \ldots, \xi(n - rk)) \ .$$

It follows from this that $\xi'(n)$ is regular and generic for the process $X_k'(\xi)$ obtained from $X(\xi)$ by setting

$$x_n' = x_{nk} \ .$$

Thus the problem we originally posed will be completely solved in this case. The main result of this section will provide us with conditions under which this case actually occurs. What we have just shown is that under such conditions the periodic subsequences of $\xi(n)$ will again be regular and generic for a well-determined process.

Before proceeding to our main theorem we give an example showing that if no conditions other than regularity are imposed on $\xi(n)$ then the periodic subsequences of $\xi(n)$ need not be regular. For this we consider a sequence $\xi^{(0)}(n)$ defined as follows: the subsequence $\xi^{(0)}(2n)$ is defined by $\xi^{(0)}(2n) = \eta_1(n)$ where $\eta_1(n)$ is a random $\{1, -1\}$-sequence and $\xi^{(0)}(2n + 1) = \eta_2(n)$ where $\eta_2(n)$ is another random $\{1, -1\}$-sequence, but where $\eta_1(n)$ and $\eta_2(n)$ are not generic for the same process. That is, the frequency of 1's and -1's in $\eta_1(n)$ and $\eta_2(n)$ are different. If we choose $\eta_1(n)$ and $\eta_2(n)$ such

that (η_1, η_2) is generic for a random $\{(\pm 1, \pm 1)$-process then the
resulting sequence $\xi^{(o)}(n)$ will be regular, as is easily verified.

Now let $\{r_k, k \leq 0\}$ and $\{s_k, k \leq 0\}$ be two sequences of
negative integers with upper density 0 satisfying $r_k < s_k < r_{k+1}$ for
each k. Define the sequence $\xi(n)$ by setting

$$
\xi(n) = \begin{cases} \xi^{(o)}(n) & \text{for } r_k \leq n < s_k \\[2ex] \xi^{(o)}(n+1) & \text{for } s_k \leq n < r_{k+1} \end{cases} \quad .
$$

Since we assume that the $\{r_k\}$ and $\{s_k\}$ have density 0 it can be
seen that the transformation from $\xi^{(o)}(n)$ to $\xi(n)$ does not alter the
average of the derived sequences and $\xi(n)$ will again be regular. (The
regularity of both $\xi^{(o)}(n)$ and $\xi(n)$ follows, granting that they are
stochastic, from the fact that $\Omega^-(\xi^{(o)}) = \Omega^-(\xi) = \{-1, 1\}^-_\infty$.) On the
other hand $\xi(2n)$ is alternately chosen from η_1 and η_2 according
as $r_k \leq 2n < s_k$ or $s_k \leq 2n < r_{k+1}$. Since $\eta_1^{(n)}$ and $\eta_2^{(n)}$ are not
generic for the same process it follows that by choosing the sequences
$\{r_k\}$ and $\{s_k\}$ sufficiently thin we can cause the averages of $\xi(2n)$
to oscillate between different limits and thereby arrange that $\xi'(n) =$
$\xi(2n)$ will not be stochastic.

27.2. <u>Regularity of σ-Sequences</u>. We have seen that a sufficient
condition for the regularity of periodic subsequences of a regular sequence is the regularity of the σ-sequences over the given sequence and
their representing the same process, where σ is a root of unity. The
condition for this is given in the next theorem.

> THEOREM 27.1. Let $\xi(n)$ be a regular left-infinite
> sequence and let $\sigma \in G$, a compact abelian group
> such that σ generates G (i.e., no proper closed
> subgroup of G contains σ). Let $L(\sigma)$ be the
> set of λ, $0 < \lambda < 1$ such that $\chi(\sigma) = e^{2\pi i\lambda}$ for
> some $\chi \in \hat{G}$ the character group of G. If $\xi(n)$
> is ergodic and λ-ergodic for every $\lambda \in L(\sigma)$ then
> there will be just one σ-process over $X(\xi)$ and
> every σ-sequence over $\xi(n)$ will be regular and
> generic for this σ-process.

PROOF. Let us write X for $X(\xi)$ so that $\xi(n)$ is a generic
point of X. Let Ω be the set of all σ-sequences associated with sequences of Ω^-_X so that if $\omega(n)$ is a sequence in Ω^-_X then

$$\ldots \ (\omega(-2),\ \sigma^{-2}\tau),\ (\omega(-1),\ \sigma^{-1}\tau),\ (\omega(0),\ \tau)$$

is a sequence in Ω. Ω is a T-space and if Φ is the natural projection
from Ω to $\Omega_{\bar{X}}$ then Φ is a T-map. Clearly any σ-process over X
must exist on Ω and must satisfy $X = \Phi(Z)$ so that our theorem asserts
that there is only one solution on Ω to $X = \Phi(Z)$.

For each $\tau \in G$ let $\bar{\tau}$ denote the operation on Ω sending
the sequence $(\omega(n),\ \eta(n))$ into the sequence $(\omega(n),\ \tau\eta(n))$ and let $\bar{\tau}$
also denote the induced transformation of $C(\Omega)$. Suppose now that it
has been shown that if Z is an ergodic solution to $X = \Phi(Z)$ then Z
has the property that $E_Z(\bar{\tau}f) = E_Z(f)$ for every $\tau \in G$, and $f \in C(\Omega)$,
where E_Z is the expectation functional of Z. We claim that the theo-
rem will follow from this. For first of all this implies that there is
a unique ergodic solution to $X = \Phi(Z)$. Namely if Z and Z' are two
ergodic solutions we may find, by the ergodic theorem, a pair of points
$\bar{\zeta}$ and $\bar{\zeta}'$ in Ω with $\Phi(\bar{\zeta}) = \Phi(\bar{\zeta}')$ and $\bar{\zeta}$ generic for Z and $\bar{\zeta}'$
generic for Z'. But since $\Phi(\bar{\zeta}') = \Phi(\zeta)$ it follows that $\bar{\zeta}' = \bar{\tau}\bar{\zeta}$
for some $\tau \in G$ so that

$$E_{Z'}(f) = \lim_{N \to \infty} \frac{1}{N+1} \sum_{n=0}^{N} f(T^n\bar{\zeta}') = \lim_{N \to \infty} \frac{1}{N+1} \sum_{n=0}^{N} \bar{\tau}f(T^n\bar{\zeta}) =$$

$$= E_Z(\bar{\tau}f) = E_Z(f)$$

and $Z = Z'$. Alternatively we may argue that $E_Z(f) = E_Z(\bar{\tau}f)$ for all
$\tau \in G$ implies that

$$E_Z(f) = E_Z \left(\int_G \bar{\tau}f \ d\tau \right)$$

where the integral has the obvious meaning. But

$$\int_G \bar{\tau}f \ d\tau$$

is invariant under all the operations $\bar{\rho}$ for $\rho \in G$ and it is easy to
see that this implies that

$$\int_G \bar{\tau}f \ d\tau$$

has the form $g \circ \Phi$ for $g \in A_X^-$. But then $E_Z(f) = E(g)$ so that Z is uniquely determined.

Since there exists a unique ergodic solution to $X = \Phi(Z)$ by Lemma 25.1 there will exist a unique solution. Thus the first assertion of the theorem will be proven. Moreover by Lemma 25.2 any sequence $\zeta(n)$ with $\Phi(\zeta) = \xi$ and $\zeta \in \Omega_Z^-$ will be generic for Z. However since $E_Z(\bar{\tau}f) = E_Z(f)$ it follows that $\bar{\tau}\Omega_Z^- = \Omega_Z^-$ so that if any preimage of ξ under Φ is in Ω_Z^- all preimages will be. Since Φ takes Ω_Z^- onto Ω_X^- it follows that there is some ζ satisfying $\Phi(\zeta) = \xi$ and $\zeta \in \Omega_Z^-$; hence all ζ in $\Phi^{-1}(\xi)$ are in Ω_Z^- and by Lemma 25.2 all are generic for Z. This would then complete the proof of the theorem.

We need therefore only establish that if Z is an ergodic solution to $X = \Phi(Z)$ where X satisfies the hypotheses of the theorem, then $E_Z(f) = E_Z(\bar{\tau}f)$ for $\tau \in G$ and $f \in C(\Omega)$. Assume this is not the case and let H be the closed subgroup of G given by

$$H = \{\tau \in G : E_Z(\bar{\tau}f) = E_Z(f) \text{ for all } f \in C(\Omega)\} \quad .$$

The transformations $\bar{\tau}$ for $\tau \in H$ act on A_Z^- since $A_Z^- = C(\Omega)/J_Z$ where $J_Z = \{f \in C(\Omega) : E_Z(ff^*) = 0\}$ and clearly f is invariant under $\bar{\tau}$ for $\tau \in H$. We may therefore define a subalgebra of A_Z^- by

$$A_Z^H = \{f \in A_Z^- : \bar{\tau}f = f \text{ for all } \tau \in H\} \quad .$$

The algebra A_X^- is isomorphically imbedded in A_Z^- by the mapping $g \longrightarrow g \circ \Phi \longrightarrow g \circ \Phi + J_Z$; we denote this imbedding by j so that $j(A_X^-) \subset A_Z^H$. Clearly we have $j(A_X^-) \subset A_Z^H \subset A_Z^-$. By the same method used in Theorem 26.1 we may show that $A_Z^H \subset L(j(A_X^-))$. Namely by Lemma 26.1 it suffices to show that for a set S of measure 1 of homomorphisms of A_Z^H no two homomorphisms of S restrict to the same homomorphism of $j(A_X^-)$. We take the set S to be the restrictions of generic points of A_Z^-. If two such points ζ_1 and ζ_2 restrict to the same homomorphism of $j(A_X^-)$ then as points of Ω, $\zeta_2 = \bar{\tau}\zeta$, for some τ in G. But using the assumption that ζ_1 and ζ_2 are generic this implies that $E_Z(\bar{\tau}f) = E_Z(f)$ for all $f \in C(\Omega)$ so that $\tau \in H$. But if $\tau \in H$ and $f \in A_Z^H$ then $f(\bar{\tau}\zeta) = f(\zeta)$; hence $f(\zeta_2) = f(\zeta_1)$ for all $f \in A_Z^H$ so that ζ_1 and ζ_2 restrict to the same homomorphism of A_Z^H. The set S therefore satisfies the conditions of Lemma 26.1 and so $A_Z^H \subset L(j(A_X^-))$.

Under the assumption that H is a proper subgroup of G we can find a character χ on G that is not identically 1 but which takes on the value 1 everywhere on H. In particular $\chi(\sigma) \neq 1$ since σ generates G. Define the function $\psi : \Omega \longrightarrow G$ by $\psi(\omega, \eta) = \eta(0)$

where $\eta(n)$ is the G-sequence corresponding to the point of Ω in question. Composing X with ψ we obtain a function in $C(\Omega)$ which corresponds to an element in A_Z^-; we denote the latter as $X \circ \psi$. We shall show that actually $X \circ \psi \in A_Z^H$. To see this suppose that $\tau \in H$; then $X(\tau) = 1$ so that

$$\bar{\tau}X \circ \psi(\omega, \eta) = X \circ \psi(\omega, \tau\eta) = X(\tau\eta(0)) = X(\tau)X(\eta(0)) = X(\eta(0)) =$$

$$X \circ \psi(\omega, \eta)$$

and $\bar{\tau}X \circ \psi = X \circ \psi$. By what we have shown this implies that $X \circ \psi \in L(j(A_X^-))$.

We next evaluate $T(X \circ \psi)$. For this we recall the definition of the shift operator on Ω which gives $T(\omega, \eta) = (T\omega, T\eta) = (T\omega, \sigma^{-1}\eta)$. Hence

$$T(X \circ \psi)(\omega, \eta) = X \circ \psi(T\omega, \sigma^{-1}\eta) = X(\sigma^{-1}\eta(0)) = X(\sigma^{-1})X(\eta(0)) =$$

$$X(\sigma^{-1})X \circ \psi(\omega, \eta)$$

so that

$$T(X \circ \psi) = X(\sigma^{-1})(X \circ \psi) \quad .$$

Since $|X \circ \psi| = 1$ so that $X \circ \psi \neq 0$ and $X(\sigma^{-1}) \neq 1$ it follows that there is a λ in $L(\sigma)$ such that X is not λ-ergodic. This contradicts the hypothesis of the theorem and so our assumption that H is a proper subgroup of G is false. We therefore have $H = G$ and as we have seen this implies the result of the theorem.

It might be remarked that the assumption that σ generates G is not an essential one. If σ generates a subgroup G_1 of G then any σ-sequence will have its "G-component" lying in a coset of $G \bmod G_1$ and from the theorem just proven it is quite easy to deduce that with the same hypothesis a σ-process over X with the G-values being taken in a single coset of $G \bmod G_1$ is unique and any such σ-sequence over a generic point of X is generic for the unique σ-process.

27.3. <u>Existence of Fourier Coefficients</u>. We have seen in §27.1 that the foregoing theorem may be applied to the question of regularity of the periodic subsequences of a regular sequence. A more obvious application is to the existence of "Fourier coefficients" of certain numerically valued sequences. If $\xi(n)$ is a numerically valued sequence, we say that C_λ is the λth Fourier coefficient of $\xi(n)$ if

$$(27.6) \qquad C_\lambda = \lim_{N \to \infty} \frac{1}{N+1} \sum_{N=-N}^{0} \xi(n) e^{-2\pi i n\lambda}$$

For a regular sequence $\xi(n)$ the limit in (27.6) need not exist in general, however it will exist if the composite sequence $\zeta(n) = (\xi(n), e^{2\pi i n\lambda})$ is regular. Now, the sequence $\zeta(n)$ is an $e^{2\pi i n\lambda}$-sequence over $\xi(n)$, in the sense of §27.1, with G the circle group if λ is irrational or a group of kth roots of unity if $\lambda = \ell/k$. Hence we may apply Theorem 27.1 which asserts that $\zeta(n)$ will be regular if $\xi(n)$ is λ'-ergodic for a certain set of λ'. Specifically, the condition is that $\xi(n)$ be λ'-ergodic for $\lambda' \in L(e^{2\pi i\lambda})$ where $L(e^{2\pi i\lambda}) = \{\chi(e^{2\pi i\lambda}),$ $\{\chi\epsilon$ character group of $G\}$. It is easily seen that both in the case that G is the circle group or the case that G is the group of kth roots of unity, $L(e^{2\pi i\lambda}) = \{e^{2\pi i m\lambda}, m = 0, \pm 1, \pm 2, \ldots\}$. Thus $\zeta(n)$ is regular if $\xi(n)$ is λ'-ergodic for all $\lambda' \equiv m\lambda \pmod{1}$. Moreover if this is the case then by Theorem 27.1 the sequences $\zeta_\tau(n) = (\xi(n), \tau e^{2\pi i n\lambda})$ for $\tau \in G$ are also regular and are generic for the same process $X(\zeta)$ for which ζ is generic. But this implies that the limit C_λ in (27.6) must be the same as τC_λ for each τ in G and so C_λ must vanish. Now we have observed in the proof of Theorem 26.1 that a process X will be λ'-ergodic for all but a denumerable number of λ'. It follows from this and the condition $\lambda' \equiv m\lambda \pmod{1}$ that for an arbitrary regular sequence $\xi(n)$, the Fourier coefficients C_λ exist and are 0 for all but a denumerable set of λ.

From this we may deduce a theorem of Wiener and Wintner ([16]) which asserts that if f is an L^1 function on a probability space Ω with an ergodic measure-preserving transformation T then for almost all ω in Ω __all__ the Fourier coefficients

$$C_\lambda = \lim_{N \to \infty} \frac{1}{N+1} \sum_{n=0}^{N} f(T^n\omega) e^{-2\pi i n\lambda}$$

exist simultaneously. For it may be easily shown that it suffices to know this for continuous functions f on compact Hausdorff spaces Ω and in this case the sequence $f(T^n\omega)$ may be taken as a sample sequence of a process X. Since we assume X is ergodic it suffices to establish that all the C_λ exist for almost all generic points of X. However for all but a denumerable set of λ, the C_λ exist for all generic points of X. For each of the exceptional λ, C_λ still exists by virtue of the ergodic theorem ([3], p. 469). It follows that all the C_λ exist for almost all generic points of X and this implies the aforementioned result.

It might be remarked that the theorem of Wiener and Wintner might also be deduced from Theorem 7.3 which gives a necessary and sufficient condition for λ-ergodicity.

CHAPTER 8. INDUCTIVE FUNCTIONS AND MARKOFF PROCESSES

In the preceding chapter we considered inductive functions of
a special kind, namely, the function ψ determining the relationship of
the two processes is given in terms of group multiplication. In the
present chapter a more general class of ψ will be admitted but the
ground process X will be assumed to be a Markoff process. Our discussion will be broken up into two parts; we shall first study the
processes that can arise as inductive functions of Markoff processes and
we then treat the relationship between such processes and the inductive
functions of individual Markoff sequences.

§28. Inductive Functions of Markoff Processes

28.1. _Preliminaries_. We shall continue using X to denote the
original process; Z will denote an inductive function of X. The range
of X will be Π and that of Z will be Λ. We shall assume throughout this chapter that the function ψ of the relationship $z_{n+1} =
\psi(x_{n+1}, z_n)$ is defined on all of $\Pi \times \Lambda$. As a result each $\pi \in \Pi$ de-
fines a mapping $\psi(\pi) : \Lambda \longrightarrow \Lambda$ satisfying

$$(28.1) \qquad\qquad \psi(\pi)(\lambda) = \psi(\pi, \lambda) \quad .$$

It will be found that the behavior of the inductive function Z will, in
the case treated in this chapter, depend primarily upon the nature of the
functions $\psi(\pi)$.

When we write $z_{n+1} = \psi(x_{n+1}, z_n)$ the variables x_n and z_n
are assumed to be defined on a common sample space which means that we
have in mind a composite process $Z^* = (X, Z)$ with variables (x_n, z_n).
Although the processes X and Z together do not determine the process
Z^* of which they are subprocesses (see the remark following Theorem 24.2)
when we consider a particular inductive function Z we shall suppose
that Z^* is fixed.

The main problem of this section will be to determine the nature
of the process Z (or, more generally, of Z^*) under the assumption that
X is a finitely-valued Markoff process. To give a more precise formulation of the problem we require the following basic definition:

DEFINITION 28.1. Let Π be a finite set, $\Pi = \{\pi_1, \pi_2, \ldots, \pi_m\}$; Λ a compact Hausdorff space, and let ψ be a map of $\Pi \times \Lambda$ into Λ. If X is a Π-valued Markoff process with transition probability matrix $(p_{ij} = P(x_{n+1} = \pi_j \mid x_n = \pi_i))$ then a process Z is a <u>standard ψ-inductive function of</u> X if the composite process (X, Z) is a Markoff process with transition probabilities

$$(28.2)\ P(x_{n+1} = \pi_j,\ z_{n+1} \in \Delta \mid x_n = \pi_i,\ z_n = \lambda) = \begin{cases} p_{ij} & \text{if } \psi(\pi_j, \lambda) \in \Delta \\ 0 & \text{if } \psi(\pi_j, \lambda) \notin \Delta \end{cases}$$

for $\Delta \subset \Lambda$.

The meaning of (28.2) is that the transition probabilities of (X, Z) are to be determined from those of X as follows: for a particular value (π_i, λ) of (x_n, z_n), the only choice that need be made is that of x_{n+1}, the value of z_{n+1} being obtained from $z_{n+1} = \psi(x_{n+1}, z_n)$. The point to be emphasized here is that although the relationship $z_{n+1} = \psi(x_{n+1}, z_n)$ is satisfied for any ψ-inductive function Z, it need not always be the case that (X, Z) is a Markoff process. We shall find however that under certain assumptions regarding ψ, (X, Z) will have to be a Markoff process and Z a standard ψ-inductive function, if Z is at all a ψ-inductive function of X. In this case the process (X, Z) and hence Z will be essentially determined by the process X in the same sense that a stationary Markoff process is determined by its transition probabilities, i.e., determined up to the stationary measure on the state space. For this reason we shall find it useful to have conditions under which a ψ-inductive function of X is a standard ψ-inductive function.

It will be necessary for us on several occasions to know that if (X, Z) has transition probabilities of the form (28.2) then one or the other of X and Z is also a Markoff process with certain transition probabilities. (In our case (28.2) implies that X is a Markoff process with transition probabilities (p_{ij}). For this we use the following lemma.

LEMMA 28.1. Let (X, Y) be a Markoff process with transition probabilities

$$p(\lambda, \lambda';\ \Delta, \Delta') = P((x_{n+1}, y_{n+1}) \in \Delta \times \Delta' \mid (x_n, y_n) = (\lambda, \lambda'))$$

where Λ and Λ' are the ranges of X and Y and $\lambda \in \Lambda,\ \Delta \subset \Lambda,\ \lambda' \in \Lambda',\ \Delta' \subset \Lambda'$. Then X is a Markoff process if

$$(28.3) \qquad \tilde{p}(\lambda, \lambda'; \Delta) = p(\lambda, \lambda'; \Delta, \Lambda')$$

is independent of λ'. In that case

$$P(x_{n+1} \in \Delta \mid x_n = \lambda) = \tilde{p}(\lambda, \lambda'; \Delta) \quad .$$

PROOF:

$$P(x_{n+1} \in \Delta \mid x_n, x_{n-1}, x_{n-2}, \dots) =$$

$$E(P(x_{n+1} \in \Delta \mid x_n, x_{n-1}, x_{n-2}, \dots ; y_n, y_{n-1}, y_{n-2}, \dots) \mid x_n, x_{n-1}, x_{n-2} \dots) =$$

$$E(P(x_{n+1} \in \Delta, y_{n+1} \in \Lambda' \mid x_n, x_{n-1}, x_{n-2}, \dots ; y_n, y_{n-1}, y_{n-2}, \dots) \mid x_n, x_{n-1},$$
$$x_{n-2}, \dots) =$$

$$E(\tilde{p}(x_n, y_n; \Delta) \mid x_n, x_{n-1}, x_{n-2}, \dots)$$

by (28.3). Hence if $\tilde{p}(\lambda, \lambda'; \Delta)$ does not depend upon λ' then $\tilde{p}(x_n, y_n; \Delta)$ is a function of x_n alone for a fixed Δ. Therefore

$$P(x_{n+1} \in \Delta \mid x_n, x_{n-1}, x_{n-2}, \dots) = P(x_{n+1} \in \Delta \mid x_n) = \tilde{p}(x_n, \lambda', \Delta) \quad .$$

This proves the lemma.

We note that in general the subprocess X of a Markoff process (X, Y) need not itself be a Markoff process. Thus the process Z in (28.2) will in general not be a Markoff process itself. A sufficient condition for Z to be a Markoff process is that

$$(28.4) \qquad \tilde{p}(i, \lambda; \Delta) = \sum_{\{j: \psi(\pi_j, \lambda) \in \Delta\}} p_{ij}$$

be independent of i, according to the lemma.

Using the lemma we may prove

THEOREM 28.1. Every ergodic finitely-valued Markoff process is an inductive function of a random process with finite range.

PROOF. Let Z be an ergodic Markoff process with range $\Lambda = \{a_1, a_2, \dots, a_r\}$ with (p_{ij}) $i, j = 1, 2, \dots, r$ the transition probability matrix. We suppose that each of the a_i is taken on with positive probability by the variables of Z. Since Z is ergodic it will follow that it is the unique stationary Markoff process with the given transition probabilities ([3], Chapter V, §2). Denote by P_i the set of intervals $\{I_{ij}\}$ where I_{ij} is the interval $[p_{i1} + p_{i2} + \cdots + p_{i,j-1},\ p_{i1} + p_{i2} + \cdots + p_{ij}]$ with $I_{i0} = [0,\ p_{i1}]$ so that each P_i is a partition of $[0, 1]$. Take a partition P refining all the partitions P_i, $i = 1, 2, \ldots, r$, so that each I_{ij} is a union of $I_\alpha \in P$.

Now take (X', Z') to be an ergodic Markoff process with range $P \times \Lambda$ and with transition probabilities defined by

$$(28.5) \quad P((x'_{n+1}, z'_{n+1}) = (I_\alpha, a_j) \mid (x'_n, z'_n) = (I_\beta, a_i)) = \begin{cases} 0 & \text{if } I_\alpha \not\subset I_{ij} \\ |I_\alpha| & \text{if } I_\alpha \subset I_{ij} \end{cases}$$

where $|I_\alpha|$ denotes the length of the interval I_α. We define a map $\psi : P \times \Lambda \longrightarrow \Lambda$ by setting $\psi(I_\alpha, a_i) = a_j$ where j is the unique index for which $I_\alpha \subset I_{ij}$ for the given index i. (28.5) then becomes

$$(28.6) \quad P((x'_{n+1}, z'_{n+1}) = (I_\alpha, a_j) \mid (x'_n, z'_n) = (I_\beta, a_i)) = \begin{cases} 0 & \text{if } \psi(I_\alpha, a_i) \neq a_j \\ |I_\alpha| & \text{if } \psi(I_\alpha, a_i) = a_j \end{cases}$$

By (28.6) we have

$$\sum_j P((x'_{n+1}, z'_{n+1}) = (I_\alpha, a_j) \mid (x'_n, z'_n) = (I_\beta, a_i)) = |I_\alpha|$$

which is independent of a_i. It follows by Lemma 28.1 that X' is a Markoff process with transition probabilities $P(x'_{n+1} = I_\alpha \mid x'_n = I_\beta) = |I_\alpha|$. Since this is independent of I_β as well it follows that X' is a random process with $P(x'_n = I_\alpha) = |I_\alpha|$. Moreover comparing (28.6) with (28.2) it follows that Z' is, in fact, a ψ-inductive function of X'. The proof of the theorem will be complete once it is shown that Z and Z' are identical. This too follows from Lemma 28.1 since

$$(28.7) \quad \sum_\alpha P((x'_{n+1}, z'_{n+1}) = (I_\alpha, a_j) \mid (x'_n, z'_n) = (I_\beta, a_i)) = \sum_{\psi(I_\alpha, a_i) = a_j} |I_\alpha|$$

and since $\psi(I_\alpha, a_i) = a_j$ for $I_\alpha \subset I_{ij}$ we find that the sum on the right of (28.7) is $|I_{ij}| = p_{ij}$. Since this is independent of β it follows by Lemma 28.1 that Z' is a Markoff process with transition probabilities

p_{ij}. Since Z has the same transition probabilities and Z is ergodic it follows that Z = Z'. This proves the theorem.

The above theorem will be useful in reducing certain problems regarding Markoff processes to the analogous problems for random processes.

28.2. <u>A Counterexample</u>. We shall present an example showing that a ψ-inductive function of a Markoff process need not necessarily be a standard ψ-inductive function. We first observe that if X in Definition 28.1 is a random process, i.e., if $p_{ij} = p_j$ is independent of i, then if Z is a standard ψ-inductive function of X, Z must itself be Markoff. Namely in that case

$$(28.8) \quad \sum_j P((x_{n+j} = \pi_j, \ z_{n+1} \in \Delta \mid x_n = \pi_i, \ z_n = \lambda) = \sum_{\psi(\pi_j, \lambda) \in \Delta} p_j$$

which is independent of i. Hence Z is a Markoff process with transition probabilities

$$p(\lambda, \ \Delta) = \sum_{\psi(\pi_j, \lambda) \in \Delta} p_j \quad .$$

To construct our counterexample we shall find a ψ-inductive function of a random process which is not a Markoff process.

For the random process X we take a $\{0, 1\}$-valued random process so that $\Pi = \{0, 1\}$, and we let Λ be the group of reals mod 1. For each n, let z_n be the element in Λ corresponding to the real number

$$\frac{1}{2} x_n + \frac{1}{2} x_{n+1} + \frac{1}{2^2} x_{n+2} + \frac{1}{2^3} x_{n+3} + \cdots$$

the x_n being the variables of X. We shall write this

$$(28.9) \quad z_n \sim \frac{1}{2} x_{n+1} + \sum_{k=1}^{\infty} \frac{1}{2^k} x_{n+k} \quad .$$

From this we have

$$z_{n+1} \sim \frac{1}{2} x_{n+1} + \sum_{k=1}^{\infty} \frac{1}{2^k} x_{n+k+1}$$

so that $z_{n+1} - 2z_n \sim 1/2 \, x_{n+1}$ since each integer ~ 0 in Λ. Thus if we define $\psi(\pi, \lambda) \sim 2\lambda + 1/2 \, \pi$ then Z will be a ψ-inductive function of X.

Suppose now that Z were a Markoff process so that the conditional distribution of z_{n+1} given $z_n, z_{n-1}, z_{n-2}, \cdots$ depends only on z_n. Now from (28.9) we have

$$2z_n \sim \sum_{k=1}^{\infty} \frac{1}{2^k} \, x_{n+k+1}$$

so that z_n determines the variables $x_{n+2}, x_{n+3}, x_{n+4}, \cdots$ almost everywhere. Since the latter in turn determine z_{n+2} it follows that z_{n+2} is determined by z_n, or z_{n+1} is determined by z_{n-1}. But if Z were a Markoff process it would follow that z_{n+1} is determined by z_n. But then z_n would also determine x_{n+1} since $z_{n+1} - 2z_n \sim 1/2 \, x_{n+1}$ and x_{n+1} can take on only the values 0 and 1. Finally by (28.9) since z_n would determine $x_{n+1}, x_{n+2}, x_{n+3}, \cdots$ it would also determine x_n. This however is impossible since z_n has the same value when $x_n = 0$, $x_{n+1} = 0$ as when $x_n = 1$, $x_{n+1} = 1$. This shows that the assumption that Z is a Markoff process and hence the assumption that Z is a ψ-inductive function of X is false.

28.3. <u>A Law of Large Numbers for Markoff Processes</u>. The result we wish to obtain regarding standard ψ-inductive functions of Markoff processes will be based on a "law of large numbers" that we wish to establish for a certain class of Markoff processes.

DEFINITION 28.2. If W is a Markoff process (not necessarily stationary) with transition probabilities $p(\lambda, \Delta)$ then a left- or right-infinite sequence $\xi(n)$ <u>is generic for</u> W if $\xi(n)$ is a stochastic sequence and the process $X(\xi)$ it generates is a Markoff process with transition probabilities $p(\lambda, \Delta)$.

The process W is here relevant only insofar as it determines the transition probabilities $p(\lambda, \Delta)$. If W is stationary and $\xi(n)$ is a left-infinite sequence then the notion of $\xi(n)$ being generic for W given in §5.3 is not the same as that given here since for one thing $\xi(n)$ would have to be regular and for another all the probability relations of W would have to be taken into account. We shall use the present notion only in this section and there should be no confusion in its usage. We remark that if $\xi(n)$ is a right-infinite sequence then the process $X(\xi)$ is constructed in analogy with the construction for a left-infinite sequence.

By a contraction of a metric space is meant a transformation for

which the distance between the images of two points does not exceed the
distance between the original points. We now have

THEOREM 28.2. Let M be a compact, metric space and
let τ_i, $i = 1, 2, \ldots, m$ be a set of contractions of
M. Suppose that W is a Markoff process with transition probabilities given by

$$(28.10) \qquad p(\lambda, \triangle) = \sum_{\tau_i(\lambda)\epsilon\triangle} p_i$$

where the p_i, $i = 1, 2, \ldots, m$ are a set of non-
negative numbers with

$$\sum_{i=1}^{m} p_i = 1 \quad ,$$

and $\lambda \epsilon M$, $\triangle \subset M$. We assume moreover that W is
one-sided in the positive direction so that the
variables of W are w_1, w_2, w_3, $\ldots$. Then almost
every sample sequence of W is generic for W.

For example if M is finite then any transformation
is a contraction if the distance is taken as a constant between any two distinct points. Moreover by
Theorem 28.1 it follows that any M-valued Markoff
process has transition probabilities of the form
(28.10). Hence the theorem applies at least to
finitely-valued Markoff processes.

PROOF. Let Ω be the space of all right-infinite M-sequences
so that any M-valued process is determined by a measure on Ω. We wish to
examine under what conditions such a measure corresponds to a Markoff
process with transition probabilities $p(\lambda, \triangle)$. Let w_1, w_2, w_3, $\ldots$ be
the coordinate functions on Ω. If E_W denotes the expectation functional
on $C(\Omega)$ induced by the process W and W is a Markoff process with
transition probabilities $p(\lambda, \triangle)$ then

$$E_W(\varphi(w_{k+1}, w_{k+2}, \ldots, w_{k+\ell})|w_k) =$$

$$\int_M \cdots \int_M \varphi(\lambda_1, \lambda_2, \ldots, \lambda_\ell)p(w_k, d\lambda_1)p(\lambda_1, d\lambda_2)\cdots p(\lambda_{\ell-1}, d\lambda_\ell) =$$

$$= \sum_{i_1, i_2, \ldots, i_\ell} p_{i_1} p_{i_2} \cdots p_{i_\ell} \varphi(\tau_{i_1}(w_k), \ \tau_{i_2}\tau_{i_1}(w_k), \ldots, \tau_{i_\ell}\tau_{i_{\ell-1}}\cdots\tau_{i_1}(w_k)) = \bar{\varphi}(w_k)$$

for any continuous function ψ of ℓ variables. We note that the right hand side of (28.11) is again a continuous function of w_k. (28.11) assigns thereby to any continuous $\varphi(w_{k+1}, \ w_{k+2}, \ \ldots, \ w_{k+\ell})$ a continuous function $\bar{\varphi}(w_k)$. Suppose now that W' is an M-valued process with the same transition probabilities as W. Then

$$(28.12)\quad\begin{aligned}&E_{W'}(\varphi(w_{k+1}, \ w_{k+2}, \ \ldots, \ w_{k+\ell})\theta(w_1, \ w_2, \ \ldots, \ w_k)) = \\[4pt] &E_{W'}(E_{W'}(\varphi(w_{k+1}, \ w_{k+2}, \ \ldots, \ w_{k+\ell}) \mid w_k)\theta(w_1, \ w_2, \ \ldots, \ w_k)) = \\[4pt] &E_{W'}(\bar{\varphi}(w_k)\theta(w_1, \ w_2, \ \ldots, \ w_k))\end{aligned}$$

for all $\varphi(w_{k+1}, \ w_{k+2}, \ \ldots, \ w_{k+\ell})$ and $\theta(w_1, \ w_2, \ \ldots, \ w_k)$. Conversely if W' is a process on Ω satisfying

$$(28.13)\quad\begin{aligned}&E_{W'}(\varphi(w_{k+1}, \ w_{k+2}, \ \ldots, \ w_{k+\ell})\theta(w_1, \ w_2, \ \ldots, \ w_k)) = \\[4pt] &\qquad\qquad\qquad E_{W'}(\bar{\varphi}(w_k)\theta(w_1, \ w_2, \ \ldots, \ w_k))\end{aligned}$$

then W' is a Markoff process with transition probabilities $p(\lambda, \ \Delta)$. For in order that W' be a Markoff process with the transition probabilities as prescribed it is sufficient that for every $\varphi(w_{k+1}, \ w_{k+2}, \ \ldots, \ w_{k+\ell})$ we have

$$E_{W'}(\varphi(w_{k+1}, \ w_{k+2}, \ \ldots, \ w_{k+\ell}) \mid w_1, \ w_2, \ \ldots, \ w_k) = \bar{\varphi}(w_k)$$

and this will hold if (28.13) is true for all $\theta(w_1, \ w_2, \ \ldots, \ w_k)$. Thus (28.13) is the criterion for W' being a Markoff process with the transition probabilities $p(\lambda, \ \Delta)$ where $\bar{\varphi}$ is defined by (28.11).

Two conclusions that may be drawn from this are the following. First, if a set of measures on Ω gives the prescribed transition probabilities so will any convex linear combination of them. Secondly, if (28.13) holds for W' it will also hold for the shifted process SW' where

$$(28.14)\quad E_{SW'}(\varphi(w_1, \ w_2, \ \ldots, \ w_\ell)) = E_{W'}(\varphi(w_2, \ w_3, \ \ldots, \ w_{\ell+1})) \ .$$

Consider now the special case where W is stationary. Then by the ergodic theorem almost all sample sequences of W will be stochastic sequences; we wish to prove that they generate processes with the same transition probabilities as W. By the separability of M it can be seen that (28.13) will hold for all φ and θ if it holds for an appropriate denumerable subset of such pairs of functions. Hence if we show that the set of points for which (28.13) does not hold with a particular φ and θ has measure 0 it will follows that the non-generic points for W form a set of measure 0. Here we speak of (28.13) holding for a point ω in Ω with W' referring to the process $X(\omega)$ generated by the stochastic sample sequence $w_n(\omega)$. (28.13) then becomes

$$\lim_{N \to \infty} \sum_{n=1}^{N} [\psi(w_{k+n+1}(\omega), \ldots, w_{k+n+}(\omega)) -$$

(28.15)

$$- \bar{\psi}(w_{k+n}(\omega))] \theta(w_{n+1}(\omega), \ldots, w_{n+k}(\omega)) = 0 \quad .$$

Hence we must show that the limit in (28.15) is 0 almost everywhere. In any case we can write

(28.16) $$\lim_{N \to \infty} \frac{1}{N} \sum_{n=1}^{N} [\psi(w_{k+n+1}, \ldots, w_{k+n+\ell}) - \bar{\psi}(w_{k+n})] \theta(w_{n+1}, \ldots, w_{m+k}) = u$$

almost everywhere where u is a measurable function on Ω invariant under the shift $w_n \longrightarrow w_{n+1}$, the limit existing almost everywhere with respect to the measure of the process W. Since W is a Markoff process, by ([3], p. 460) u will be measurable with respect to w_1 and hence with respect to each w_n. For every n, $\theta(w_{n+1}, \ldots, w_{n+k})\bar{u}$ is a measurable function of $w_{n+1}, \ldots, w_{n+k}$: $\theta(w_{n+1}, \ldots, w_{n+k})\bar{u} = \theta'(w_{n+1}, \ldots, w_{n+k})$. Replacing θ in (28.16) by θ' will give $|u|^2$ on the right. However the expectation of the quantity on the left of (28.16) vanishes for any θ since W has transition probabilities $p(\lambda, \Delta)$. Hence $E_W(|u|^2) = 0$ so that $u = 0$ a.e. This proves our theorem for the case that W is stationary.

To prove the theorem in the general case it suffices to consider those processes, denoted W_ρ, where the first variable w_1 equals ρ with probability one with ρ a fixed point of M, and which are Markoff processes with transition probabilities $p(\lambda, \Delta)$. **The theorem** will then follow for an arbitrary initial distribution by the Fubini theorem so we consider only the processes W_ρ. We let Q be the set of all ρ such that the theorem is true for W_ρ, i.e., almost all sample sequences of

W_ρ are generic for W_ρ, or equivalently, generic for W. We wish to
show that Q = M.

Let C_ε be the set of points of M at a distance $\geq \varepsilon$ from Q.
We will first show that for any W_ρ, the probability that some w_n lies
outside the set C_ε is 1, or that the probability of remaining indefinitely in C_ε is 0. Consider the processes $S_k W_\rho$ where for a
process W', SW' is defined by (28.14). If we identify a process with
the corresponding measure on Ω then we may speak of the averaged process

$$W_\rho^N = \frac{1}{N + 1} (W_\rho + SW_\rho + \ldots + S^N W_\rho) \ .$$

By the remakrs made previously each W_ρ^N is again a Markoff process with
transition probabilities $p(\lambda, \Delta)$ and evidently so also, a limit of such
processes is a Markoff process with the prescribed transition probabilities.
Suppose then that for some sequence N_k, $W_\rho^{N_k} \longrightarrow \bar{W}$. Since $S\bar{W} = \bar{W}$, $\bar{W}$
is a stationary Markoff process and our theorem holds for $\bar{W}$. Now let
P_ρ^N denote the probability measure corresponding to W_ρ^N and $\bar{P}$ the
probability measure for $\bar{W}$. Also Let E_ρ^N and $\bar{E}$ denote the corresponding
expectation functionals. Since the theorem holds for $\bar{W}$ we may say that
with $\bar{P}$ measure 1 a sequence of $\bar{W}$ begins outside of C_ε for each
$\varepsilon > 0$.

Now choose a function g, continuous on M, equal to 1 on
C_ε, to 0 outside of $C_{\varepsilon/2}$ and between 0 and 1 elsewhere. Then
since almost all the sample **paths** of $\bar{W}$ begin outside of $C_{\varepsilon/2}$ it
follows that $\bar{E}(g(w_1)) = 0$ and hence

$$\lim_{k \to \infty} E_\rho^{N_k}(w_1)) = 0 \ .$$

But this implies that

$$\lim_{k \to \infty} \frac{1}{N_k + 1} \sum_{n=1}^{N_{k+1}} E_\rho^O(g(w_n)) = 0$$

so that we must have $E_\rho^O(g(w_n)) < \delta$ for arbitrarily small δ and infinitely many n. This means that $P_\rho^O(w_n \in C_\varepsilon) < \delta$ for infinitely many n
and this implies that the probability of all w_n belonging to C_ε is 0.
Thus we find that starting from any point of M the probability is 1
that the path comes arbitrarily close to Q.

Now if we knew that any path must eventually hit Q we would
be through since the property of being generic for W only depends on the

"tail end" of a sequence. We shall see however that it suffices to know
that for each initial point almost all paths come arbitrarily close to Q.

A sample path in Ω will be generic for W if first of all it
is stochastic, i.e., certain averages exist, and if these averages satisfy
(28.15). By the separability of M both of these conditions need be
verified only for a denumerable set of functions. Therefore it suffices
to show that the sample paths not satisfying one of these conditions for
a particular function have measure 0. Consider, say, the former; we
must show that for a real-valued function $\varphi(\lambda_0, \lambda_1, \ldots, \lambda_k)$

$$
(28.17) \quad
\begin{aligned}
&\limsup_{N \to \infty} \frac{1}{N} \sum_{n=1}^{N} \varphi(w_n(\omega),\, w_{n+1}(\omega),\, \ldots,\, w_{n+k}(\omega)) = \\[2mm]
&\liminf_{N \to \infty} \frac{1}{N} \sum_{n=1}^{N} \varphi(w_n(\omega),\, w_{n+1}(\omega),\, \ldots,\, w_{n+k}(\omega)) \,,
\end{aligned}
$$

for almost all sample paths of W_ρ. We shall show that for an arbitrary
$\varepsilon > 0$ the difference between the two sides of (28.17) is less than ε
for almost all sample paths of W_ρ.

For this choose a $\delta > 0$ such that

$$
|\varphi(\lambda_0', \lambda_1', \ldots, \lambda_k') - \varphi(\lambda_0, \lambda_1, \ldots, \lambda_k)| < \varepsilon/2
$$

if $d(\lambda_0', \lambda_0), d(\lambda_1', \lambda_1), \ldots, d(\lambda_k', \lambda_k) < \delta$, where $d(\cdot,\, \cdot)$ is the metric
on M. Suppose now that ρ is at a distance $< \delta$ from a point of Q. We
shall deduce from this that the two sides of (28.17) differ by less than
ε for almost all sample paths of W_ρ. First of all, let ρ be the
point of Q for which $d(\rho, \rho') < \delta$, and consider the processes W_ρ and
$W_{\rho'}$. If we consider the random process Y with range $\{\tau_1, \tau_2, \ldots, \tau_m\}$
and with $P(y_n = \tau_1) = p_i$ then we can find measure preserving maps from
the space of right-infinite sample sequences of Y to the sample spaces of
W_ρ and $W_{\rho'}$. For each sequence of W_ρ is determined by a sequence of τ_1
and the probability of a set of sequences in both W_ρ and $W_{\rho'}$ is exactly the probability of the corresponding set of τ_1-sequences.

Moreover if ω and ω' are W_ρ and $W_{\rho'}$ sequences respectively
corresponding to the same Y sequence then each $w_n(\omega) = \tau_1 w_{n-1}(\omega)$ and
$w_n(\omega') = \tau_1 w_{n-1}(\omega')$ for the same τ_1. Since the τ_1 are contractions
this implies that $d(w_n(\omega), w_n(\omega')) \leq d(w_{n-1}(\omega), w_{n-1}(\omega')) \leq \cdots \leq d(\rho, \rho')$.
Hence each $d(w_n(\omega), w_n(\omega')) < \delta$. But then by (24.18)

$$
|\varphi(w_n(\omega), w_{n+1}(\omega), \ldots, w_{n+k}(\omega)) - \varphi(w_n(\omega'), w_{n+1}(\omega'), \ldots, w_{n+k}(\omega'))| < \frac{\varepsilon}{2}
$$

and so the difference of the two sides of (28.17) for ω differs from the corresponding difference for ω' by less than ϵ. Since for almost all Y-sequences the corresponding sequence $w_n(\omega')$ are stochastic since $\omega' \in Q$ it follows that for almost all sequences in W the difference between the two sides of (28.17) is less than ϵ as asserted.

The above was shown on the hypothesis that ρ is within δ of Q. However we have already shown that with probability 1 a successor of ρ comes within δ of Q and this establishes the same result for all ρ. It thus follows that for all ρ almost all sample sequence of W_ρ are stochastic sequences. An argument of exactly the same kind may be used to show that almost all sequences of W_ρ satisfy (28.15) since we may "uniformly" approximate the sequences of W_ρ by those of an $W_{\rho'}$ with ρ' in Q. This completes the proof of the theorem.

We might point out that the hypothesis that M is compact was used in just one place in the proof of this theorem, namely, it enabled us to take for granted that the limit of the processes W_ρ^N was again a process $\bar{W}$ and that the measures P_ρ^N do not "dissipate to ∞". It is therefore possible to prove this theorem under weaker assumptions, for example, if M is taken to be locally compact and there is some stationary measure on M for the transition probabilities $p(\lambda, \Delta)$.

It is quite easy to show that one cannot obtain the same results without any assumptions regarding the transition probabilities $p(\lambda, \Delta)$. For example let M be the space of all doubly infinite $(0, 1)$-sequences, and define a Markoff process W on M by setting $w_1 = \lambda_0$, a fixed sequence of M and $P(w_{n+1} = \tau\lambda \mid w_n = \lambda) = 1$ where τ is the shift to the right defined in M. The Markoff process in question is, in fact, of the type described in the theorem since $P(w_{n+1} \in \Delta \mid w_n = \lambda) = 1$ or 0 according as $\tau\lambda \mid \Delta$ on $\tau\lambda \not\in \Delta$. The reason the hypotheses of the theorem are not fulfilled is that τ will not be a contraction in any metric for which M will be compact. For this process it is easily seen that the conclusion of the theorem is, in general, false, since the sequence $\{\tau^n\lambda_0\}$ will be a stochastic sequence if and only if λ_0 is a stochastic $(0, 1)$-sequence and λ_0 is arbitrary.

28.4.   <u>Compact Inductive Functions of Random Processes</u>. We shall next apply Theorem 28.2 to obtain conditions under which a ψ-inductive function of a random process is a standard ψ-inductive function. As we have seen in §28.2 a necessary condition for this is that the inductive function Z itself be a Markoff process with transition probabilities (see (28.8)):

$$(28.19) \qquad P(z_{n+1} \in \Delta \mid z_n = \lambda) = \sum_{\psi(\pi_j, \lambda) \in \Delta} p_j \quad .$$

If, conversely, (28.19) holds, but not only for Z but for the composite process (X, Z) then Z is a standard ψ-inductive function of X for then (28.19) includes (28.2).

What we shall show now is that if Z is a "compact inductive function" of a random process X, the definition of which will be given shortly, then Z is a Markoff process with transition probabilities as in (28.19). Moreover we shall see that if Z is a compact inductive function of X then so is (X, Z). Hence (28.19) will hold for (X, Z) and Z will be a standard ψ-inductive functions of X.

DEFINITION 28.3. A mapping $\psi : \Pi \times \Lambda \longrightarrow \Lambda$ will be called <u>compact</u> if there is a metric on Λ such that each $\psi(\pi)$ (see (28.1)) is a contraction of Λ in this metric. A ψ-inductive function of a Π-valued process will be termed a <u>compact inductive function</u> if ψ is compact.

In terms of an arbitrary metric on Λ compatible with the given topology, the condition that ψ be compact is equivalent to the requirement that the semigroup of mappings of Λ into itself generated by the $\psi(\pi)$ forms an equicontinuous family with respect to the given metric.

Suppose now that ψ is a compact map of $\Pi \times \Lambda \longrightarrow \Lambda$ and that Π is finite. Let $d(\ ,\ )$ be the metric in Λ. We may define a metric $d'(\ ,\ )$ on $\Pi \times \Lambda$ by $d'(\pi_1, \lambda;\ \pi_j, \lambda') = d(\lambda, \lambda') + \delta(\pi_1, \pi_j)$ where $\delta(\pi_1, \pi_j) = 1$ if $\pi_1 \neq \pi_j$ and $\delta(\pi_1, \pi_1) = 0$. ψ defines a map ψ^* of $\Pi \times (\Pi \times \Lambda) \longrightarrow (\Pi \times \Lambda)$ by

$$\psi^*(\pi_1,\ \pi_j,\ \lambda) = (\pi_1,\ \psi(\pi_1,\ \lambda))$$

so that $\psi^*(\pi_1) : \Pi \times \Lambda \longrightarrow \Pi \times \Lambda$ acts as ψ on Λ and sends all of Π onto π_1. It follows from this that the $\psi^*(\pi_1)$ are contractions of $\Pi \times \Lambda$ in the metric d'. Since (X, Z) is a ψ^*-inductive function of X if Z is a ψ-inductive function of X it follows that if Z is a compact-inductive function of X, so is (X, Z).

THEOREM 28.3. If X is a finitely valued random process
and Z is a compact inductive function of X then Z .
is a Markoff process with transition probabilities

(28.19)
$$P(z_{n+1} \in \Delta)\ |\ z_n = \lambda) = \sum_{\psi(\pi_j, \lambda) \in \Delta} p_j$$

where the π_j lie in the range of X, $z_{n+1} = \psi(x_{n+1}, z_n)$ and $p_j = P(x_n = \pi_j)$.

PROOF. If we write $\psi(\pi_j) = \tau_j$ then we recognize that the conclusion of the theorem asserts that Z is a Markoff process of the kind discussed in Theorem 28.2, namely one with transition probabilities

$$p(\lambda, \Delta) = \sum_{\tau_j(\lambda)\epsilon\Delta} p_j \quad .$$

As we have shown in the proof of Theorem 28.2, the condition for this is that

$$E(\varphi(z_{k+1}, z_{k+2}, \ldots, z_{k+\ell})\theta(z_1, z_2, \ldots, z_k)) =$$

$$E(\bar{\varphi}(z_k)\theta(z_1, z_2, \ldots, z_k))$$

where $\bar{\varphi}$ is defined as in (28.11). To show that this is so we shall show that there is a generic sequence $\zeta(n)$ of Z for which

$$\lim_{N \to \infty} \frac{1}{N} \sum_{n=1}^{N} \varphi(\zeta(n+k+1),\zeta(n+k+2),\ldots,\zeta(n+k+\ell))\theta(\zeta(n+1),\zeta(n+2),\ldots,\zeta(n+k))$$

$$(28.20)$$

$$= \lim_{N \to \infty} \frac{1}{N} \sum_{n=1}^{N} \bar{\varphi}(\zeta(n+k))\theta(\zeta(n+1),\zeta(n+2),\ldots,\zeta(n+k)) \quad .$$

Here the sequence $\zeta(n)$ is right-infinite and by "generic" we mean in the sense of §5.3 but for right-infinite rather than left-infinite sequences.

In order to prove this we shall have to assume that (X, Z) is an ergodic process. However it is possible to reduce the general case to that of an ergodic extension of X which we note is ergodic since it is random. For as in the proof of Lemma 25.1, the set of extensions (X, Z) for which Z is a ψ-inductive functions forms a convex set which is compact when identified with the corresponding measures and the extremals are just the ergodic extensions. It follows then by the Krein-Milman theorem that once the theorem is established for ergodic extensions it will be true for arbitrary extensions (X, Z).

Let M now be the semigroup of all contractions of the space Λ, the metric being that in which the $\tau_i = \psi(\pi_i)$ are contractions. M will have an identity element e and will be compact in the metric

$$D(\sigma_1, \sigma_2) = \sup_{\lambda\epsilon\Lambda} d(\sigma_1\lambda, \sigma_2\lambda)$$

d(,) being the metric of Λ. On M the elements $\tau_i \in$ M operate by left multiplication and are again contractions since

$$\sup_{\lambda \in \Lambda} d(\tau_1\sigma_1\lambda,\ \tau_1\sigma_2\lambda) \leq \sup_{\lambda \in \Lambda} d(\sigma_1\lambda,\ \sigma_2\lambda)\ .$$

We now define a Markoff process W in M by setting

$$P(w_o = e) = 1,\ P(w_{n+1} \in \Delta \mid w_n = \lambda) = \sum_{\tau_j\lambda\in\Delta} p_j,\ \lambda \in M,\ \Delta \subset M\ .$$

This is a Markoff process to which Theorem 28.2 applies and so almost all samples paths of W will be generic for W in the sense of Definition 28.2.

Now it is possible to define a measure preserving map from the sample space of X to the sample space of W. Namely if $\xi(n)$ is a right-infinite Π-sequence we define $\Psi(\xi)$ as the right-infinite M-sequence

$$\Psi(\xi) = (e,\psi(\xi(1))e,\psi(\xi(2))\psi(\xi(1))e,\ldots,\psi(\xi(n))\psi(\xi(n-1))\ldots\psi(\xi(1))e,\ldots)\ .$$

It is clear from the definition of W that Ψ is measure preserving. We thus infer that for almost all sample sequences ξ of X the sequence $\Psi(\xi)$ is generic for W. Let us set

$$\zeta*(n) = \psi(\xi(n))\psi(\xi(n-1))\ldots\psi(\xi(1))e$$

so that $\Psi(\xi) = (e,\ \zeta*(1),\ \zeta*(2),\ \ldots,\ \zeta*(n),\ \ldots)$. Thus for almost all ξ we may say that

$$\lim_{N\to\infty} \frac{1}{N} \sum_{n=1}^{N} \varphi*(\zeta*(n+k+1),\ \zeta*(n+k+2),\ \ldots,\ \zeta*(n+k+\ell))$$
$$\theta*(\zeta*(n+1),\ \zeta*(n+2),\ \ldots,\ \zeta*(n+k)) =$$

(28.21)

$$\lim_{N\to\infty} \frac{1}{N} \sum_{n=1}^{N} \bar{\varphi}*(\zeta*(n+k))\theta*(\zeta*(n+1),\ \zeta*(n+2),\ \ldots,\ \zeta*(n+k))$$

for all pairs $\varphi*$, $\theta*$ defined on product spaces M^ℓ and M^k. Here $\bar{\varphi}*$ as given by (28.11) is

$$
\bar{\varphi}*(\zeta*(n+k)) = \sum_{i_1, i_2, \ldots, i_\ell} \varphi(\tau_{i_1}\zeta*(n+k),\ \tau_{i_2}\tau_{i_1}\zeta*(n+k),
$$

(28.22)

$$
\ldots,\ \tau_{i_\ell}\tau_{i_{\ell-1}}\cdots\tau_{i_1}\zeta*(n+k))\ .
$$

We now suppose that we have a ψ-inductive function Z of X for which (X, Z) is ergodic. Then by the ergodic theorem almost all sample sequences, say, right-infinite sample sequences, are generic for (X, Z). It follows that there is a composite sequence $(\xi(n),\ \zeta(n))$ generic for (X, Z) such that simultaneously (28.21) holds for $\Psi(\xi)$ and $\zeta(n)$ is generic for Z. We now choose $\varphi*(\sigma_1,\ \sigma_2,\ \ldots,\ \sigma_\ell) = \varphi(\sigma_1\zeta(0),\ \sigma_2\zeta(0),\ \ldots,\ \sigma_\ell\zeta(0))$ where φ is a function on Λ^ℓ and since $\sigma_1 \in M$, $\zeta(0) \in \Lambda$ the products $\sigma_1\zeta(0) \in \Lambda$. Similarly we choose $\theta*(\sigma_1,\ \sigma_2,\ \ldots,\ \sigma_k) = \theta(\sigma_1\zeta(0),\ \sigma_2\zeta(0),\ \ldots,\ \sigma_k\zeta(0))$ for θ defined on Λ^k. Since Z is a ψ-inductive function of X we find that $\zeta(n) = \psi(\xi(n),\ \zeta(n-1)) = \psi(\xi(n))\zeta(n-1)$ where $\psi(\xi(n)) \in M$. Iterating this we obtain $\zeta(n) = \psi(\xi(n))\psi(\xi(n-1))\ldots\psi(\xi(1))\zeta(0) = \zeta*(n)\zeta(0)$. But then $\varphi*(\zeta*(n+k+1),\ \zeta*(n+k+2),\ \ldots,\ \zeta*(n+k+\ell)) = \varphi(\zeta*(n+k+1)\zeta(0),\ \zeta*(n+k+2)\zeta(0),\ \ldots,\ \zeta*(n+k+\ell)\zeta(0)) = \varphi(\zeta(n+k+\ell),\ \zeta(n+k+2),\ \ldots,\ \zeta(n+k+\ell))$. It follows that (28.21) implies (28.20) for a generic sequence $\zeta(n)$ of Z. But then Z is a Markoff process with the prescribed transition probabilities and this proves the theorem.

28.5. <u>Compact Inductive Functions of Markoff Processes</u>. We shall need the following lemma quite frequently.

LEMMA 28.2. If Z_1 and Z_2 are two generic extensions of a process X, then there exists an ergodic extension Z of X extending both Z_1 and Z_2.

PROOF. Let $\emptyset_1$ and $\emptyset_2$ be the induced T-maps of $\Omega_{Z_1}^-$ and $\Omega_{Z_2}^-$ respectively onto Ω_X^-. Since Z_1 and Z_2 are ergodic there will exist a pair of points $\zeta_1 \in \Omega_{Z_1}^-$ and $\zeta_2 \in \Omega_{Z_2}^-$, generic for Z_1 and Z_2 respectively with the property that $\emptyset_1(\zeta_1) = \emptyset_2(\zeta_2)$. Consider the T-space $\Omega_{Z_1}^- \times \Omega_{Z_2}^-$ and let Ω be the closure in the latter space of the translates of the point $(\zeta_1,\ \zeta_2)$. If $(\omega_1,\ \omega_2) \in \Omega$ then we must have $\emptyset_1(\omega_1) = \emptyset_2(\omega_2)$ since this holds for all translates of $(\zeta_1,\ \zeta_2)$. Hence we have a T-map defined from Ω onto Ω_X^- by setting $\emptyset(\omega_1,\ \omega_2) = \emptyset_1(\omega_1) = \emptyset_2(\omega_2)$. We also have $\psi_1 : \Omega \longrightarrow \Omega_{Z_1}^-$ and $\psi_2 : \Omega \longrightarrow \Omega_{Z_2}^-$ defined by $\psi_1(\omega_1,\ \omega_2) = \omega_1$. All these maps are onto because their images are dense since ζ_1 and ζ_2 are generic for Z_1 and Z_2. Letting Λ_1 and Λ_2

represent the ranges of Z_1 and Z_2 we can define a $\Lambda_1 \times \Lambda_2$-valued
process Z on Ω whose expectation functional is given as some limit

$$(28.23) \qquad E_Z(f) = \lim_{k \to \infty} \frac{1}{N_k + 1} \sum_{n=0}^{N_k} f(T^k \zeta)$$

where $\zeta = (\zeta_1, \zeta_2) \in \Omega$. The sequence N_k may be chosen so that the limit
in (28.23) exists for every $f \in C(\Omega)$. Now it follows directly from
(28.23) and the fact that ζ_1 is generic for Z_1 that Z satisfies
$\psi_1(Z) = Z_1$ and $\emptyset(Z) = X$. Namely

$$E_Z(g \circ \psi_1) = \lim_{k \to \infty} \frac{1}{N_k + 1} \sum_{n=0}^{N_k} g(T^k \zeta_1) = E_{Z_1}(g)$$

so that $\psi_1(Z) = Z_1$. Since $\emptyset_1 \circ \psi_1 = \emptyset$ it follows that $\emptyset(Z) = \emptyset_1(Z_1) = Z$.
This shows that there exists a process Z simultaneously extending both
Z_1 and Z_2. (This is non-trivial because the Z_1 have a common subprocess X.) That Z may be chosen ergodic follows in the manner of
Lemma 25.1, by applying the Krein-Milman theorem to the set of measures
on Ω corresponding to possible processes Z satisfying the requirements
of the theorem.

 The purpose of this lemma is to enable us to speak of the
composite process (Z_1, Z_2) when only the individual processes are given.
If the Z_1 were unrelated (Z_1, Z_2) could be chosen in an arbitrary
manner, e.g., the variables of Z_1 and Z_2 might be taken independent.
Since however Z_1 and Z_2 have a common subprocess it is not obvious
that any composite process (Z_1, Z_2) exists.

 We may now prove the following theorem.

 THEOREM 28.4. If a ψ-inductive function Z of a
finitely-valued ergodic Markoff process X is
compact then Z is a standard ψ-inductive function
of X.

 PROOF. As in the proof of Theorem 28.3 we may suppose that
(X, Z) is an ergodic extension of X since the general case reduces to
this by the Krein-Milman theorem. By Theorem 28.1 the Markoff process X
is an inductive function of a random process Y with finite range. Moreover in giving the proof of that theorem we chose the process (X, Y) as

ergodic. Thus we have two ergodic extensions of X, (X, Y) and (X, Z)
and by the preceding lemma we may find a common extension (X, Y, Z) of
all these processes.

Let Π, P, and Λ respectively denote the ranges of X, Y,
and Z; let $\varphi : P \times \Pi \longrightarrow \Pi$ determine X as an inductive function
of Y and $\psi : \Pi \times \Lambda \longrightarrow \Lambda$ determine Z as an inductive function of
X. Π and P are finite sets whereas Λ is a compact metric space. The
mappings $\psi(\pi)$, $\pi \in \Pi$ are contractions of Λ, and $\varphi(\rho)$ for $\rho \in P$ are
mappings of Π into itself. Let (p_{ij}) be the transition probability
matrix for X, and let (p_k) be the set of probabilities for the random
process Y. Since X is a φ-inductive function of Y we have by (28.19)

$$p_{ij} = P(x_{n+1} = \pi_j \mid x_n = \pi_i) = \sum_{\varphi(\rho_k, \pi_i) = \pi_j} p_k \quad .$$

We now observe that in the composite process (X, Y, Z), (X, Z)
is an inductive function of the random process Y. For

$$(x_{n+1}, \ z_{n+1}) = (\varphi(y_{n+1}, \ x_n), \ \psi(x_{n+1}, \ z_n)) =$$

(28.25)

$$(\varphi(y_{n+1}, \ x_n), \ \psi(\varphi(y_{n+1}, \ x_n), \ z_n)) = \psi^*(y_{n+1}, \ (x_n, \ z_n)) \quad .$$

Moreover since Π and P are discrete it can easily be verified that
ψ^* is a compact map of $P \times (\Pi \times \Lambda)$ into $\Pi \times \Lambda$. It follows then by
Theorem 28.3 that (X, Z) is a Markoff process with transition prob-
abilities

$$P(x_{n+1} = \pi_j, \ z_{n+1} \in \Delta \mid x_n = \pi_i, \ z_n = \lambda) = {\sum}' p_k$$

the latter summation taken over all k such that $\psi^*(\rho_k, \ \pi_i, \ \lambda) \in (\pi_j, \ \Delta)$.
By (28.25) this means that k is restricted by the conditions $\varphi(\rho_k, \ \pi_i) =$
π_j and $\psi(\varphi(\rho_k, \ \pi_i), \ \lambda) \in \Delta$. Thus the sum will be empty if $\psi(\pi_j, \ \lambda) \notin \Delta$
and if $\psi(\pi_j, \ \lambda) \in \Delta$ it will be

$$\sum_{\varphi(\rho_k, \pi_i) = \pi_j} p_k \quad .$$

By (28.24) the latter is p_{ij} so that

$$P(x_{n+1} = \pi_j,\ z_{n+1} \in \Delta \mid x_n = \pi_j,\ z_n = \lambda) = \begin{cases} 0 & \text{if } \psi(\pi_j,\ \lambda) \notin \Delta \\ p_{ij} & \text{if } \psi(\pi_j,\ \lambda) \in \Delta \end{cases}$$

and this is just (28.2). This proves the theorem.

§29. Compact Inductive Functions of Markoff Sequences

29.1. <u>Uniqueness of Inductive Functions</u>. In the foregoing section we found that if the mapping ψ defining a process as an inductive function of a Markoff process is compact then the inductive function had to be a process of a special type. More specifically, the composite process (X, Z) would have to be a Markoff process with given transition probabilities. In this section we shall apply this to the problems raised at the beginning of the last chapter regarding the correspondence between the inductive functions of a sequence and the inductive functions of a process.

Our first step is to show that subject to certain restrictions a process Z satisfying $z_{n+1} = \psi(x_{n+1},\ z_n)$ will be uniquely determined thereby.

> THEOREM 29.1. If X is a random Π-valued process and ψ is a compact map of $\Pi \times \Lambda$ into Λ, then two ergodic ψ-inductive functions of X whose ranges overlap are identical. (The range of a process, or the range of the variables of a process, is the least closed set in which the variables take their values with probability 1.)

PROOF. If Z is a ψ-inductive function of X then by Theorem 28.3 it is a Markoff process with prescribed transition probabilities and so Z is determined by giving the stationary measure on Λ corresponding to it. The theorem will thus follow if it is shown that the supports in Λ of the stationary measures for distinct ergodic processes are disjoint.

To do this we shall identify the elements π_i of Π with the contractions $\psi(\pi_i)$ of Λ and define the operator R on $C(\Lambda)$ to $C(\Lambda)$ by

$$(29.1) \qquad\qquad R(f)(\lambda) = \Sigma_i p_i f(\pi_i(\lambda))$$

where the p_i are the probabilities in X of the $\pi_i \in \Pi$. Since $\Sigma p_i = 1$ it follows that $R(f) = f$ when f is a constant. What we will show is that the only functions satisfying $R(f) = f$ are those that are constant over any set $\Delta \subset \Lambda$ where Δ is the support of the stationary measure for an ergodic process Z.

Let us first observe how our theorem follows from this. Suppose that μ_1 and μ_2 are two stationary measures on Λ and that the supports of μ_1 and μ_2 overlap. Now a measure μ is stationary for the transition probabilities $p(\lambda, \Delta)$ if

$$\mu(\Delta) = \int p(\lambda', \Delta)d\mu(\lambda')$$

or if

$$(29.2) \qquad \int f(\lambda)d\mu(\lambda) = \int \int f(\lambda)p(\lambda', d\lambda)d\mu(\lambda')$$

for all continuous f as follows from $E(f(x_1)) = E(E(f(x_1) \mid x_0))$. In our case

$$p(\lambda, \Delta) = \sum_{\pi_1(\lambda) \in \Delta} p_1$$

so that

$$\int f(\lambda)p(\lambda', d\lambda) = \Sigma p_1 f(\pi_1(\lambda')) = R(f)(\lambda') \quad .$$

Thus condition (29.2) is that $\mu(f) = \mu(Rf)$ for all f in $C(\Lambda)$. Now it is easy to see that, because the π_1 are contractions of Λ and $\Sigma p_1 = 1$, for a continuous f on Λ the set of functions $R^n f$ is an equicontinuous family. The same will be true of the averages

$$(29.3) \qquad R_N(f) = \frac{1}{N+1} \sum_{0}^{N} R^n f \quad .$$

Hence a sequence N_k exists such that $R_{N_k}(f)$ converges to a function $\bar{f}$ on Λ. Clearly $\bar{f}$ satisfies $R\bar{f} = \bar{f}$. Moreover by 29.2 we must have for any stationary measure μ, $\mu(f) = \mu(\bar{f})$. Hence for our two measures μ_1 and μ_2, $\mu_1(f) = \mu_1(\bar{f})$ and $\mu_2(f) = \mu_2(\bar{f})$. To prove then that $\mu_1 = \mu_2$ it suffices to show that $\mu_1(\bar{f}) = \mu_2(\bar{f})$ for the functions $\bar{f}$. Now if it is shown that any function satisfying $R(f) = f$ is constant over the support of μ_1 and μ_2 then since these supports overlap and $R(\bar{f}) = \bar{f}$ it follows that $\bar{f}$ is constant over the combined support of μ_1 and μ_2. But then since the μ_1 are probability measures $\mu_1(\bar{f}) = \mu_2(\bar{f})$.

Thus our theorem depends upon proving the foregoing assertion, namely that a function satisfying $R(f) = f$ is constant over the ranges of ergodic processes Z (equivalently, over supports of the stationary measures on Λ corresponding to ergodic Z). Suppose that Δ is the range of an ergodic Z and that $f \in C(\Lambda)$ satisfies $R(f) = f$. We shall show that an f which is non-negative in Δ satisfying $R(f) = f$ either vanishes nowhere in Δ or everywhere in Δ. Once this is shown then if $R(f) = f$ and f is arbitrary, then setting

$$\gamma = \inf_{\lambda \in \Delta} \varphi(\lambda) \ ,$$

we will have $R(f - \gamma) = f - \gamma$, and since $f - \gamma \geq 0$ in Δ and $f - \gamma$ does vanish somewhere in Δ it will have to vanish everywhere.

We suppose then that $Rf = f$ and that $f(\lambda) \geq 0$ for $\lambda \in \Delta$, and $f \not\equiv 0$. Let

$$\beta = \sup_{\lambda \in \Delta} f(\lambda)$$

so that $\beta > 0$ and let

$$Q_1 = \{\lambda \in \Delta : f(\lambda) > 0\}, \quad Q_2 = \{\lambda \in \Delta : f(\lambda) > \beta/2\} \ .$$

We shall show that for any $\lambda \in \Delta$ there is a sequence $\pi_{i_1}, \pi_{i_2}, \ldots, \pi_{i_k}$ with $\pi_{i_k} \pi_{i_{k-1}} \cdots \pi_{i_1}(\lambda) \in Q_1$. Let μ be the stationary measure on Δ for the process Z. By definition of the range of Z, every open subset of Δ must have positive μ measure. Consequently by the fact that almost every sample sequence of Z is generic for Z it follows that almost every sample sequence of Z enters into every non-empty open subset of Δ. In particular, for almost all $\lambda \in \Delta$ (with respect to μ) there will be a sequence $\pi_{i_1}, \pi_{i_2}, \ldots, \pi_{i_k}$ with $\pi_{i_k} \pi_{i_{k-1}} \cdots \pi_{i_1}(\lambda) \in Q_2$. It follows that this holds for a dense set of λ (since the set of λ for which this is not true has μ-measure 0 it cannot contain an open neighborhood). Hence for every $\lambda \in \Delta$ there exists a $\lambda' \in \Delta$ arbitrarily close to λ with some successor $\pi_{i_k} \pi_{i_{k-1}} \cdots \pi_{i_1}(\lambda) \in Q_2$. Now there exists a $\delta > 0$ such that any point within δ of Q_2 lies in Q_1. Hence since for every λ there exists a λ' within δ of λ and the π_i are contractions it follows that $\pi_{i_k} \pi_{i_{k-1}} \cdots \pi_{i_1}(\lambda) \in Q_1$ for some sequence of π's.

From this we see that for every λ one of the functions $f(\pi_{i_k} \pi_{i_{k-1}} \cdots \pi_{i_1}(\lambda))$ will be positive. Now it is clear from the fact that Δ is the range of Z that each π_i leaves Δ invariant. Hence since f is non-negative in Δ, each $f(\pi_{i_\ell} \pi_{i_{\ell-1}} \cdots \pi_{i_1}(\lambda))$ is also

non-negative. Combining this with the foregoing remark we find that for
any λ and some k, $R^k f(\lambda) > 0$. Since $R(f) = f$ it follows that
$f(\lambda) > 0$ for each $\lambda \in \Delta$ and so f cannot vanish anywhere in Δ. This
proves our assertion and completes the proof of the theorem.

COROLLARY. If Z is a compact inductive function of
X where X is an ergodic Markoff process with finite
range, then Z is uniquely determined by the range of
the composite process (X, Z).

PROOF. As in the proof of Theorem 28.4, (X, Z) is a compact
inductive function of a random process Y and so the preceding theorem
applies.

29.2. <u>Application to Markoff Sequences</u>. The principal result
of this chapter is the following:

THEOREM 29.2. Let X be a finitely valued Markoff
process and let $\xi(n)$ be a left-infinite sequence
generic for X. Let Π be the range of X and suppose
that for a compact metric space Λ, ψ is a compact
map of $\Pi \times \Lambda$ into Λ. Then every solution to

$$(29.4) \qquad\qquad \zeta(n + 1) = \psi(\xi(n + 1),\ \zeta(n))$$

is regular and (hence) generic for some Z satisfying
$z_{n+1} = \psi(x_{n+1},\ z_n)$ and for every ergodic solution Z
to the latter equation there exists a sequence $\zeta(n)$
generic for Z and satisfying (29.4).

PROOF. It will be convenient to consider the composite process
(X, Z) as an inductive function of X so we set $\Lambda^* = \Pi \times \Lambda$ and define
$\psi^* : \Pi \times \Lambda^* \longrightarrow \Lambda^*$ by $\psi^*(\pi, (\pi', \lambda)) = (\pi, \psi(\pi, \lambda))$. Since Π is
discrete we can take as metric in Λ^* $D(\pi, \lambda;\ \pi', \lambda') = \delta(\pi, \pi') +$
$d(\lambda, \lambda')$ where $\delta(\pi, \pi') = 0$ or 1 according as $\pi = \pi'$ or $\pi \neq \pi'$ and
$d(\lambda, \lambda')$ is the metric in Λ. $Z^* = (X, Z)$ is a compact inductive function of X with this metric and $z^*_{n+1} = \psi^*(x_{n+1},\ z^*_n)$.

We point out that since X has a generic sequence, it must be
an ergodic Markoff process. For a finite state Markoff process is ergodic
if for every two states that occur with positive probability there is a
possible passage from one to the other and this is evidently the case if
the two states occur in a generic sequence. As a result of this, Theorem
28.4 and the Corollary to Theorem 29.1 apply so that all the solutions Z^*

to $z^*_{n+1} = \psi^*(x_{n+1}, z^*_n)$ are Markoff processes with prescribed transition probabilities and also the ergodic solutions Z^*_α are determined by their ranges in Λ^* which we denote as Δ_α.

Our first step will be to show that if $\zeta(n)$ satisfies (29.4) and if the composite sequence $\zeta^*(n) = (\xi(n), \zeta(n))$ has all its entries in a single Δ_α, then $\zeta^*(n)$ is regular and (therefore) generic for the process Z^*_α. In particular, $\zeta(n)$ will be regular and generic for a ψ-inductive function of X.

For this consider the space Ω of all left-infinite Λ^*-sequences whose entries are in Δ_α and are such that the Λ component of the sequence is a ψ-inductive function of the Π-component. That is, an $\omega^*(n)$ in Ω has the form $\omega^*(n) = (\eta(n), \omega(n))$ where $\omega^*(n) \in \Delta_\alpha$ and $\omega(n + 1) = \psi(\eta(n + 1), \omega(n))$. Also we specify that the Π-component lies in $\Omega^-_X : \eta \in \Omega^-_X$. In particular the sequence $\xi^*(n) = (\xi(n), \zeta(n))$ referred to previously belongs to Ω. There is then a projection $\emptyset$ of Ω into Ω^-_X where $\emptyset(\omega^*) = \eta$. We first observe that $\emptyset$ is onto. That is tantamount to saying that each sequence in Ω^-_X has a ψ^*-inductive function of it with values in Δ_α. Now this is a consequence of the following remark. If $(\pi_i, \lambda) \in \Delta_\alpha$ and the transition from π_i to π_j is possible in X, i.e., $p_{ij} > 0$, then $\psi^*(\pi_j)(\pi_i, \lambda) = (\pi_j, \psi(\pi_j, \lambda)) \in \Delta_\alpha$. This follows quite readily from the definition of the range of a process, for if every neighborhood of (π_i, λ) has positive probability, and $p_{ij} > 0$, then every neighborhood of $\psi^*(\pi_j)(\pi_i, \lambda)$ will have positive probability by the definition of the transition probabilities of Z^*_α. As a result of this we have $\omega^*(m) \in \Delta_\alpha$ for some $m < 0$ and $\omega^*(m) = (\eta(m), \omega(m))$ where $\eta(n)$ is a sequence in Ω^-_X, then defining $\omega^*(m + 1)$ as $\psi^*(\eta(m + 1)\omega^*(m)$ we will have $\omega^*(m + 1) \in \Delta_\alpha$ as well. For since $\eta(n)$ is a sequence in Ω^-_X each transition $\eta(n) \longrightarrow \eta(n + 1)$ has positive probability. We may then define $\omega^*(m + 2)$ as $\psi^*(\eta(m + 2))\omega^*(m + 1)$, etc., so that we obtain a sequence of $\omega^*(n)$, $n \geq m$ with $\omega^*(n) \in \Delta_\alpha$ and satisfying the functional equation. However m is arbitrary here since for every $\eta(m) \in \Pi$ we may find an $\omega(m) \in \Lambda$ such that $(\eta(m), \omega(m)) \in \Delta_\alpha$ (since $\eta(m)$ occurs in Z^*_α) and therefore we can find arbitrarily long Δ_α sequences over $\eta(n)$ satisfying the functional equation. By the compactness of Δ_α it follows that there is a left-infinite Δ_α-sequence satisfying the functional equation over $\eta(n)$. This shows that each $\eta \in \Omega^-_X$ has an inverse image in Ω under $\emptyset$.

Now the space Ω with the shift to the right defined on Ω as a sequence space is a T-space (Definition 25.4) and $\emptyset$ is a T-map. Moreover, any process W on Ω satisfying $\emptyset(W) = X$ is clearly a ψ^*-inductive function of X with range in Δ_α. However by the Corollary to Theorem 29.1 there exists a unique ergodic process with range in Δ_α that is

a ψ^*-inductive function of X and hence there exists a unique ergodic
solution Z_α^* to $\emptyset(W) = X$. By Lemma 25.1 there exists then a unique
solution altogether to $\emptyset(W) = X$. We may therefore apply Lemma 25.2
which asserts that in this case if a point of Ω restricts to a generic
point of X and if it lies in the sample space of the unique solution
to $\emptyset(W) = X$ then it is generic for this solution process. Now in our
case Z_α^* is the unique process on Ω satisfying $\emptyset(W) = X$ and $\xi(n)$
is a generic point in Ω_X^- so that the sequence $\zeta^*(n) = (\xi(n), \zeta(n))$
is the preimage under $\emptyset$ of a generic point of X. If we show that it
belongs to the sample space of Z_α^* then it will follow that $\zeta^*(n)$ is
regular and generic for Z_α^* as claimed. However Z_α^* is a Markoff
process and in order for a sequence to belong to the sample space of Z_α^*
it suffices that every entry belong to the range of the process and that
every transition have positive probability. But this is the case for
every sequence in Ω since the entries belong to Δ_α the range of Z_α^*
and by the definition of the transition probability in Z_α^*, the transition from $(\eta(n), \omega(n))$ to $(\eta(n + 1), \psi(\eta(n + 1), \omega(n)))$ has the
same probability as the transition in X from $\eta(n)$ to $\eta(n + 1)$.
Hence the sequence $\zeta^*(n)$ belongs to the sample space of Z_α^* and is
therefore generic for Z_α^*. We have thus shown that a ψ^*-inductive function of X with values in one of the Δ_α is regular and generic for
the corresponding Z_α^*.

Suppose now that $\zeta(n)$ is an arbitrary solution to (29.4) and
let $\zeta^*(n)$ be the composite sequence $\zeta^*(n) = (\xi(n), \zeta(n))$. We shall see
next that there must be an α for which all the $\zeta^*(n) \in \Delta_\alpha$. To see this,
construct a T-space from the sequence $\zeta^*(n)$ by considering all its translates in Λ_∞^{*-} and the weak limits of such translates. This space which
we call Ω will again be a T-space and there will exist a T-map $\emptyset$ from
Ω onto Ω_X^-. $\emptyset$ will be onto since $\emptyset(\zeta^*) = \xi$ and the translates of ξ
are dense in Ω_X^-. As before the solutions on Ω to $X = \emptyset(W)$ represent
ψ^*-inductive functions of X and by the first part of Lemma 25.2 there
exist such solutions on Ω. By Lemma 25.1 it follows that there exist
ergodic solutions on Ω. In other words there is some α such that Z_α^*
exists on Ω. But this implies that the $\zeta^*(n)$ must come arbitrarily
close to Δ_α as $n \longrightarrow -\infty$. Now once this is established it follows
that all the $\zeta^*(n)$ in fact belong to Δ_α. For suppose that
$D(\zeta^*(m), \lambda^*) < \epsilon < 1$ for $\lambda^* \in \Delta_\alpha$. By the definition of the metric D
it follows that $\zeta^*(m)$ and λ^* have the same Π-component $\xi(m)$. Now
the transition from $\xi(m)$ to $\xi(m + 1)$ has positive probability in X;
hence since $\lambda^* \in \Delta_\alpha$ it follows that $\psi^*(\xi(m + 1))\lambda^* \in \Delta_\alpha$. However
$\psi^*(\xi(m + 1))$ is a contraction of Λ^* and also $\zeta^*(m + 1) = \psi^*(\xi(m + 1))$
$\zeta^*(m)$. Hence $\zeta^*(m + 1)$ will also be at a distance $< \epsilon$ from Δ_α.
Proceeding in this way it follows that all $\zeta^*(m + j)$ are at a distance

$< \epsilon$ from Δ_α once this is so for $\zeta*(m)$. Since in our case the $\zeta*(m)$ come arbitrarily close to Δ_α it follows immediately that all the $\zeta*(m)$ are in fact in Δ_α.

With this we have proven the first part of the theorem. For the second assertion of the theorem we point out that if Z is an ergodic solution to $z_{n+1} = \psi(x_{n+1}, z_n)$ then it may be taken as the Λ-component of an ergodic $\psi*$-inductive function Z_α^* of X. Namely let Ω be the sample space of (X, Z) and consider on Ω the projections $\emptyset_1$ and $\emptyset_2$ onto the sample spaces Ω_X^- and Ω_Z^- respectively. (X, Z) is then a solution of $\emptyset_1(W) = X$, $\emptyset_2(W) = Z$. As with Lemma 25.1 the set of processes W satisfying these equations form a compact convex set when identified with the corresponding measures on Ω. Again the set of extremals of this convex set will correspond to the ergodic solutions W to the above pair of equations. It follows that there exists an ergodic solution W on Ω. This however means that Z is the Λ-component of a Z_α^* as asserted since the Z_α^* are just the ergodic $\psi*$-inductive functions of X and a solution W to $\emptyset(W) = X$ is a $\psi*$-inductive function of X.

To complete the proof of the theorem we have then to prove that for each Z_α^* there exists a solution $\zeta*(n)$ to $\zeta*(n + 1) = \psi*(\xi(n + 1), \zeta*(n))$ generic for Z_α^*. However by what has been shown already it suffices for this to show that there exists a solution $\zeta*(n)$ to the above equation with values in Δ_α. But this too was already demonstrated when we showed in the first part of the proof, that the map $\emptyset$ from Ω to Ω_X^- is onto. This completes the proof of the theorem.

We remark that the foregoing theorem does not hold if "left-infinite" is replaced by "right-infinite". The reason for this is that in proving that the entries of a solution to (29.4) satisfy $(\xi(n), \zeta(n)) \epsilon \Delta_\alpha$ for some α we made use of the fact that for m sufficiently far to the <u>left</u> of n, $(\xi(m), \zeta(m))$ is within ϵ of Δ_α for arbitrarily small $\epsilon > 0$. This argument would therefore break down for a right-infinite sequence and we could only conclude that the $\zeta*(n)$ tend to some Δ_α. As a result the sequence $\zeta*(n)$ would not have to be regular; it can be shown however that it would still be a stochastic sequence.

Before closing this chapter we mention the connection of the foregoing results to the random ergodic theorem of Ulam and von Neumann. For simplicity assume that both Π and Λ are finite spaces so that the $\psi(\pi)$ will necessarily be contractions and ψ will be compact. Now take X to be a random Π-valued process. In this case the random ergodic theorem states that for any function f in $C(\Lambda)$ and almost all right-infinite sample sequences of $X : \xi(1), \xi(2), \xi(3), \ldots,$ the sequence

$$(29.5) \qquad f(\lambda), \; f(\psi(\xi(1))\lambda), \; f(\psi(\xi(2))\psi(\xi(1))\lambda), \; \dots$$

possesses an average. Now we notice that the sequence $\psi(\xi(n))\psi(\xi(n-1))$ $\dots\psi(\xi(1))\lambda$ is a ψ-inductive function of the sequence $\xi(n)$. What we have just proven gives then the more precise result that for all generic right-infinite sequences of X, the resulting sequence

$$(29.6) \qquad \lambda, \; \psi(\xi(1))\lambda, \; \psi(\xi(2))\psi(\xi(1))\lambda, \; \dots$$

is a stochastic sequence. Thus we can specify a definite set of sequences $\xi(n)$ of probability 1 for which the assertion of the random ergodic theorem is true.

CHAPTER 9.   PROJECTIVE INDUCTIVE FUNCTIONS AND PREDICTION

In this chapter we shall treat yet another case of processes obtained by inductive functions; this time however the results obtained will have a direct bearing on our problem of statistical predictability. As in the last chapter we consider inductive functions Z of X where X is a finite state Markoff process. The range Λ of Z will be a subset of a projective space and the operation $\psi(\pi)$ associated with the variables of X will be projective transformations of Λ. The results of the preceding chapter, in particular Theorem 29.2, will not cover this situation because the transformations $\psi(\pi)$ cannot be taken as contractions in a metric compatible with the topology of Λ. It will in fact be possible to find sequential solutions $\zeta(n)$ to $\zeta(n + 1) = \psi(\xi(n + 1), \zeta(n))$ with $\xi(n)$ generic for X but where $\zeta(n)$ is not regular. Nevertheless the analogue of the second part of Theorem 29.2 does hold; namely, for every ergodic solution Z to $z_{n+1} = \psi(x_{n+1}, z_n)$ there exists for any $\xi(n)$ generic for X a $\zeta(n)$ generic for Z with $\zeta(n + 1) = \psi(\xi(n + 1), \zeta(n))$. The proof of this assertion will occupy us for most of this chapter. In giving this proof we shall draw partly on the results but mostly on the methods of the preceding two chapters. Finally in the last section of the chapter we connect the present results with the problem of prediction which we first posed in §22.

<h3 style="text-align:center">§30. Projective Inductive Functions</h3>

30.1. <u>Notation</u>. Throughout §§30-32 X will denote a Π-valued Markoff process where Π is a finite set $\{\pi_1, \pi_2, \ldots, \pi_s\}$. The transition probabilities $P(x_{n+1} = \pi_j \mid x_n = \pi_i)$ will be denoted p_{ij}. The process X will be supposed ergodic and moreover the term "process" will be used in this chapter only in reference to an "ergodic process" unless it is otherwise indicated.

We shall also use the term "cone" in a restricted sense; namely, a cone V will be the set of all vectors with non-negative components in an m-dimensional Euclidean space for some integer m. We shall be considering linear transformations of one cone into another and such a transformation

207

may be represented by a matrix with non-negative entries. When a cone is
denoted by V, V_n, etc. we shall suppose its elements represented as row
vectors and so the linear transformations will be given by right multiplication with matrices. For a cone $V \subset R^m$ we denote by $L(V)$ the cone
of positive linear functionals on V, i.e., the set of m-dimensional
column vectors with non-negative components. Here we denote by vv' the
inner product of two vectors $v \in V$ and $v' \in L(V)$. For $V \subset R^m$ we
introduce the functional $\| \cdot \|$ defined by $\|v\| = \Sigma_1^m v_1$ where
$v = (v_1, v_2, \ldots, v_m)$. The subset of V of vectors v with $\|v\| = 1$
will be denoted $\tilde{V}$ and referred to as the simplex of V. We shall also
speak of the projective space $\bar{V}$ of V with reference to the set of
equivalence classes of V for the equivalence relation: $v_1 \sim v_2$ if
$v_1 = \lambda v_2$ for $\lambda > 0$. With the exception of the 0-element the points of
$\bar{V}$ are in $1 - 1$ correspondence with the points of $\tilde{V}$; there is thus
induced on $\bar{V}$ a compact topology. The interior of the cone V will be
denoted V^0, its boundary $V' = V - V^0$ and the corresponding subsets of
$\bar{V}$ and $\tilde{V}$ will be denoted $\bar{V^0}, \bar{V}', \tilde{V}^0, \tilde{V}'$.

We have seen in §15 that defining $D(v_1, v_2) = - \log \Gamma_V(v_1, v_2)$
for $v_1, v_2 \in V$ then $D(v_1, v_2)$ depends only upon the images of v_1, v_2
in $\bar{V}$. Hence $D(u_1, u_2)$ is defined for pairs u_1, u_2 in $\bar{V}$. Moreover,
we have seen that $D(v_1, v_2)$ defines an ordinary metric in $\bar{V}$ (which
may take on the value $+ \infty$) if we further stipulate that $D(o, v) = \infty$
for any $v \neq 0$. If we compare the topology induced by the metric D with
the original topology of $\bar{V}$ we find that the former is stronger than the
latter. Namely if we have $D(v_n, v) \longrightarrow 0$ then $\Gamma_V(v_n, v) \longrightarrow 1$ or
$K_V(v_n, v)K_V(v, v_n) \longrightarrow 1$, where v, v_n are taken as elements of $\tilde{V}$ so
that K_V is defined (we may assume that neither v nor v_n is 0 since
$D(v_n, v) \longrightarrow 0$). Since $v_n - K_V(v_n, v)v \in V$ we have $\|v_n\| \geq K_V(v_n, v)\|v\|$
or $K_V(v_n, v) \leq 1$ since $v_n, v \in \tilde{V}$. Similarly $K_V(v, v_n) \leq 1$. If the
product $\Gamma_V(v_n, v) \longrightarrow 1$ we must then have separately $K_V(v_n, v) \longrightarrow 1$
and $K_V(v, v_n) \longrightarrow 1$. It follows from this that the ratio of each pair
of corresponding components tends to 1 and so $v_n \longrightarrow v$ in the Euclidean
metric. The converse however is not true since in the Euclidean metric
$\tilde{V}$, and therefore $\bar{V}$, is compact, but in the topology of D, $\bar{V}$ cannot be
compact as follows from the fact that $D(v_n, v) \longrightarrow \infty$ as $v_n \longrightarrow \bar{V}'$ if
$v \in \bar{V}^0$. On the other hand it is easily seen that a set of bounded
D-diameter is conditionally compact in the topology of the metric D and
hence on such a   the two topologies must agree. In particular for a
closed subset of $\bar{V}^0$ or more generally for a closed subset contained in
any single "face" of $\bar{V}$, the two topologies agree. This last observation
will be applied on a number of occasions.

We have used the expression "face" which we now define precisely.

It is recognized that the relation $u \sim v$ if $D(u, v) < \infty$ is an equivalence relation; the equivalence classes will be referred to as the _faces_ of $\bar{V}$ (or of V). The _principal face_ is the equivalence class consisting of the interior $\bar{V}^O$. The set of faces of V will be denoted $\theta(V)$, and the face containing a particular element v of $\bar{V}$ or V will be denoted $\theta(v)$. We note that $\theta(V)$ forms a lattice under the order relationship

(30.1) $\theta_1 \leq \theta_2$ if for some $v_1 \in \theta_1$, $v_2 \in \theta_2$, $K_V(v_2, v_1) > 0$.

The meaning of (30.1) is that θ_1 forms a subspace of the cone determined by θ_2. When $V \subset R^m$ we observe that $\theta(V)$ may be identified with the subsets of (1, 2, ..., m), the image $\theta(v)$ of a vector $v = (v_1, v_2, ..., v_m)$ being the set of indices for which $v_1 > 0$. The ordering $\theta_1 \leq \theta_2$ then is simply the inclusion ordering of the subsets of (1, 2, ..., m). We remark that the functional $\|\cdot\|$, as well as the sets $\bar{V}$, $\tilde{V}$, V^O, V', $\theta(V)$ etc. may all be defined for the dual cones L(V) as well as for V.

Together with the latter spaces we shall also associate to V the space of "fractional linear" functionals on V, namely, the functionals $\ell(v)$ defined by

$$(30.2) \qquad\qquad \ell(v) = \frac{vv_1'}{vv_2'}$$

with $v_1', v_2' \in L(V)$. It will be convenient to represent ℓ simply as v_1'/v_2' and then we write $\ell(v)$ as $v\ell$ or $v(v_1'/v_2')$. We note that F(V) may be represented as the subset of the projective space $\overline{L(V)} \times \overline{L(V)}$ for which the second component does not vanish. We also note that $\ell(v) = \ell(\lambda v)$ and so the fractional linear functionals on V are functions on $\bar{V}$.

If σ is an $m_1 \times m_2$ matrix it defines a map of $V_1 \subset R^{m_1}$ into $V_2 \subset R^{m_2}$ and a map of $L(V_2)$ into $L(V_1)$, the former denoted by $v \longrightarrow v\sigma$ and the latter by $v' \longrightarrow \sigma v'$. Since σ is linear it follows that these transformations induce transformations from $\bar{V}_1$ to $\bar{V}_2$ and $\overline{L(V_2)}$ to $\overline{L(V_1)}$. We shall use the same σ to denote these transformations which will be termed _projective transformations_. σ also defines a map of $F(V_2)$ into $F(V_1)$ by $v_1'/v_2' \longrightarrow \sigma v_1'/\sigma v_2'$. We then have the associativity relationship $(v\sigma)\ell = v(\sigma\ell)$ if we write $\sigma v_1'/\sigma v_2'$ as $\sigma(v_1'/v_2')$.

An important observation is that as a projective transformation σ is a contraction in the metric D. Namely by Lemma 15.1 (iv), $\Gamma_{V_2}(v_1\sigma, v_2\sigma) \geq \Gamma_{V_1}(v_1, v_2)$ and so $D(v_1\sigma, v_2\sigma) \leq D(v_1, v_2)$. In particular

if $\theta(v_1) = \theta(v_2)$ so that $D(v_1, v_2) < \infty$ then the same holds for $v_1\sigma$
and $v_2\sigma$ so that $\theta(v_1\sigma) = \theta(v_2\sigma)$. Hence σ induces a mapping again denoted by σ, of $\theta(V_1)$ into $\theta(V_2)$, $\theta \longrightarrow \theta\sigma$. Similarly for
$\theta \in \theta(L(V_2))$, $\sigma\theta$ will be defined in $\theta(L(V_1))$. We point out that whereas
a projective transformation is a contraction and hence certainly continu-
ous in the topology of the metric D, such a transformation need not be
continuous in the original topology. However it is easy to see that a
projective transformation σ of $\bar{V}$ is continuous in the neighborhood of
a point of $\bar{V}$ whose preimage in V is not mapped onto 0 by σ. If
$v\sigma = 0$ with $v \neq 0$ then the induced transformation will not be continuous
unless all of V is mapped into 0 since $\bar{v}\sigma$ is mapped onto the isolated
point 0 of $(\overline{V\sigma})$.

30.2. _Preliminary Lemmas_. We shall reformulate the definition
of a standard inductive function (Definition 28.1) to cover more general
situations and to enable us to prove two lemmas that we shall need.

DEFINITION 30.1. If Y is a Markoff process with state space Λ
and transition probabilities $p(\lambda, \Delta)$, $\lambda \in \Lambda$, $\Delta \subset \Lambda$, then Z is a
standard ψ-inductive function (st. ψ-ind. fn.) of Y, if the composite process
(Y, Z) is a Markoff process with transition probabilities defined by

$$(30.3) \qquad E(f(y_{n+1}, z_{n+1}) | y_n, z_n) = \int_\Lambda f(\lambda, \psi(\lambda, z_n)) p(y_n, d\lambda) \ .$$

From (30.3) it follows that

$$E(f(y_{n+1}, z_{n+1}) g(y_n, z_n) | y_n, z_n) = \int_\Lambda f(\lambda, \psi(\lambda, z_n)) g(y_n, z_n) p(y_n, d\lambda) \ ;$$

by taking linear combinations of the products fg and taking limits we find

$$(30.4) \ E(f(y_{n+1}, z_{n+1}, y_n, z_n) | y_n, z_n) = \int_\Lambda f(\lambda, \psi(\lambda, z_n) y_n, z_n) p(y_n, d\lambda) \ .$$

Applying the above to $f = |z_{n+1} - \psi(y_{n+1}, z_n)|^2$ we find that $E(f) = 0$
and so $z_{n+1} = \psi(y_{n+1}, z_n)$. Thus (30.3) implies in particular that Z is
a ψ-inductive function of Y. The significance of a process being a st.
ψ-ind. fn. of a Markoff process is that in that case the essential probability relations are determined by the old process and the function ψ.

LEMMA 30.1. If Z is a ψ-ind. fn. of Y then Z is
a standard ψ-ind. fn. if and only if for all $f \in C(\Lambda)$

(30.5) $E(f(y_{n+1})|z_n,\ z_{n-1},\ \ldots;\ y_n,\ y_{n-1},\ \ldots) = E(f(y_{n+1})|y_n)$.

PROOF. For by (30.5),

$$E(f(y_{n+1})g(z_n)|z_n,\ z_{n-1},\ \ldots;\ y_n,\ y_{n-1},\ \ldots) =$$

$$E(f(y_{n+1})|y_n)g(z_n) = \int_\Lambda f(\lambda)g(z_n)p(y_n,\ d\lambda)$$

so that by taking linear combinations of fg and limits of the latter we
have

$$E(f'(y_{n+1},\ z_n)|z_n,\ z_{n-1},\ \ldots;\ y_n,\ y_{n-1},\ \ldots) = \int_\Lambda f'(\lambda,\ z_n)p(y_n,\ d\lambda)\ .$$

Setting $f'(y_{n+1},\ z_n) = f(y_{n+1},\ z_{n+1}) = f(y_{n+1},\ \psi(y_{n+1},\ z_n))$ we find

$$E(f(y_{n+1},\ z_{n+1})|z_n,\ z_{n-1},\ \ldots;\ y_n,\ y_{n-1},\ \ldots) = \int_\Lambda f(\lambda,\ \psi(\lambda,\ z_n))p(y_n,\ d\lambda)$$

which depends only on $(y_n,\ z_n)$. Hence $(y_n,\ z_n)$ forms a Markoff process
and (30.3) holds. The converse follows directly from the definition.

The meaning of (30.5) is that the past of Z should add no more
information as regards the future of Y than what is known from the past
of Y itself. For example, if Y is a random process, the condition for
Z to be a st. ind. fn. of Y is that the future of Y be independent of
the past of Z as well as the past of Y.

LEMMA 30.2. If Z is a ψ-ind. fn. of Y such that
(Y, Z) is an L-extension of Y (Definition 2.6) then
Z is a standard ψ-ind. fn. of Y.

PROOF. For, since (Y, Z) is an L-extension of Y, by (1.4) (a)

$$E(f(y_{n+1})|z_n,\ z_{n-1},\ \ldots;\ y_n,\ y_{n-1},\ \ldots) = E(f(y_{n+1})|y_n,\ y_{n-1},\ \ldots)$$

and since Y is a Markoff process (30.5) follows.

LEMMA 30.3. If Z and W are inductive functions of
a Markoff process Y and W is a st. ind. fn. of Y
and Z is a st. ind. fn. of (Y, W) (which is a
Markoff process since W is a st. ind. fn. of Y) then

Z is a st. ind. fn. of Y.

PROOF. $E(f(y_{n+1}) \mid z_n,\ z_{n-1},\ \ldots;\ y_n,\ y_{n-1},\ \ldots) =$

$$E\{E(f(y_{n+1}) \mid z_n,\ z_{n-1},\ \ldots;\ y_n,\ y_{n-1},\ \ldots;\ w_n,\ w_{n-1},\ \ldots) \mid z_n,\ z_{n-1},\ \ldots;\ y_n,\ y_{n-1},\ \ldots\} =$$

$$E\{E(f(y_{n+1}) \mid y_n,\ y_{n-1},\ldots;\ w_n,\ w_{n-1},\ \ldots) \mid z_n,\ z_{n-1},\ \ldots;\ y_n,\ y_{n-1},\ \ldots\} =$$

$$E\{E(f(y_{n+1}) \mid y_n,\ y_{n-1},\ \ldots) \mid z_n,\ z_{n-1},\ \ldots;\ y_n,\ y_{n-1},\ \ldots\} =$$

$$E(f(y_{n+1}) \mid y_n,\ y_{n-1},\ \ldots)$$

using successively the fact that Z is a st. ind. fn. of (Y, W) and that
W is a st. ind. fn. of Y.

Before stating the next definition there is a matter that needs
clarification and this is the notion of the identity of two processes. We
point out that there are two senses in which two processes are considered
identical depending upon whether the processes are considered individually
or jointly. When two processes are considered individually their identity
depends upon their being isomorphic as processes. Thus the random
$(1 - 1)$-process X with variables x_n and $P(x_n = 1) = P(x_n = - 1)$ is
the "same" as the process Y with variables $y_n = - x_n$. However when
taken jointly as processes on the sample space of X these processes are
evidently distinct. Thus in the latter sense the notion of identity is
taken in its strict sense whereas in the former it refers to an equivalence.

In the same way we shall distinguish between the set of all
solutions to a functional equation and what will be termed the "complete
solution". Namely, a set $\{Z_\alpha\}$ of processes may be set to furnish the
set of all solutions to a stochastic functional equation $Y = \emptyset(Z)$ if
every solution Z is isomorphic to some Z_α. Hence the joint processes
(Z_α, Z_β) are of no concern to us and need not be defined. On the other
hand we have

DEFINITION 30.2. A composite process $Z^* = (Z_\alpha)$ is the <u>complete
solution</u> to a functional equation $Y = \emptyset(Z)$ if each of the component
processes Z_α is a solution and the Z_α are distinct (but possible isomorphic to one another) and if any other process with the same property is
a subprocess of Z^* with its component processes corresponding to a subset of the Z_α.

Here it is quite possible that among the Z_α there are some isomorphic to one another. Thus suppose that Z is an (ergodic) $(1, - 1)$-

process equidistributed in the group $\{1, -1\}$ (§26) and let Y be the sub-process defined by $y_n = z_n z_{n-1}^{-1}$. By Theorem 26.3 the solution Z to this equation is unique but this means "unique up to a isomorphism". The complete solution, on the other hand, is given by $z_n^* = (z_n, -z_n)$. In general it will be the case that the components of the complete solution consist of the set of all solution each taken with a certain multiplicity. Here we reiterate the important observation that the individual components of a composite process do not determine the process. Hence knowledge of the complete solution comprises more than just a knowledge of all the solutions together with the respective multiplicities.

It is not entirely obvious that a complete solution exists. This is established in the next lemma.

> LEMMA 30.4. Let $\emptyset$ be a T-map of a T-space Ω onto the sample space of a process Y. Then there exists an (ergodic) process Z^* representing the complete solution to $Y = \emptyset(Z)$. (§25).

PROOF. Consider the set of composite processes (Z_β) with all components satisfying the given functional equation. These may be partially ordered as follows: if $Z_\sigma^* = (Z_\alpha^\sigma)_{\alpha \in I}$ and $Z_\tau^* = (Z_\beta^\tau)_{\beta \in J}$ then $Z_\sigma^* \leq Z_\tau^*$ if there is an injection $\varphi : I \longrightarrow J$ such that the process $(Z_{\alpha_1}^\sigma, Z_{\alpha_2}^\sigma, \ldots, Z_{\alpha_r}^\sigma)$ and $Z_{\varphi(\alpha_1)}^\tau, Z_{\varphi(\alpha_2)}^\tau, \ldots, Z_{\varphi(\alpha_r)}^\tau)$ are isomorphic for all $(\alpha_1, \alpha_2, \ldots, \alpha_r)$. By Zorn's lemma there exists a maximal process Z^* which we claim satisfies the conditions of the definition. For if Z^{**} also has the properties in question then applying Lemma 28.2 to Z^* and Z^{**} which both extend Y (and are ergodic since we assume all our processes are ergodic), we can find a process extending simultaneously Z^* and Z^{**}. Then (Z^*, Z^{**}) has the same properties and extends Z^*, so by the maximality of Z^*, (Z^*, Z^{**}), and hence Z^{**}, is a subprocess of Z^*.

For arbitrary stochastic functional equations $F(Y, Z) = 0$ we may also apeak of the complete solution by reducing it to the case con-sidered. Namely if we take the T-space of all pairs of sequences (η, ζ) satisfying $F(\eta, \zeta) = 0$ and let $\emptyset$ be the projection of (η, ζ) onto η then the solutions to $F(Y, Z) = 0$ correspond to the solutions to $\emptyset(Z') = Y$ with Z' as the composite process (Y, Z).

One more notion that we shall need is given in the next definition.

DEFINITION 30.3. If Λ is compact separable and $\zeta(n)$ is a left-infinite Λ-sequence then a process Z is a <u>limit process</u> of $\zeta(n)$ if Z exists on Λ_∞^- (Definition 25.6) and there exists a sequence

$N_k \longrightarrow \infty$ such that for all $\varphi \in C(\Lambda_\infty^-)$

$$(30.6) \qquad \lim_{k \to \infty} \frac{1}{N_k + 1} \sum_{n=0}^{N_k} \varphi(\ldots, \zeta(-n-1), \zeta(-n))$$

exists and equals $E_Z(\varphi)$. Here the process Z is not required to be ergodic.

That a limit process of $\zeta(n)$ exists follows from an often repeated argument showing that the limit in (30.6) exists for some $\{N_k\}$. Moreover the following is evident:

LEMMA 30.5. If a sequence $\zeta(n)$ has a unique limit process Z then $\zeta(n)$ is stochastic and $Z = X(\zeta)$.

30.3. Projective Inductive Functions of a Markoff Process

DEFINITION 30.4. Let X be a Markoff process with finite range Π. Suppose that to each $\pi \in \Pi$ there is associated a pair of cones $V_0(\pi)$ and $V_1(\pi)$ and a linear transformation $\bar{\pi}$ of $V_0(\pi)$ into $V_1(\pi)$. Suppose furthermore that if $p_{ij} > 0$ then $V_1(\pi_i) = V_0(\pi_j)$. Then a process Z whose range is the set $U_\pi \overline{V_1(\pi)}$ with variables z_n satisfying $z_n \in \overline{V_1(x_n)}$ and almost everywhere

$$(30.7) \qquad\qquad z_{n+1} = z_n \, \overline{x_{n+1}}$$

is said to be a <u>projective inductive function of</u> X.

The condition that $z_n \in \overline{V_1(x_n)}$ is necessary in order that (30.7) be meaningful. Namely since the probability of transition from x_n to x_{n+1} is positive we have $V_1(x_n) = V_0(x_{n+1})$ and so $\overline{x_{n+1}}$ operates on z_n taking it into $\overline{V_1(x_{n+1})}$. In the simplest case the cones $V_0(\pi)$ and $V_1(\pi)$ are all equal to a single cone V and the transformations $\bar{\pi}$ are all defined on the same set $\bar{V}$. Although this is the case in which we are primarily interested we will in any case be led to the more general situation of Definition 30.4.

We observe that (30.7) and the requirement of ergodicity imply that z_n is either almost always the 0 element or almost never the 0 element. Since there is little to be said regarding the former case we shall always assume that we are in the latter situation. Since the 0 elements of the $\overline{V_1(\pi)}$ are isolated, if they occur with probability 0 then they do not belong to the range of the process in question. As a result we may take as the range the union of the $\overline{V_1(\pi)}$ without their 0

elements. This would amount to replacing the z_n by variables with values in $\tilde{V}_1(x_n)$ rather than in $\overline{V_1(x_n)}$. Equation (30.7) is then to be replaced by

$$(30.7')\qquad\qquad \|\tilde{z}_n\,\overline{x_{n+1}}\|\,\tilde{z}_{n+1} = \tilde{z}_n\,\overline{x_{n+1}}\ .$$

In this last form all the operations are continuous and so the relationship must hold everywhere rather than almost everywhere. Moreover if (30.7') holds for a pair of sequences $(\xi(n),\ \tilde{\zeta}(n))$ it will hold for any limit process of the sequences. On the other hand neither of these statements need be true for (30.7) since the $\bar{\pi}$ are not continuous transformations. In particular if $(\xi(n),\ \zeta(n))$ is an arbitrary sample sequence of $(X,\ Z)$, it need not be the case that $\zeta(n+1) = \zeta(n)\overline{\xi(n+1)}$. However Using (30.7') we can say the following:

> LEMMA 30.6. If $(\xi(n),\ \zeta(n))$ is generic for $(X,\ Z)$
> where X is a finite state Markoff process and Z
> is a projective inductive function of X, then

$$(30.8)\qquad\qquad \zeta(n+1) = \zeta(n)\overline{\xi(n+1)}\ .$$

PROOF. Suppose we have shown that $\zeta(n)$ never belongs to the null space of $\overline{\xi(n+1)}$. Then $\zeta(n)\overline{\xi(n+1)}$ is not the 0 element of $\overline{V_1(\xi(n+1))}$ and so it corresponds to a vector in $\tilde{V}_1(\xi(n+1))$. The $\zeta(n)$ can never correspond to the 0 element of $\overline{V_1(\xi(n))}$ since $\zeta(n)$ is generic for Z and the 0 elements are not in the range of Z. Suppose that $\zeta'(n)$ is the vector in $\tilde{V}_1(\xi(n))$ corresponding to $\zeta(n) \in V_1(\xi(n))$. The vector in $\tilde{V}_1(\xi(n+1))$ corresponding to $\zeta(n)\overline{\xi(n+1)}$ will then be

$$\frac{\zeta'(n)\overline{\xi(n+1)}}{\|\zeta'(n)\overline{\xi(n+1)}\|}$$

and (30.8) reduces to

$$\|\zeta'(n)\xi(n+1)\|\zeta(n+1) = \zeta'(n)\xi(n+1)$$

Now the latter equation involves only continuous functions of the $\zeta(n)$; hence it will hold for any sample sequence of $(X,\ Z)$ since it holds almost everywhere for the variables of $(X,\ Z)$. Moreover, (30.8) in any case holds for those n for which $\zeta(n)$ is not in the null space of $\overline{\xi(n+1)}$. Thus the lemma depends upon showing that $\zeta(n)$ is never in the null space of $\overline{\xi(n+1)}$. The argument for this depends upon the assumption that X is a finite state Markoff process. We consider the process W defined together with Z by $w_n = \theta(z_n)$. W is an inductive function of X since $\theta(z_{n+1}) = \theta(z_n)\,\overline{x_{n+1}}$, and since it is finitely valued it is a compact inductive function. By Theorem 28.4, W is a standard ind. fn. of X so that (X, W) is a Markoff process with specified transition probabilities. We use this to conclude that if (π, θ) is in the range of (X, W) and if $\pi, \pi_{i_1}, \pi_{i_2}, \ldots, \pi_{i_k}$ is a succession of elements of Π such that the transitions between successive elements have positive probabilities then $(\pi_{i_k}, \theta\bar{\pi}_{i_1}\bar{\pi}_{i_2}\ldots\bar{\pi}_{i_k})$ is also in the range of (X, W). For the transition probability from (π, θ) to $(\pi_{i_1}, \theta\bar{\pi}_{i_1})$ is the same as that from π to π_{i_1}, hence positive, hence $(\pi_{i_1}, \theta\bar{\pi}_{i_1})$ is in the range of (X, W) and continuing in this way we find that $(\pi_{i_k}, \theta\bar{\pi}_{i_1}\bar{\pi}_{i_2}\ldots\bar{\pi}_{i_k})$ is in the range of (X, W).

Now suppose that $\theta_o \in \theta(V(\pi))$ is a maximal face of that set with the property that z_n passes through θ_o with positive probability. That is to say, θ_o is chosen so that no face $> \theta_o$ has the same property. Since $\zeta(n)$ is generic for Z there must be infinitely many n for which $\xi(n) = \pi$ and $\theta(\zeta(n)) = \theta_o$. The reason for this is that $(\xi(n), \zeta(n))$ comes arbitrarily close to come (π, λ) in the range of (X, Z) with $\theta(\lambda) = \theta_o$. But if $\zeta(n)$ is sufficiently close to λ then $\theta(\zeta(n)) \geq \theta(\lambda)$ and since strict inequality is impossible we have $\theta(\zeta(n)) = \theta_o$. Thus we find that infinitely often $(\xi(n), \theta(\zeta(n)))$ is in the range of (X, W). Let n_1 be a value of n for which this is the case and suppose that n_2 is the first value of $n \geq n_1$ for which $\zeta(n)$ is in the null space of $\overline{\xi(n+1)}$. If $n_1 \leq n < n_2$ we will have $\zeta(n+1) = \zeta(n)\overline{\xi(n+1)}$ and so $\theta(\zeta(n+1)) = \theta(\zeta(n))\overline{\xi(n+1)}$. By the preceding remarks it follows using induction that $(\xi(n+1), \theta(\zeta(n+1)))$ is in the range of (X, W) for each such value of n. We thus find that $(\xi(n_2), \theta(\zeta(n_2)))$ is in the range of (X, W). But then it would be impossible that $\zeta(n_2)\xi(n_2 + 1) = 0$ for this would then imply that $(\xi(n_2 + 1), 0)$ is in the range of (X, W) and since the 0 element of $\theta(V_o(\pi))$ is always carried into the 0 element of $\theta(V_1(\pi))$ it follows that $(\pi, 0)$ cannot belong to the range of any (ergodic) (X, W) for which W is not identically 0. As a result n_2 does not exist and since there are infinitely many n_1 it follows that $\zeta(n)$ is never in the null space of $\xi(n+1)$. This proves the lemma.

Suppose we start with a composite sequence $(\xi(n), \zeta(n))$ satisfying (30.8) and suppose that (X, Z) is a limit process of this sequence. Passing to the vector-valued process $\tilde{Z}$, we find that (30.7') is satisfied. The process Z will be a projective inductive function of X if (30.7) holds, i.e., if $\tilde{z}_n\overline{x_{n+1}}$ can vanish only with probability 0. Under certain circumstances this may be determined from the sequence

$(\xi(n), \zeta(n))$; for example, if $\|\widetilde{\zeta(n)}\overline{\xi(n+1)}\|$ is bounded away from 0. This observation will be useful later.

The next lemma is of the same type, asserting that a functional relationship between the random variables is retained for certain of the sample sequences. Note that this lemma would be trivial if $\theta : \overline{V} \longrightarrow \theta(V)$ were a continuous map.

LEMMA 30.7. Let Z be a projective inductive function of X and let W be given by $w_n = \theta(z_n)$. Suppose that $(\xi(n), \zeta(n))$ is generic for (X, Z) and that $(\xi(n), \omega(n), \zeta(n))$ is a sample sequence of (X, W, Z). Then $\omega(n) = \theta(\zeta(n))$.

PROOF. We first show that for all points $(\omega(n), \zeta(n))$ in the sample space of (W, Z), $\omega(n) \geq \theta(\zeta(n))$ in the ordering of $\theta(V_n)$ where $V_n = V_1(\xi(n))$. To see this we observe that $\omega(n) = \theta(\zeta(n))$ for almost all (ω, ζ) since $w_n = \theta(z_n)$; thus we have equality for a dense set of (ω, ζ) in the sample space. Now suppose that $(\omega_k, \zeta_k) \longrightarrow (\omega, \zeta)$ where $\omega_k(n) = \theta(\zeta_k(n))$. Since the $\omega_k(n)$ form a discrete set we have for sufficiently large k, $\omega_k(n) = \omega(n)$. On the other hand if $\zeta_k(n) \longrightarrow \zeta(n)$ then eventually the $\zeta_k(n)$ must lie either in the same face as $\zeta(n)$ or in a larger one. Hence $\omega(n) = \omega_k(n) = \theta(\zeta_k(n)) \geq \theta(\zeta(n))$ for k sufficiently large. Now suppose that $\zeta(n)$ is generic for Z and suppose that $\theta_o \in \theta(V(\pi))$ is a maximal face of $V(\pi)$ with the property that z_n passes through θ_o. Since $\zeta(n)$ is generic for Z we must have infinitely often $\theta(\zeta(n)) = \theta_o$ (see the proof of the preceding lemma). On the other hand, $\omega(n)$ cannot be $\geq \theta_o$ unless it equals θ_o by the maximal property of θ_o. Hence we have infinitely often $\omega(n) = \theta(\zeta(n))$. However $\omega(n)$ satisfies $\omega(n+1) = \omega(n)\overline{\xi(n+1)}$ and $\zeta(n)$ satisfies $\zeta(n+1) = \zeta(n)\overline{\xi(n+1)}$, the first because W is finitely valued so that the $\bar{\pi}$ act continuously on the range of W and the second by the preceding lemma. Hence $\theta(\zeta(n+1)) = \theta(\zeta(n))\overline{\xi(n+1)}$ as well and so if $\omega(n) = \theta(\zeta(n))$ infinitely often the two must be equal for all n.

The following lemma will be needed both in this and in a later section.

LEMMA 30.8. Let $\xi(n)$ be a Π-valued sequence and suppose that $\zeta'(n)$ and $\zeta''(n)$ are regular with $\zeta'(n)$, $\zeta''(n) \in \overline{V_1(\xi(n))}$ and that they satisfy (30.8). Then if for some metric $d(., .)$ on $\cup_\pi \overline{V_1(\pi)}$ (compatible with the given topology) the sequence $d(\zeta'(n), \zeta''(n))$ forms a null sequence (Definition 4.2) then the sequences $\zeta'(n)$ and $\zeta''(n)$ are identical.

PROOF. Let V be any one of the cones $V(\pi)$ and let $\theta_0 \in \theta(V)$ be a maximal element of $\theta(V)$ for which some $\theta(\zeta'(n)) = \theta_0$ or for which $\theta(\zeta''(n)) = \theta_0$. In other words the face θ_0 is occupied by some $\zeta'(n)$ or $\zeta''(n)$ and no strictly larger face of V has this property. Suppose that $\theta(\zeta'(m)) = \theta_0$. Now it is easily seen that given any point in $\bar{V}$ there is a neighborhood containing the point that intersects only faces $\geq$ the face containing the given point. Let U_k be a sequence of such neighborhoods corresponding to the point $\zeta'(m)$ in $\bar{V}$ such that the U_k shrink to the point $\zeta'(m)$. By our discussion regarding the two topologies on $\bar{V}$ we see that for k sufficiently large the two topologies must agree on $U_k \cap \theta_0$ and hence the diameter of $U_k \cap \theta_0$ in the metric D of V becomes arbitrarily small.

Now the sequence $\zeta'(n)$ is by hypothesis regular and since $\zeta'(m) \in U_k$ for each k, there will be a set of n of positive density for which $\zeta'(n) \in U_k$. Since the U_k intersect only faces $\geq \theta_0$ the $\zeta'(n)$ for which this is so either belong to θ_0 or to a strictly larger face. The latter however is excluded by our choice of θ_0. Hence we find that for a set of n of positive density, $\zeta'(n) \in U_k \cap \theta_0$. The latter now implies that $D(\zeta'(n), \zeta'(m)) < \varepsilon$ for k sufficiently large. Let S_k be the set of n in question. Now by hypothesis the set of n for which $d(\zeta'(n), \zeta''(n)) > \delta$ has density 0. Also for any k there is a k' such that any point within a distance $< \delta$ (in the metric d) of $U_{k'}$ is inside U_k if δ is sufficiently small. Hence for $n \in S_{k'}$ and n outside a set of density 0, $\zeta''(n) \in U_k$. Thus for a set of positive density both $\zeta'(n)$ and $\zeta''(n)$ are within D distance $\leq \varepsilon$ of $\zeta'(m)$. We thus have

(30.9)
$$\lim_{n \to -\infty} D(\zeta'(n), \zeta''(n)) = 0 \ .$$

On the other hand, the mappings $\bar{\pi}$ are contractions in the metric D. Thus $D(\zeta'(n+1), \zeta''(n+1)) = D(\zeta'(n)\xi(n+1), \zeta''(n)\xi(n+1)) \leq D(\zeta'(n), \zeta''(n))$. As a result (30.9) implies that $D(\zeta'(n), \zeta''(n)) = 0$ for all n and $\zeta'(n)$ and $\zeta''(n)$ must be identical.

We are now in a position to prove

THEOREM 30.1. A projective inductive function of a finite state Markoff process X is a standard inductive function of X.

PROOF. We assume fixed the process X, the cones $V_o(\pi)$ and $V_1(\pi)$, and the transformations $\bar{\pi}$ for $\pi \in \Pi$ and suppose that Z is a solution to (30.7). We shall say that Z is an <u>extremal</u> solution if whenever (X, Z, Z') is a composite process with Z' also a solution to (30.7) such that $\theta(z_n) = \theta(z_n')$ for the variables z_n and z_n' of Z and Z', then Z and Z' are identical (not only isomorphic).

We point out here that it does not matter whether we speak of (X, Z, Z') as an arbitrary process or only as an ergodic process. That is, if whenever (X, Z, Z') is an ergodic process satisfying $\theta(z_n) = \theta(z_n')$ with Z' a solution to (30.7) then Z and Z' are identical, then for an arbitrary composite process (X, Z, Z') with $\theta(z_n) = \theta(z_n')$ and Z' a solution to (30.7) we will have $Z = Z'$. To see this we let Ω be the set of all triples of sequences $(\xi(n), \zeta(n), \zeta'(n))$ where $(\xi(n), \zeta(n))$ is a sample sequence of (X, Z) and $\zeta'(n + 1) = \zeta'(n)\overline{\xi(n + 1)}$, $\theta(\zeta'(n)) = \theta(\zeta(n))$. Thus any triple (X, Z, Z') of the kind described exists on Ω (§25). Now let $\Phi(\xi, \zeta, \zeta') = (\xi, \zeta)$; the process in question then satisfies $\Phi(X, Z, Z') = (X, Z)$. Our assertion is that if (X, Z, Z) is the only ergodic solution then it is also the only solution. However this follows directly by Lemma 25.1. The theorem will now first be proven in the case that the projective inductive function Z in question is extremal and afterwards we treat the general case.

Suppose that Z is an extremal solution, and let W be the process whose variables are given by $w_n = \theta(z_n)$. W is not a subprocess of Z because θ is not a continuous function. However it is easily seen that (Z, W) is an L-extension of Z so that the composite process (X, Z, W) is well defined as an L-extension of (X, Z). We shall now show that under the assumption that Z is extremal, (X, Z, W) will be an L-extension of (X, W) as well (Definition 2.6).

In order to do this we shall show that if η_1 and η_2 are two generic points for (X, Z, W) which have the same restriction to (X, W) then η_1 and η_2 are identical. Thus two distinct generic points of (X, Z, W) must correspond to distinct points of (X, W) and by Lemma 26.1, taking S to be the set of generic points of (X, Z, W), it follows that (X, Z, W) is an L-extension of (X, W).

Suppose that $\eta_1 = (\xi, \zeta_1, \omega)$ and $\eta_2 = (\xi, \zeta_2, \omega)$; we wish to show that ζ_1 and ζ_2 must be equal if η_1 and η_2 are generic for (X, Z, W). Note that since $\theta(z_n) = w_n$ we have by Lemma 30.7, $\theta(\zeta_1(n)) = \omega(n)$ and since z_n satisfies (30.7) we will have, by Lemma 30.6, $\zeta_1(n + 1) = \zeta_1(n)\overline{\xi(n + 1)}$. Now consider the sequence $\rho(n) = (\xi(n), \zeta_1(n), \zeta_2(n))$. Although the individual components of $\rho(n)$ are regular it does not follow that $\rho(n)$ is regular or even stochastic. Thus it is <u>a priori</u>

possible that the sequence $\rho(n)$ have more than one limit process (Definition 30.3). Suppose that Y is some limit process of $\rho(n)$. We point out that as a limit process Y is not necessarily ergodic. Since the components of $\rho(n)$ are regular it follows that Y has the form $(X,\ Z^{(1)},\ Z^{(2)})$ where $Z^{(1)}$ and $Z^{(2)}$ are isomorphic to Z and the variables of the latter satisfy (30.7). We also have $\theta(z_n^{(1)}) = \theta(z_n^{(2)})$ almost everywhere. To see this we consider the compound sequence $(\zeta(n),\ \zeta_1(n),\ \zeta_2(n),\ \omega(n))$. It is clear that a limit process of $\rho(n)$ may be extended to a limit process of the above sequence (by choosing a subsequence of the N_k in (30.6)). The new process has the form $(X,\ Z^{(1)},\ Z^{(2)},\ W)$ where $w_n = \theta(z_n^{(1)}) = \theta(z_n^{(2)})$. Hence originally we must have $\theta(z_n^{(1)}) = \theta(z_n^{(2)})$. However by the hypothesis that Z is an extremal solution it follows that $Z^{(2)} = Z^{(1)}$ and so the variables $z_n^{(1)}$ and $z_n^{(2)}$ are identical almost everywhere. If $d(\cdot,\ \cdot)$ is a metric on the range of Z we will have $d(z_n^{(1)},\ z_n^{(2)}) = 0$ and as a result

$$\lim_{N \to \infty} \frac{1}{N+1} \sum_{n=0}^{N} d(\zeta_1(-n),\ \zeta_2(-n)) = E(d(z_n^{(1)},\ z_n^{(2)})) = 0$$

from which it follows that $d(\zeta_1(n),\ \zeta_2(n))$ is a null sequence. We now apply Lemma 30.8 which asserts that under these circumstances the sequence $\zeta_1(n)$ and $\zeta_2(n)$ must actually be identical. This shows that $(X,\ Z,\ W)$ is an L-extension of $(X,\ W)$.

Turning to the process W we note that each w_n takes its values in the finite set $\theta(V_1(x_n)) = \theta(V_0(x_{n+1}))$. The operator $\overline{x_{n+1}}$ taking $V_0(x_{n+1})$ into $V_1(x_{n+1})$ induces a transformation of $\theta(V_0(x_{n+1}))$ into $\theta(V_1(x_{n+1}))$. Thus in view of (30.7) we find

$$(30.10) \qquad\qquad\qquad w_{n+1} = w_n\,\overline{x_{n+1}}\ .$$

It follows that W is a finitely-valued inductive function of X; hence it is a compact inductive function and by Theorem 28.4, W is a standard inductive function of X. On the other hand since Z is an ind. fn. of X, is it an ind. fn. of $(X,\ W)$. Since $(X,\ Z,\ W)$ is an L-extension of $(X,\ W)$ it follows by Lemma 30.2 that Z is a st. ind. fn. of $(X,\ W)$. If we now apply Lemma 30.3 we find that Z is a st. ind. fn. of X as was to be shown.

Actually our method of proof gives slightly more. Namely, suppose $(Z_1,\ Z_2,\ \ldots,\ Z_r)$ is a composite process with each component an extremal solution to (30.7). Then the entire composite process is a st. ind. fn. of X. For, we can find for each i a finitely-valued inductive function

W_1 of X such that (X, W_1, Z_1) is an L-extension of (X, W_1). By
repeated application of Lemma 28.2 we may find a process $(X, W_1, \ldots,$
$W_r, Z_1, \ldots, Z_r)$ extending simultaneously each of the (X, W_1, Z_1) and
$(Z_1, \ldots, Z_r)$. It is clear that this process is an L-extension of
$(X, W_1, \ldots, W_r)$. But $(W_1, W_2, \ldots, W_r)$ is a finitely-valued inductive
function of Z, hence a st. ind. fn. $(Z_1, Z_2, \ldots, Z_r)$ is an ind. fn.
of X and hence of $(X, W_1, \ldots, W_r)$ and by Lemma 30.2 it is a st. ind.
fn. of $(X, W_1, \ldots, W_r)$. Finally by Lemma 30.3, $(Z_1, Z_2, \ldots, Z_r)$ is
a st. ind. fn. of X. To prove the theorem in the general case we shall
show that for any solution Z to (30.7) there exists a compositum of
extremal solutions $(Z_1, Z_2, \ldots, Z_r)$ such that $(Z, Z_1, Z_2, \ldots, Z_r)$
is an L-extension of $(Z_1, Z_2, \ldots, Z_r)$. The theorem will then follow by
another application of Lemmas 30.2 and 30.3. For by the first of these Z
will be a st. ind. fn. of $(X, Z_1, Z_2, \ldots, Z_r)$. Since we have shown that
$(Z_1, Z_2, \ldots, Z_r)$ is a st. ind. fn. of X it follows by Lemma 30.3 that
Z is a st. ind. fn. of X. We have therefore to prove that for each
solution Z there exists a set of extremal solutions as described.

     For this it will be convenient to introduce the complete solution
Z^* to the functional relationship (30.7). Z^* consists of one component
isomorphic to X and a number of components Z^α with (X, Z^α) satisfying
(30.7). By the definition of a complete solution the given (X, Z) is
isomorphic to some (X, Z^α) where the X and Z^α refer to components
of Z^*. Our procedure will now be the following. We will show that if
$Z = Z^\alpha$ is not an extremal solution then there exist two components $Z^{(1)}$
and $Z^{(2)}$ of Z^* which are in some sense "smaller than" Z^α and such
that $(Z, Z^{(1)}, Z^{(2)})$ is an L-extension of $(Z^{(1)}, Z^{(2)})$. If the $Z^{(1)}$
are not already both extremal and say $Z^{(1)}$ is not extremal, then by re-
peating this argument we may find components $Z^{(11)}$ and $Z^{(12)}$ strictly
smaller than $Z^{(1)}$ with $(Z^{(1)}, Z^{(11)}, Z^{(12)})$ an L-extension of $(Z^{(11)},$
$Z^{(12)})$ from which it follows that $(Z, Z^{(11)}, Z^{(12)}, Z^{(2)})$ is an
L-extension of $(Z^{(11)}, Z^{(12)}, Z^{(2)})$. We will also see that this procedure
must come to an end after finitely many steps so that eventually we arrive
at the set $(Z_1, Z_2, \ldots, Z_r)$ of extremal solutions sought after.

     We shall require for this the notion of a solution Z^α being
a <u>combination</u> of two other solutions Z^β and Z^γ. The solutions we speak
of here are all to be thought of as components of Z^* so that the relationship between them is fixed. To begin with we may suppose that the variables
z_n^α are almost never 0 since otherwise by (30.7) and the requirement of
ergodicity they would be identically 0 so that Z^α would be a trivial
(constant) process. As a result each z_n^α in $\overline{V_1(x_n)}$ corresponds to vector
u_n^α in $\widetilde{V}_1(x_n)$ with probability 1 (see §30.1). The u_n^α satisfy a
modified form of (30.7), namely:

$$(30.11) \qquad u_{n+1} = \frac{u_n \, \overline{x_{n+1}}}{\|u_n \, \overline{x_{n+1}}\|}$$

the denominator supplying the required normalizing factor. Now for the
solutions Z^α, Z^β, Z^γ we shall say that Z^α is a <u>combination</u> of Z^β
and Z^γ if each u_n^α is a convex linear combination of u_n^β and u_n^γ :
$u_n^\alpha = t_n u_n^\beta + (1 - t_n) u_n^\gamma$, $0 \le t_n \le 1$. The coefficient t_n need not be
constant and may be a $[0, 1]$-valued random variable. We shall now show
that if Z^α is not extremal then there exist two components $Z^{(1)}$ and
$Z^{(2)}$ with Z^α a combination of $Z^{(1)}$ and $Z^{(2)}$ such that $\theta(z_n^{(1)}) <$
$\theta(z_n^\alpha)$ and $\theta(z_n^{(2)}) < \theta(z_n^\alpha)$ the inequality being taken in the ordering
of $\theta(V_1(x_n))$ with "$<$" meaning "$\le$ but $\ne$". Since the dimensions of
the $z_n^{(i)}$ are strictly less than that of z_n it follows that at most a
finite number of steps are required to decompose the Z^α into extremal
components.

Suppose then that Z^α is not an extremal solution to (30.7).
Then Z^α forms a part of a triple (X, Z^α, Z') with Z' also satisfying
(30.7), and with $\theta(z_n') = \theta(z_n)$. Now (X, Z^α) is a subprocess of both
Z^* and (X, Z^α, Z') so that by Lemma 28.2 there exists a process extending both Z^* and (X, Z^α, Z'). It is easy to see that in this process
Z' must be identical with some component Z^β of Z^* by the maximal
property of Z^*. Thus for Z^α non-extremal there exists another component
Z^β in Z^* with $\theta(z_n^\alpha) = \theta(z_n^\beta)$. Denote the corresponding $\tilde{V}_1(x_n)$
valued variables by u_n^α and u_n^β. We have $\Gamma_{V_1(x_n)}(u_n^\alpha, u_n^\beta) > 0$ and also,
by Lemma 15.1 (iv) and (30.11), $\Gamma_{V_1(x_{n+1})}(u_{n+1}^\alpha, u_{n+1}^\beta) \ge \Gamma_{V_1(x_n)}(u_n^\alpha, u_n^\beta)$.
Consequently for each λ, the set of sample points for Z^* for which
$\Gamma_{V_1(x_n)}(u_n^\alpha, u_n^\beta) < \lambda$ is invariant under the transformation T. By ergodicity it follows that $\Gamma_{V_1(x_n)}(u_n^\alpha, u_n^\beta)$ is constant.

Moreover by (30.11) we have

$$K_{V_1(x_{n+1})}(u_{n+1}^\alpha, u_{n+1}^\beta) = K_{V_1(x_{n+1})} \left(\frac{u_n^\alpha \, \overline{x_{n+1}}}{\|u_n^\alpha \, \overline{x_{n+1}}\|} , \frac{u_n^\beta \, \overline{x_{n+1}}}{\|u_n^\beta \, \overline{x_{n+1}}\|} \right) \ge$$

$$K_{V_1(x_n)} \left(\frac{u_n^\alpha}{\|u_n^\alpha \, \overline{x_{n+1}}\|} , \frac{u_n^\beta}{\|u_n^\beta \, x_{n+1}\|} \right) = \frac{\|u_n^\beta \, \overline{x_{n+1}}\|}{\|u_n^\alpha \, \overline{x_{n+1}}\|} K_{V_1(x_n)}(u_n^\alpha, u_n^\beta)$$

and similarly

$$K_{V_1}(x_{n+1})(u_{n+1}^\beta, u_{n+1}^\alpha) \geq \frac{\|u_n^\alpha \; \overline{x_{n+1}}\|}{\|u_n^\beta \; \overline{x_{n+1}}\|} K_{V_1}(x_n)(u_n^\beta, u_n^\alpha)$$

by §15.1. Since however $\Gamma_{V_1}(x_n)(u_n^\alpha, u_n^\beta) = K_{V_1}(x_n)(u_n^\alpha, u_n^\beta) K_{V_1}(x_n)(u_n^\beta, u_n^\alpha)$ and the product is not zero we must have

$$K_{V_1}(x_{n+1})(u_{n+1}^\alpha, u_{n+1}^\beta) = \frac{\|u_n^\beta \; \overline{x_{n+1}}\|}{\|u_n^\alpha \; \overline{x_{n+1}}\|} K_{V_1}(x_n)(u_n^\alpha, u_n^\beta)$$

(30.12)

$$K_{V_1}(x_{n+1})(u_{n+1}^\beta, u_{n+1}^\alpha) = \frac{\|u_n^\alpha \; \overline{x_{n+1}}\|}{\|u_n^\beta \; \overline{x_{n+1}}\|} K_{V_1}(x_n)(u_n^\beta, u_n^\alpha)$$

We now define the processes $Z^{(1)}$ and $Z^{(2)}$ by

$$(30.13) \quad u_n^{(1)} = \frac{u_n^\alpha - K_{V_1}(x_n)(u_n^\alpha, u_n^\beta) u_n^\beta}{\|u_n^\alpha - K_{V_1}(x_n)(u_n^\alpha, u_n^\beta) u_n^\beta\|}, \quad u_n^{(2)} \frac{u_n^\beta - K_{V_1}(x_n)(u_n^\beta, u_n^\alpha) u_n^\alpha}{\|u_n^\beta - K_{V_1}(x_n)(u_n^\beta, u_n^\alpha) u_n^\alpha\|} \quad .$$

The denominators of these expressions are > 0 with probability 1 since otherwise they would vanish identically with probability 1 and the processes Z^α and Z^β would be identical contrary to hypothesis; hence the processes of (30.13) are well-defined. The vectors $u_n^{(1)}$ and $u_n^{(2)}$ are simply the endpoints of the line segment joining u_n^α to u_n^β and cut off by the bounding faces of the cone $V_1(x_n)$. From this it is geometrically clear that $\theta(u_n^{(1)}) < \theta(u_n^\alpha)$, but we may deduce this also from the definition (30.1) using (30.13). The processes $Z^{(1)}$ and $Z^{(2)}$ are defined by setting $z_n^{(1)}$ equal to the image in $\overline{V_1(x_n)}$ of the vector $u_n^{(1)}$ in $V_1(x_n)$. To show that the $Z^{(1)}$ are components of Z^* we must show that the variables $z_n^{(1)}$ satisfy (30.7) or equivalently that the $u_n^{(1)}$ satisfy (30.11). This however follows by (30.12):

$$u_{n+1}^{(1)} = \frac{u_{n+1}^{\alpha} - K_{V_1}(x_{n+1})(u_{n+1}^{\alpha}, u_{n+1}^{\beta})u_{n+1}^{\beta}}{\|u_{n+1}^{\alpha} - K_{V_1}(x_{n+1})(u_{n+1}^{\alpha}, u_{n+1}^{\beta})u_{n+1}^{\beta}\|} =$$

$$(30.14) \qquad \frac{\dfrac{u_n^{\alpha}\,\overline{x_{n+1}}}{\|u_n^{\alpha}\overline{x_{n+1}}\|} - \dfrac{\|u_n^{\beta}\,\overline{x_{n+1}}\|}{\|u_n^{\alpha}\,\overline{x_{n+1}}\|}\,K_{V_1}(x_n)(u_n^{\alpha},\ u_n^{\beta})\,\dfrac{u_n^{\beta}\,\overline{x_{n+1}}}{\|u_n^{\beta}\,\overline{x_{n+1}}\|}}{\left\|\dfrac{u_n^{\alpha}\,\overline{x_{n+1}}}{\|u_n^{\alpha}\,\overline{x_{n+1}}\|} - \dfrac{\|u_n^{\beta}\,\overline{x_{n+1}}\|}{\|u_n^{\alpha}\,\overline{x_{n+1}}\|}\,K_{V_1}(x_n)(u_n^{\alpha},\ u_n^{\beta})\,\dfrac{u_n^{\beta}\,\overline{x_{n+1}}}{\|u_n^{\beta}\,\overline{x_{n+1}}\|}\right\|} =$$

$$\frac{(u_n^{\alpha} - K_{V_1}(x_n)(u_n^{\alpha}, u_n^{\beta})u_n^{\beta})\,\overline{x_{n+1}}}{\|(u_n^{\alpha} - K_{V_1}(x_n)(u_n^{\alpha}, u_n^{\beta})u_n^{\beta})\,\overline{x_{n+1}}\|} = \frac{u_n^{(1)}\,\overline{x_{n+1}}}{\|u_n^{(1)}\,\overline{x_{n+1}}\|}$$

with a similar expression for $u_{n+1}^{(2)}$. Finally the fact that Z^{α} is a combination of $Z^{(1)}$ and $Z^{(2)}$ follows from (30.13) using the inequality

$$\begin{vmatrix} 1 & -K_{V_1}(x_n)(u_n^{\alpha},\ u_n^{\beta}) \\ -K_{V_1}(x_n)(u_n^{\beta},\ u_n^{\alpha}) & 1 \end{vmatrix} = 1 - \Gamma_{V_1}(x_n)(u_n^{\alpha},\ u_n^{\beta}) > 0 \ .$$

To complete the proof of the theorem we need only show that $(Z^{\alpha}, Z^{(1)}, Z^{(2)})$ is an L-extension of $(Z^{(1)}, Z^{(2)})$. As we have pointed out an iteration of the above procedure will then yield a finite set $(Z_1, Z_2, \ldots, Z_r)$ of components of Z^* that are extremal with $(Z^{\alpha}, Z_1, Z_2, \ldots, Z_r)$ an L-extension of $(Z_1, Z_2, \ldots, Z_r)$, and as we have seen this implies that Z^{α} is a st. ind. fn. of X.

If we write

$$(30.15) \qquad u_n^{\alpha} = t_n u_n^{(1)} + (1 - t_n)u_n^{(2)}$$

then u_n^{α} is determined by $u_n^{(1)}$, $u_n^{(2)}$, and t_n so that it suffices to prove that $(S, Z^{(1)}, Z^{(2)})$ is an L-extension of $(Z^{(1)}, Z^{(2)})$ where S is the process with variables t_n. Now by (30.11) and (30.15) we find that t_n satisfies

$$(30.16) \qquad t_{n+1} = \frac{t_n \|u_n^{(1)} \overline{x_{n+1}}\|}{t_n \|u_n^{(1)} \overline{x_{n+1}}\| + (1-t_n)\|u_n^{(2)} \overline{x_{n+1}}\|}$$

so that

$$(30.17) \qquad \frac{t_{n+1}}{1 - t_{n+1}} = \frac{t_n}{1 - t_n} \frac{\|u_n^{(1)} \overline{x_{n+1}}\|}{\|u_n^{(2)} \overline{x_{n+1}}\|} \ .$$

To show that $(S, Z^{(1)}, Z^{(2)})$ is an L-extension of $(Z^{(1)}, Z^{(2)})$ we apply Lemma 26.1 by showing that no two distinct generic points of $(S, Z^{(1)}, Z^{(2)})$ restrict to the same generic point of $(Z^{(1)}, Z^{(2)})$. Namely if η_1 and η_2 are two generic points of $(S, Z^{(1)}, Z^{(2)})$ restricting to the same point of $(Z^{(1)}, Z^{(2)})$ then we may write $\eta_1(n) = (\sigma_1(n), \zeta^{(1)}(n), \zeta^{(2)}(n))$ and $\eta_2(n) = (\sigma_2(n), \zeta^{(1)}(n), \zeta^{(2)}(n))$. By (30.17) we find that

$$\frac{\sigma_1(n+1)}{1 - \sigma_1(n+1)} \cdot \frac{1 - \sigma_1(n)}{\sigma_1(n)} = \frac{\sigma_2(n+1)}{1 - \sigma_2(n+1)} \cdot \frac{1 - \sigma_2(n)}{\sigma_2(n)}$$

and hence

$$(30.18) \qquad \frac{\sigma_2(n)}{1 - \sigma_2(n)} = \gamma \frac{\sigma_1(n)}{1 - \sigma_1(n)}$$

for some constant γ. Now $\sigma_1(n)$ and $\sigma_2(n)$ are both generic for the same process S. If we could conclude from this that $\sigma_1(n)/(1 - \sigma_1(n))$ and $\sigma_2(n)/(1 - \sigma_2(n))$ are generic for the same process we would have a real valued process invariant under multiplication by γ. Now if a non-zero random variable Y has the same distribution as γY it is easy to conclude that $\gamma = 1$ (consider $E(Y^{1t})$). Thus we would find that $\sigma_1(n) = \sigma_2(n)$ as desired. We cannot use this argument directly since the sequences $\sigma_1(n)/(1 - \sigma_1(n))$ are generally unbounded. However we deduce from (30.18) that

$$\sigma_2(n) = \frac{\gamma \sigma_1(n)}{(\gamma-1)\sigma_1(n) + 1}$$

so that the variable t_n has the same distribution as

$$\frac{\gamma t_n}{(\gamma-1)\, t_n + 1}$$

and hence $t_n/(1 - t_n)$ has the same distribution as $\gamma t_n/(1 - t_n)$. From this we may legitimately conclude that $\gamma = 1$ and that $\sigma_2(n) = \sigma_1(n)$ as required. This completes the proof of the theorem.

   30.4. <u>The Range of a Projective Inductive Function</u>. In Theorem 29.1 and its corollary it was shown that for a compact inductive function ψ, the ranges of distinct processes (X, Z^α) were disjoint where the Z^α are the various ψ-inductive functions of X. In our case they need not necessarily be disjoint but the points common to two ranges must lie on the "boundary" of the ranges. This is made more precise in the following theorem.

   THEOREM 30.2. Let $\Lambda = U_{\pi \epsilon \Pi}(\pi, \overline{V_1(\pi)})$ and $\Lambda' = U(\pi, \overline{V_1'(\pi)})$ so that Λ' forms the boundary of Λ. For each solution Z^α to (30.7) let $\Delta^\alpha \subset \Lambda$ be the range of the composite process (X, Z^α). Then if (X, Z^α) and (X, Z^β) are non-isomorphic, $\Delta^\alpha \cap \Delta^\beta \subset \Lambda'$.

   PROOF. By Theorem 30.1 each (X, Z^α) is a Markoff process with preassigned transition probabilities and so the various solutions are determined by the corresponding stationary measures on Λ. We therefore need only show that a stationary measure μ for a solution (X, Z^α) is uniquely determined given a point of the range in $\Lambda - \Lambda'$. To do so we show that to every function $f \epsilon C(\Lambda)$ we may assign a function $\bar{f}$ such that (i) $\bar{f}$ is continuous in $\Lambda - \Lambda'$, (ii) $\mu(f) = \mu(\bar{f})$, (iii) $\bar{f}$ is constant almost everywhere with respect to μ. From this our assertion will follow, for if $\lambda \epsilon \Lambda - \Lambda'$ and λ is in the range Δ^α of (X, Z^α) then every neighborhood of λ in $\Lambda - \Lambda'$ will have positive μ measure so that by (i) and (iii) $\bar{f}$ must be constant everywhere in $\Delta^\alpha - \Delta^\alpha \cap \Lambda'$. If this set is non-empty it has positive μ measure and so $\mu(\bar{f}) = \bar{f}(\lambda)$ for $\lambda \epsilon \Delta^\alpha - \Delta^\alpha \cap \Lambda'$ and so by (ii) μ is determined by the point λ in $\Delta^\alpha - \Delta^\alpha \cap \Lambda'$.

   The construction of the function $\bar{f}$ is carried out as follows. We first define the operator R on $C(\Lambda)$ by setting

$$(30.19) \qquad Rf(\pi_1, u) = \sum_{j=1}^{S} p_{1j}\, f(\pi_j, u\bar{\pi}_j)$$

where $u \in \overline{V_1(\pi_1)}$ so that if $p_{ij} > 0, u \in \overline{V_0(\pi_j)}$ and $u\overline{\pi_j}$ is well-defined. Because the $\overline{\pi}_j$ are not continuous the function Rf need no longer belong to $C(\Lambda)$. Comparison with (30.3) shows that

$$(30.20) \qquad E(f(x_{n+1}, z_{n+1})|x_n, z_n) = Rf(x_n, z_n)$$

if Z satisfying $z_{n+1} = z_n \overline{x_{n+1}}$ is a st. ind. fn. of X which by Theorem 30.1 is the case. Thus for a stationary measure μ of a solution (X, Z^α) we will have $\mu(f) = \mu(Rf)$.

We next introduce a metric $D^*(\lambda, \lambda')$ into Λ where for $\lambda = (\pi, u), \lambda' = (\pi', u'),$

$$D^*(\lambda, \lambda') = \begin{cases} \infty & \text{if } \pi \neq \pi' \\ D(u, u') & \text{if } \pi = \pi' \end{cases}$$

where $D(u, u') = - \log \Gamma_{V_1(\pi)}(u, u')$ as in §30.1. By the remarks of §30.1 the topology of D^* agrees with the original topology of Λ in the interior $\Lambda - \Lambda'$. Now we can show that if f has the property that $|f(\lambda) - f(\lambda')| \leq \epsilon$ whenever $D^*(\lambda, \lambda') < \delta$ then Rf has the same property. For if $\lambda = (\pi_1, u), \lambda' = (\pi', u')$ and $D^*(\lambda, \lambda') < \infty$ then $\pi' = \pi_1$ and so $Rf(\lambda') = Rf(\pi_1, u') = \Sigma p_{ij}f(\pi_j, u'\overline{\pi}_j)$ and

$$|Rf(\lambda') - Rf(\lambda)| \leq \Sigma p_{ij} |f(\pi_j, u'\overline{\pi}_j) - f(\pi_j, u\overline{\pi}_j)|$$

and since $D(u'\overline{\pi}_j, u\overline{\pi}_j) \leq D(u', u)$ as we have remarked in §30.1 it follows from $D^*(\lambda, \lambda') < \delta$ that $|f(\pi_j, u'\overline{\pi}_j) - f(\pi_j, u\overline{\pi}_j)| \leq \epsilon$ and so $|Rf(\lambda') - Rf(\lambda)| \leq \epsilon.$

Now if $f \in C(\Lambda)$ then for each ϵ there exists a δ such that if $D^*(\lambda, \lambda') < \delta$ then $|f(\lambda) - f(\lambda')| < \epsilon.$ This is a consequence of the uniform continuity of f in the original topology and the relationship between the two topologies. From this and the foregoing it follows that the family $\{R^n f\}$ is equicontinuous in the metric of D^*; hence so is the family

$$\left\{ \frac{1}{N + 1} \sum_{n=o}^{N} R^n f \right\} .$$

It follows that a subsequence converges to a function $\overline{f}$ continuous in the topology of D^*. The limit function $\overline{f}$ clearly satisfies $\mu(f) = \mu(\overline{f})$ and $\overline{f} = R\overline{f}$. As we have observed the topology of D^* agrees with the original topology in $\Lambda - \Lambda'$ so that $\overline{f}$ is continuous in $\Lambda - \Lambda'$. There remains only (iii) and to prove this we use the fact that $\bar{f} = R\bar{f}$. Namely let $g_n = \bar{f}(x_n, z_n^\alpha)$. By (30.20) we have

$$E(g_{n+1} \mid x_n, z_n^\alpha) = R\bar{f}(x_n, z_n^\alpha) = g_n \quad .$$

But then $E(g_{n+1} \bar{g}_n) = E(E(g_{n+1} \mid x_n, z_n^\alpha)\bar{g}_n) = E(|g_n|^2)$ and similarly $E(\overline{g_{n+1}} g_n) = E(|g_n|^2)$. From this we obtain

$$E(|g_{n+1} - g_n|^2) = E(|g_{n+1}|^2) + E(|g_n|^2) - E(g_{n+1}\bar{g}_n) - E(\overline{g_{n+1}}g_n) = 0$$

so that $g_{n+1} = g_n$. By ergodicity g_n is constant and so $\bar{f}$ is constant almost everywhere with respect to the stationary measure μ. This completes the proof of the theorem.

$\S 31.$ Projective Inductive Functions of Markoff Sequences

31.1. <u>Statement of the Problem.</u> In the preceding section we have seen how some of the results of the last chapter for the case of a compact inductive function carry over to the projective case. In this section we wish to establish an analogue of Theorem 29.2 for the case of projective inductive functions. We first show that the conclusions of that theorem do not carry over entirely; in particular it is possible that a sequence $\xi(n)$ be generic for a Markoff process, that $\zeta(n + 1) = \zeta(n)\overline{\xi(n + 1)}$, but that $\zeta(n)$ is not regular.

We take X to be a random process on Π where Π consists of two elements π_1 and π_2 with $V_0(\pi_1) = V_1(\pi_1) = V$ where V is the cone in R^2. We let p_1 and p_2 denote the probabilities of π_1 and π_2 respectively in X. To the elements π_1 are associated the matrices

$$(31.1) \qquad \bar{\pi}_1 = \begin{pmatrix} 1 & \alpha_1 \\ 0 & \beta_1 \end{pmatrix} \qquad \bar{\pi}_2 = \begin{pmatrix} 1 & \alpha_2 \\ 0 & \beta_2 \end{pmatrix}$$

with $\alpha_1, \alpha_2 \geq 0$ and $\beta_1, \beta_2 > 0$. If we use $t = t_2/t_1$ to parametrize the projective space $\bar{V}$ of V the operators $\bar{\pi}_1$ and $\bar{\pi}_2$ become

$$(31.2) \qquad t\bar{\pi}_1 = \alpha_1 + \beta_1 t, \qquad t\bar{\pi}_2 = \alpha_2 + \beta_2 t \quad .$$

Here $\bar{V}$ is identified with the compactified half-line $[0, \infty]$.

Now consider the solutions to $z_{n+1} = z_n \overline{x_{n+1}}$. By the results of the foregoing section any such solution Z is a standard inductive function of X and since X is random Z will be a Markoff process itself with the transitions shown in (31.2) occurring with probabilities

p_1 and p_2. There is always a trivial solution to be had by taking
$z_n = \infty$ everywhere. In case $\alpha_1 = \alpha_2 = 0$ we also have the trivial
solution $z_n = 0$; if furthermore $\beta_1 = \beta_2 = 1$ then $z_n = t$, a constant,
is also a solution. It may be shown that a necessary and sufficient condition that a solution outside of these exists is that

$$\beta_1^{p_1}\beta_2^{p_2} < 1 \ .$$

We omit the proof of this assertion which will follow from our subsequent
analysis. It follows that in case

$$\beta_1^{p_1}\beta_2^{p_2} \geq 1$$

the only solutions to $z_{n+1} = z_n\,\overline{x_{n+1}}$ are constant and thus any non-
constant sequence $\zeta(n)$ satisfying $\zeta(n + 1) = \zeta(n)\overline{\xi(n + 1)}$ could not
possibly be regular since it will not be generic for any solution Z. It
is moreover not difficult to find examples of such $\zeta(n)$.

However even in the case that

$$\beta_1^{p_1}\beta_2^{p_2} < 1$$

there will exist sequential inductive functions of generic $\xi(n)$ which
are not regular. Let us define the random variables $\beta(x_n)$ by setting
$\beta(\pi_1) = \beta_1$, $\beta(\pi_2) = \beta_2$. Since $p_1 \log \beta_1 + p_2 \log \beta_2 < 0$ or
$E(\log \beta(x_n)) < 0$, it follows that for the generic sequence $\xi(n)$,

$$(31.3) \qquad \lim_{N \to \infty} \frac{1}{N + 1} \sum_{n=0}^{N} \log \beta(\xi(-n)) < 0 \ .$$

Now suppose that we have some sequence $\zeta(n)$ satisfying $\zeta(n + 1) =$
$\zeta(n)\overline{\xi(n + 1)}$. It is clear from (31.2) that if $\zeta'(n)$ is another such
solution then

$$(31.4) \qquad [\zeta'(n + 1) - \zeta(n + 1)] = [\zeta'(n) - \zeta(n)]\beta(\xi(n + 1)).$$

Conversely if $\zeta'(n) > 0$ and the sequence $\zeta'(n) - \zeta(n)$ satisfies (31.4)
it is clear that $\zeta'(n)$ will again be a solution. Now we can achieve
this by taking $\zeta'(0) > \zeta(0)$ and setting

$$(31.5) \quad \zeta'(-q) - \zeta(-q) = (\zeta'(0) - \zeta(0))\beta(\xi(0))^{-1}\beta(\xi(-1))^{-1} \ \cdots \ \beta(\xi(-q+1))^{-1}.$$

However we then have

$$\log \zeta'(-q) \geq \log (\zeta'(-q) - \zeta(-q)) =$$

$$\log (\zeta'(0) - \zeta(0)) - \sum_{0}^{q-1} \log \beta(\xi(-n))$$

and by (31.3) this tends to ∞ so that $\zeta'(-q) \longrightarrow \infty$ and $\zeta'(n)$ cannot be regular. What we have shown is that if there exists some solution in $[0, \infty)$ to $\zeta(n + 1) = \zeta(n)\overline{\xi(n + 1)}$ then there exist solutions that are not regular. Thus Theorem 29.2 and its corollary do not carry over entirely to the case of projective inductive functions.

On the other hand a closer analysis of the foregoing example shows that for the case

$$\beta_1^{p_1} \beta_2^{p_2} < 1$$

when a non-trivial solution to $z_{n+1} = z_n \overline{x_{n+1}}$ exists there also exists for every sequence $\xi(n)$ generic for X _some_ sequence $\zeta(n)$ generic for this Z with $\zeta(n + 1) = \zeta(n)\xi(n + 1)$. We observe that by the considerations of the preceding paragraph at most one such non-trivial solution can exist so that we find that "every" solution Z has a representative over any generic $\xi(n)$. The problem that we have set ourselves in this and the next section is to show that this is the situation in general. That is to say we wish to prove that if X is a finite state Markoff process, Z is a projective inductive function of X and $\xi(n)$ is generic for X then there exists a $\zeta(n)$ generic for Z with $\zeta(n + 1) = \zeta(n)\overline{\xi(n + 1)}$, or more precisely, with $(\xi(n), \zeta(n))$ generic for (X, Z).

31.2. _A Reformulation_. Suppose that X is a finite state Markoff process and that we have a fixed assignment of cones $V_0(\pi)$ and $V_1(\pi)$ and matrices $\bar{\pi}$ to the elements $\pi \in \Pi$, the range of X. Consider the finite sets $\theta(V_1(\pi))$ and the $U_{\pi \in \Pi} \theta(V_1(\pi))$-valued process W satisfying

$$(31.6) \qquad\qquad w_{n+1} = w_n \overline{x_{n+1}} \; .$$

Since the range of W is finite, W is a compact ind. fn. of X, hence by Theorem 28.4 it is a st. ind. fn. of X so that (X, W) is again a finite-state Markoff process. This is true not only of a particular W but also of the complete solution W^* to (31.6), (Definition 30.2) which, as is easily seen, consists of a component isomorphic to X and finitely many components W^α satisfying (31.6): $W^* = (X, W^{(1)}, \ldots, W^{(\ell)})$. $(W^{(1)}, \ldots, W^{(\ell)})$ is a compact inductive function of X so that $W^* = (X, W^{(1)}, \ldots, W^{(\ell)})$ is again a finite state Markoff process. Moreover by Theorem 29.2 for any generic sequence $\xi(n)$ there exists a generic extension of $\xi(n)$ to W^*.

We now return to the problem posed at the close of §31.1. We are thus given a generic sequence $\xi(n)$ of X which we wish to extend to (X, Z). Clearly it suffices to show that the extension $\xi^*(n)$ of $\xi(n)$ to W^* has a generic extension to (W^*, Z) (where the latter is some composite process extending W^* and Z). By renaming the process X we then have the following problem: given a finite state Markoff process X with an assignment of cones $V_0(\pi)$, $V_1(\pi)$ and matrices $\bar{\pi}$ to the elements of the range of X, and <u>such that any solution to (31.6) is a subprocess of</u> X, $w_n = g(x_n)$, then any generic sequence for X has an extension to (X, Z) where Z is a projective ind. fn. of X for the assignment $\{V_0(\pi), V_1(\pi), \bar{\pi}\}$.

We shall make one further modification in the problem and for this we need the following definition:

DEFINITION 31.1. A projective inductive function Z of X is said to be <u>non-degenerate</u> if $z_n \in \overline{V_1^0(x_n)}$ with probability 1 (the notation is that of Definition 30.4 and §30.1).

There need not always exist a non-degenerate solution Z as is the case in the example in §31.1, when

$$\beta_1^{p_1}\beta_2^{p_2} \geq 1 \ .$$

The observation that we make is that the problem just formulated need be considered only for the case that the process Z in question is non-degenerate. The reason is that we may reassign the cones $V_0(\pi)$ and $V_1(\pi)$ in such a way that the given process becomes non-degenerate. For this we replace the process X by the equivalent process X' whose variables are $x_n' = (x_{n-1}, x_n)$. X' is again a Markoff process whose range consists of the pairs (π_i, π_j) with $p_{ij} > 0$. Now the process $w_n = \theta(z_n)$ satisfies (31.6) and so is a subprocess of X of the form $w_n = g(x_n)$. That is to say, each x_n determines the face to which the element z_n belongs. We now define $V_0(\pi_i, \pi_j)$ as the closure of the face $g(\pi_i)$ in $V_1(\pi_i)$ and $V_1(\pi_i, \pi_j)$ as the closure of the face $g(\pi_j)$ in $V_1(\pi_j)$. That is, each face of a cone forms the interior of some cone, and $V_0(\pi_i, \pi_j)$ and $V_1(\pi_i, \pi_j)$ are so chosen that z_{n-1} belongs to the interior of $V_0(x_{n-1}, x_n)$ and z_n belongs to the interior of $V_1(x_{n-1}, x_n)$.

This arrangement of cones satisfies the conditions of Definition 30.4 since the transition from (π_1, π_j) to (π_k, π_ℓ) is possible only if $j = k$ and then clearly $V_1(\pi_1, \pi_j) = V_0(\pi_k, \pi_\ell)$. Finally the transformation $\overline{(\pi_1, \pi_j)}$ is defined as the restriction of $\bar{\pi}_j$ which is defined on $V_0(\pi_j) = V_1(\pi_1)$ to $V_0(\pi_1, \pi_j) \subset V_1(\pi_1)$. From $z_n \overline{x_{n+1}} = z_{n+1}$ it follows that some interior point of $V_0(\pi_1, \pi_j)$ is carried into a point of $V_1(\pi_1, \pi_j)$ and so it follows that $\overline{(\pi_1, \pi_j)}$ takes all of $V_0(\pi_1, \pi_j)$ into $V_1(\pi_1, \pi_j)$. Now given a generic sequence $\xi(n)$ of X, this is equivalent to a generic sequence $\xi'(n)$ of X'. The process Z with variables z_n defines a process Z' with variables z'_n with $z'_n \in \overline{V_1^0(x'_n)}$ and Z' satisfies $z'_{n+1} = z'_n \overline{x'_{n+1}}$. Suppose we could find a generic extension of $\xi'(n)$ to (X', Z'), say $(\xi'(n), \zeta'(n))$. Since (X, Z) and (X', Z') are evidently subprocesses of one another this would amount to having found a generic extension of $\xi(n)$ to (X, Z). Since Z' is a non-degenerate process with respect to X' our problem is thereby reduced to proving the following assertion:

(A)   If X is a finite state Markoff process and Z is
        a non-degenerate projective inductive function of X
        where X has the property that all the corresponding
        solutions to (31.6) are subprocesses of X of the
        type $w_n = g(x_n)$, then any generic point of X has
        a generic extension to (X, Z).

31.3. <u>A Sufficient Condition</u>. The following definition is needed:

DEFINITION 31.2. A left-infinite or right-infinite numerical sequence $\rho(n)$ is <u>essentially bounded</u> if for each $\varepsilon > 0$ there exists a κ such that $|\rho(n)| < \kappa$ but for a set of n of upper density $< \varepsilon$. $\rho(n)$ is <u>essentially bounded away from</u> 0 if $1/\rho(n)$ is essentially bounded.

Thus $\rho(n) = 1/n$ is not essentially bounded away from 0 whereas $\rho(n) = 1 + \cos n\alpha$ is essentially bounded away from 0 if α/π is irrational.

LEMMA 31.1. Let X and Z be as in (A) and let d
be the maximum dimension of the cones $V_1(\pi)$, and
assume that (A) holds whenever the process Z corresponds to an assignment of cones with dimension $< d$.
Let $\xi(n)$ be a generic sequence of X and denote by
$V(n)$ the cone $V_1(\xi(n))$. Suppose that there exists
a sequence $\zeta'(n)$ with $\zeta'(n) \in \overline{V(n)}$ such that the

distance from $\zeta'(n)$ to the boundary $\overline{V'(n)}$ forms
a sequence essentially bounded away from 0 and such
that $\zeta'(n + 1) = \zeta'(n)\overline{\xi(n + 1)}$. Then (A) holds for
Z so that there exists a sequence $\zeta(n)$ satisfying
the same equation with $(\xi(n),\ \zeta(n))$ generic for
$(X,\ Z)$.

PROOF. We first remark that if (A) is assumed to hold when the
maximum dimension is $< d$ then by the discussion in §31.2 a generic sequence of X will have a generic extension to any $(X,\ Z)$ where Z is
a projective inductive function of X such that that maximum dimension
of the faces $\theta(z_n)$ is $< d$. To prove the lemma we shall consider two
cases, first, that the process Z is extremal, and then the general
case (see the proof of Theorem 30.1 for the definition of "extremal" and
other notions to be used here).

Suppose then that Z is an extremal solution; we shall show
that in that case the given sequence $\zeta'(n)$ already satisfies the requirements of (A) so that $(\xi(n),\ \zeta'(n))$ is generic for $(X,\ Z)$. We
first show that $\zeta^*(n) = (\xi(n),\ \zeta'(n))$ is a stochastic sequence with
$X(\zeta^*) = (X,\ Z)$. For this it suffices to show (Lemma 30.5) that any limit
process of $\zeta^*(n)$ must be isomorphic to $(X,\ Z)$. Now if Y is a limit
process of $\zeta^*(n)$ it has the form $(X,\ Z')$ where Z' is some solution
to

$$(31.7) \qquad\qquad z_{n+1} = z_n\ \overline{x_{n+1}}$$

since $\zeta'(n)$ satisfies the analogous equation. (Strictly speaking we
may only deduce that for the corresponding vector valued variables u'_n
we have $\|u'_n\ \overline{x_{n+1}}\|u'_{n+1} = u'_n\ \overline{x_{n+1}}$ as in the proof of Lemma 30.6. This
will amount to the same as (31.7) if with probability 1, u'_n is not in
the null space of $\overline{x_{n+1}}$. This however will follow from the subsequent
discussion). By the assumption that Z is extremal it follows that any
solution Z' to (31.7) satisfying $\theta(z'_n) = \theta_0(x_n)$, where $\theta_0(x_n)$ denotes the maximal face (i.e., the interior) of the cone $V_1(x_n)$, must
have $(X,\ Z')$ isomorphic to $(X,\ Z)$. (In fact, it implies more, namely
that if Z and Z' are combined as a single process with X then
$Z = Z'$.) As remarked the same holds for solutions Z' not known to be
ergodic, by Lemma 25.1. Hence to prove that the limit process $(X,\ Z')$
of $\zeta^*(n)$ is isomorphic to $(X,\ Z)$ it suffices to show that $\theta(z'_n) = \theta_0(x_n)$.
Denoting by $\varphi(u,\ \pi)$ the distance from u which is assumed to belong
to $\overline{V'_1(\pi)}$, to the boundary $\overline{V'_1(\pi)}$, our condition reads: $\varphi(z'_n,\ x_n) > 0$
almost everywhere. (When we speak of distance here and in the statement
of the lemma it is meant with respect to any metric compatible with the

given topology.) Now since (X, Z') is a limit process of $\zeta^*(n)$ it follows that the probability that $\varphi(z_n', x_n) < \varepsilon$ does not exceed the upper density of the set of n for which $\varphi(\zeta'(n), \xi(n)) < 2\varepsilon$. But by hypothesis the latter goes to 0 as $\varepsilon \longrightarrow 0$ and so we have $\varphi(z_n', x_n) > 0$ with probability 1. (In particular, z_n' can be in the null space of $\overline{x_{n+1}}$ only with probability 0, and so the foregoing arguments show that $Z' = Z$). Thus $\zeta^*(n)$ is stochastic with the unique limit process (X, Z).

To complete the proof of the lemma in the extremal case it is only necessary to show that the sequence $\zeta^*(n)$ lies in the sample space of (X, Z). Now by Theorem 30.1, (X, Z) is a Markoff process whose transition probabilities are determined by those of X. It is easy to show that for a process of this kind a sequence belongs to the sample space of the process if and only if each transition has positive probability and if each entry of the sequence belongs to the range of the process. Now the first requirement is evidently true of $\zeta^*(n)$ since the transition probability from $\zeta^*(n)$ to $\zeta^*(n + 1)$ is the same as that from $\xi(n)$ to $\xi(n + 1)$. Thus we have only to show that each $\zeta^*(n)$ lies in the range of (X, Z). For this we introduce the metric $D^*(\lambda, \lambda')$ already used in the proof of Theorem 30.2 into $U_{\pi \in \Pi} (\pi, \overline{V_1(\pi)})$, defined for $\lambda = (\pi, u), \lambda' = (\pi', u')$ by

$$D^*(\lambda, \lambda') = \begin{cases} \infty & \text{if} \quad \pi \neq \pi' \\ D(u, u') & \text{if} \quad \pi = \pi' \end{cases}$$

where D is the metric introduced in §30.1 in the space $\overline{V(\pi)}$. Let Δ be the range of (X, Z) in $U_{\pi \in \Pi} (\pi, \overline{V_1(\pi)})$ and let $D^*(\lambda, \Delta)$ denote the distance from λ to Δ. We wish to show that $D^*(\zeta^*(n), \Delta) = 0$. We note that in any case $D^*(\zeta^*(n + 1), \Delta) \leq D^*(\zeta^*(n), \Delta)$ since if $D^*(\zeta^*(n), \Delta) < \infty$ there will be an element $(\pi, u) \in \Delta$ with $\xi(n) = \pi$ and since the transition from $\xi(n)$ to $\xi(n + 1)$ has positive probability $(\xi(n + 1), u \, \xi(n + 1)) \in \Delta$ and $D^*((\xi(n + 1), \zeta(n)\overline{\xi(n + 1)}), (\xi(n + 1), u \, \xi(n + 1)) = D(\zeta(n)\overline{\xi(n + 1)}, u \, \xi(n + 1)) \leq D(\zeta(n), u) = D^*(\xi(n), \zeta(n)),$ $(\pi, u))$. It therefore suffices to show that

$$\liminf_{n \to -\infty} D^*(\zeta^*(n), \Delta) = 0 \quad .$$

This however follows immediately from the fact that Δ is the range of a limit process of $\zeta^*(n)$ so that $\zeta^*(n)$ gets arbitrarily close to any point of Δ. Since $z_n \in V_1^O(x_n)$ with probability 1 there will be points of Δ of the form (π, u) with $u \in \overline{V_1^O(\pi)}$ and in a neighborhood of such a point the metric D^* is compatible with the original compact topology of $U_{\pi \in \Pi} (\pi, \overline{V_1(\pi)})$ which implies that

$$\liminf_{n \to -\infty} D^*(\zeta^*)n), \Delta) = 0 \ .$$

We have thereby proven the lemma in the case that Z is extremal. Actually though we have also proven more; namely by the same arguments it follows that in any case the pair $(\xi(n), \zeta'(n))$ will be generic for some non-degenerate process (X, Z'). For we have shown that the sequence $\zeta^*(n) = (\xi(n), \zeta'(n))$ has all its entries in the range of any limit process of $\zeta^*(n)$ if there are points of this range in $\Lambda - \Lambda'$ where $\Lambda = U (\pi, \overline{V_1(\pi)})$ and Λ' is the boundary. However by the hypothesis of the lemma almost all points of the range of any limit process of $\zeta^*(n)$ must belong to $\Lambda - \Lambda'$. Moreover by Theorem 30.2 the intersection of two such ranges is restricted to lie in Λ' from which it follows that $\zeta^*(n)$ has a unique limit process and lies in the sample space. Hence $\zeta^*(n)$ is generic for some (X, Z'). We remark that it is inessential here that the limit process of $\zeta^*(n)$ is <u>a priori</u> not necessarily ergodic since any point in the range of a non-ergodic solution (X, Z') is also in the range of some ergodic solution. It follows that $\zeta^*(n)$ belongs to the range of an ergodic solution and hence the limit process of $\zeta^*(n)$ is in fact ergodic.

We turn now to the general case of the lemma where Z is not necessarily extremal. It will be necessary to recall a number of results established in the course of the proof of Theorem 30.1. First of all we showed there that for an extremal solution Z, (X, W, Z) is an L-extension of (X, W) where W is given by $w_n = \theta(z_n)$. In our case by hypothesis W is a subprocess of X so that we have (X, Z) as an L-extension of X. Hence $(X, Z_1, Z_2, \ldots, Z_k)$ is an L-extension of X for any number of extremal processes. Now we also showed that if Z is an arbitrary solution then there exist extremal solutions $Z_1, Z_2, \ldots, Z_r$ with $(Z, Z_1, Z_2, \ldots, Z_r)$ an L-extension of $(Z_1, Z_2, \ldots, Z_r)$; hence $(X, Z, Z_1, Z_2, \ldots, Z_r)$ is an L-extension of $(X, Z_1, Z_2, \ldots, Z_r)$ and therefore also of X. It follows that in our case for <u>any</u> solution Z to (31.7) (X, Z) will be an L-extension of X.

Suppose that Z is not an extremal solution and consider the complete solution Z^* to (31.7), $Z^* = (X, \ldots, Z^\alpha, \ldots, Z^\beta, \ldots)$. We may identify Z with some non-degenerate component Z^α of Z^* and we may also suppose that $(\xi(n), \zeta'(n))$ is generic for another non-degenerate component (X, Z^β) where $Z^\alpha \neq Z^\beta$. From the proof to Theorem 30.1 we recall that Z^α and Z^β are combinations of two other solutions $Z^{(1)}$ and $Z^{(2)}$ in the sense that $u_n^\alpha = t_n^\alpha u_n^{(1)} + (1 - t_n^\alpha)u_n^{(2)}$ and $u_n^\beta = t_n^\beta u_n^{(1)} + (1 - t_n^\beta)u_n^{(2)}$ with u_n^α denoting the $\tilde{V}_1(x_n)$-valued counterpart of z_n^α, etc. We also have $\theta(z_n^{(i)}) < \theta(z_n^\alpha) = \theta(z_n^\beta)$ for $i = 1, 2$. We observe now that the above relationships imply that Z^α is a subprocess of $(Z^\beta, Z^{(1)}, Z^{(2)})$. For we have

$$u_n^\alpha - u_n^{(1)} = (1 - t_n^\alpha)(u_n^{(2)} - u_n^{(1)}), \quad u_n^\beta - u_n^{(1)} = (1 - t_n^\beta)(u_n^{(2)} - u_n^{(1)})$$

and

$$(31.8) \qquad u_n^\alpha = u_n^{(1)} + \frac{(1-t_n^\alpha)}{(1-t_n^\beta)} (u_n^\beta - u_n^{(1)}) \quad .$$

Now by (30.17) we have

$$\frac{t_{n+1}^\alpha}{(1-t_{n+1}^\alpha)} \cdot \frac{(1-t_n^\alpha)}{t_n^\alpha} = \frac{\|u_n^{(1)} \overline{x_{n+1}}\|}{\|u_n^{(2)} \overline{x_{n+1}}\|}$$

so that (by symmetry):

$$\frac{t_{n+1}^\alpha}{1 - t_{n+1}^\alpha} \cdot \frac{1 - t_n^\alpha}{t_n^\alpha} = \frac{t_{n+1}^\beta}{1 - t_{n+1}^\beta} \cdot \frac{1 - t_n^\beta}{t_n^\beta}$$

This in turn implies that

$$\frac{1 - t_n^\alpha}{t_n^\alpha} \cdot \frac{t_n^\beta}{1 - t_n^\beta} = \frac{1 - t_{n+1}^\alpha}{t_{n+1}^\alpha} \frac{t_{n+1}^\beta}{1 - t_{n+1}^\beta}$$

so that by ergodicity there is a constant c with

$$(31.9) \qquad \frac{1 - t_n^\alpha}{t_n^\alpha} \cdot \frac{t_n^\beta}{1 - t_n^\beta} = c \quad .$$

From this we find that

$$\frac{1 - t_n^\alpha}{1 - t_n^\beta} = \frac{c}{(1-c)t_n^\beta + c}$$

so that

$$(31.10) \qquad u_n^\alpha = u_n^{(1)} + \frac{c}{(1-c)t_n^\beta + c}\ (u_n^\beta - u_n^{(1)})\ .$$

Now the right hand side of (31.10) defines a continuous function of $u_n^{(1)}$, $u_n^{(2)}$, and u_n^β. To see this we note that t_n^β is a well-defined continuous function of these three quantities as long as $u_n^{(1)} \neq u_n^{(2)}$. However as $u_n^{(1)}$ and $u_n^{(2)}$ come closer, $u_n^\beta - u_n^{(1)}$ tends to 0 as well so it suffices to show that the coefficient

$$\frac{c}{(1-c)t_n^\beta + c}$$

remains bounded for all values of $u_n^{(1)}$, $u_n^{(2)}$, and u_n^β. But this follows from the fact that the coefficient in question may be written as

$$\frac{1 - t_n^\alpha}{1 - t_n^\beta}$$

and also as

$$c\,\frac{t_n^\alpha}{t_n^\beta}\ ,$$

the first of which is bounded for $t_n^\beta \in [0, 1/2]$ and the second for $t_n^\beta \in [1/2, 1]$. It follows that Z^α is a subprocess of $(Z^\beta, Z^{(1)}, Z^{(2)})$.

As a result of this it suffices to show that the generic sequence $\xi(n)$ for X has a generic extension to $(X, Z^\beta, Z^{(1)}, Z^{(2)})$. Now we have already seen that $\xi(n)$ has an extension $(\xi(n), \zeta'(n))$ to Z^β; that was in fact how Z^β was chosen. Now the condition that $\theta(z_n^{(1)}) < \theta(z_n^\alpha)$ implies that the dimensions of the faces containing the $z_n^{(1)}$ are all $< d$. By the hypothesis of the lemma and the remark made at the outset of the proof this implies that the sequence $\xi(n)$ has generic extensions $(\xi(n), \zeta^{(1)}(n))$ and $(\xi(n), \zeta^{(2)}(n))$ to $(X, Z^{(1)})$ and $(X, Z^{(2)})$ respectively. To complete the proof we will show that $(\xi(n), \zeta(n), \zeta^{(1)}(n), \zeta^{(2)}(n))$ is generic for $(X, Z^\beta, Z^{(1)}, Z^{(2)})$. The first step is to show that $\eta(n) = (\xi(n), \zeta(n), \zeta^{(1)}(n), \zeta^{(2)}(n))$ has as unique limit process the process $Y = (X, Z^\beta, Z^{(1)}, Z^{(2)})$ so that $\eta(n)$ will be stochastic with $X(\eta) = Y$. This follows quite readily from the fact noted before that Y is an L-extension of X. Hence the expectation of any variable in Y is the limit of expectations of a sequence of variables in X. More precisely, if f, g, h are functions on the sample space of (X, Z^β), $(X, Z^{(1)})$, and $(X, Z^{(2)})$ respectively then

$$E(fgh) = \lim_{n \to \infty} E(f_n g_n h_n)$$

with f_n, g_n, $h_n \in A_X^-$. Consequently the composite process $(X, Z^\beta, Z^{(1)}, Z^{(2)})$ is uniquely determined by the processes (X, Z^β), $(X, Z^{(1)})$ and $(X, Z^{(2)})$. Hence the process Y with the given components is unique and since any limit process of $\eta(n)$ must have these components, there is a unique limit process and $\eta(n)$ is stochastic with $X(\eta) = Y$.

To show that $\eta(n)$ is generic for Y we must show that it lies in the sample space of Y. Now since Y is an L-extension of X and $(Z^\beta, Z^{(1)}, Z^{(2)})$ is an inductive function of X it is a standard inductive function by Lemma 30.2. Hence Y is a Markoff process whose transition probabilities are derived from those of X. As a result in order for $\eta(n)$ to lie in the sample space of Y it suffices that all the transitions in $\eta(n)$ have positive probability, which they do since $\eta(n)$ has the transition probabilities of $\xi(n)$, and that each $\eta(n)$ lies in the range of Y. Now introduce a metric $D^*(\lambda, \lambda')$ into the range of Y by defining D^* for $\lambda = (\pi, u_1, u_2, u_3)$ and $\lambda' = (\pi', u_1', u_2', u_3')$ as

$$D^*(\lambda, \lambda') = \begin{cases} \infty & \text{if } \pi \neq \pi' \\ D(u_1, u_1') + D(u_2, u_2') + D(u_3, u_3') & \text{if } \pi = \pi' \end{cases}$$

where D is the metric introduced in §30.1 for $\overline{V(\pi)}$. As in the earlier part of the proof what has to be shown is that the $\liminf$ as $n \longrightarrow -\infty$ of the D^* distance from the $\eta(n)$ to the range of Y is 0. To do so we choose an element $\lambda = (\pi, u_1, u_2, u_3)$ in the range of Y which is maximal in the sense that $\lambda' = (\pi', u_1', u_2', u_3')$ is any other point in the range of Y with $\pi' = \pi$ and $\theta(u_1') \geq \theta(u_1)$, $\theta(u_2') \geq \theta(u_2)$, and $\theta(u_3') \geq \theta(u_3)$ then $\theta(u_1') = \theta(u_1)$, $\theta(u_2') = \theta(u_2)$, $\theta(u_3') = \theta(u_3)$. Since Y is a limit process of $\eta(n)$ there will be elements $\eta(n)$ arbitrarily close to λ in the ordinary topology of the set $\cup_{\pi \in \Pi}(\pi, \overline{V_1(\pi)} \times \overline{V_1(\pi)} \times \overline{V_1(\pi)})$ containing λ. Now the ordinary topology of $\overline{V(\pi)}$ agrees with that induced by D for sets interior to a single face of $\overline{V(\pi)}$. Hence to prove that there are $\eta(n)$ arbitrarily close to λ in the D^*-topology it would suffice to show that there are $\eta(n)$ arbitrarily closed to λ in the ordinary sense for which

$$(31.11) \quad \theta(\zeta'(n)) = \theta(u_1), \; \theta(\zeta^{(1)}(n)) = \theta(u_2), \; \theta(\zeta^{(2)}(n)) = \theta(u_3) \; .$$

As a matter of fact we will show that whenever $\xi(n) = \pi$ this will be the case and since $\xi(n) = \pi$ for $\eta(n)$ sufficiently close to λ our proof will be complete.

Consider the processes W^β, $W^{(1)}$, and $W^{(2)}$ with variables $w_n^\beta = \theta(z_n^\beta)$, $w_n^{(1)} = \theta(z_n^{(1)})$, $w_n^{(2)} = \theta(z_n^{(2)})$. Since these satisfy (31.6) they are subprocesses of X by hypothesis: $w_n^\beta = g^\beta(x_n)$, $w_n^{(1)} = g^{(1)}(x_n)$, and $w_n^{(2)} = g^{(2)}(x_n)$. Now let U be a neighborhood of the point λ so small that if $(\pi', u_1', u_2', u_3') \in U$, $\pi' = \pi$ and $\theta(u_1') \geq \theta(u_1)$, $\theta(u_2') \geq \theta(u_2)$, and $\theta(u_3') \geq \theta(u_3)$. With positive probability $(x_n, z_n^\beta, z_n^{(1)}, z_n^{(2)}) \in U$ and by the choice of λ as a maximal element of the range it follows that $\theta(z_n^\beta) = \theta(u_1)$, $\theta(z_n^{(1)}) = \theta(u_2)$, $\theta(z_n^{(2)}) = \theta(u_3)$, and $x_n = \pi$, with positive probability. But then we must have $\theta(u_1) = g^\beta(\pi)$, $\theta(u_2) = g^{(1)}(\pi)$, and $\theta(u_3) = g^{(2)}(\pi)$.

Since $w_n^\beta = g^\beta(x_n)$, it follows that $(g^\beta(\xi(n)), \zeta'(n))$ is a sample sequence of (W^β, Z^β). Moreover $\zeta'(n)$ is generic for Z^β. We may therefore apply Lemma 30.7 which asserts that $g^\beta(\xi(n)) = \theta(\zeta'(n))$. Hence when $\xi(n) = \pi, \theta(\zeta'(n)) = g^\beta(\pi) = \theta(u_1)$. In exactly the same way all of (31.11) may be verified in case $\xi(n) = \pi$. This completes the proof of the lemma.

31.4. __A Further Sufficient Condition__. We shall retain the notation of §31.3. In addition we will be considering sequences with values in the spaces $\overline{L(V_1(\pi))}$ and $\overline{F(V_1(\pi))}$, $i = 0, 1$. Since the $\bar\pi$ map $V_0(\pi)$ into $V_1(\pi)$ they also map $\overline{L(V_1(\pi))}$ and $\overline{F(V_1(\pi))}$ into $\overline{L(V_0(\pi))}$ and $\overline{F(V_0(\pi))}$ respectively. Thus we may consider solutions $\eta(n)$ to the equation

$$(31.12) \qquad\qquad \eta(n) = \overline{\xi(n)}\,\eta(n + 1)$$

with $\eta(n)$ either in $\overline{L(V_0(\xi(n)))}$ or $F(V_0(\xi(n)))$. We shall refer to (31.12) as the __adjoint__ equation. Note that $\eta(n)$ is an __inverse__ inductive function of $\xi(n)$, the adjective inverse referring to the reversal of the time scale. The relationship between (31.12) and the equation $\zeta(n + 1) = \zeta(n)\overline{\xi(n + 1)}$ is seen if we combine an ordinary solution $\zeta(n)$ with an ajoint solution whose terms lie in $F(V_0(\xi(n)))$. Namely since $\eta(n) \in F(V_0(\xi(n))) = F(V_1(\xi(n - 1)))$ ($\xi(n)$ being a sample sequence of X) and the elements in $F(V)$ define single valued functions on $\bar V$ we can form the products $\zeta(n - 1)\eta(n)$. By (31.12) this is $\zeta(n - 1)\overline{\xi(n)}\eta(n + 1) = \zeta(n)\eta(n + 1)$. Hence $\zeta(n - 1)\eta(n)$ is independent of n.

If $\theta_1 \in \theta(V)$ we shall mean by $c(\theta_1)$ the vector in the equivalence class $\theta_1 \subset V$ having all its non-zero components equal to 1. By $\theta_0 = \theta_0(V)$ we shall always mean the maximum element of $\theta(V)$ and $c(\theta_0)$ is the vector with all components equal to 1.

Now let X, Z, $\xi(n)$, $V(n)$ be as in Lemma 31.1. We define for each $q < 0$ and $\theta \in \theta(L(V(q)))$, $\theta \neq 0$, a sequence $\eta_{q,\theta}(n)$ with values in $F(V(n-1))$ as the unique solution to (31.12) defined for $n \leq q + 1$ and satisfying $\eta_{q,\theta}(q+1) = c(\theta)/c(\theta_O)$ where $\theta_O = \theta_O(L(V(q)))$. The sequences $\eta_{q,\theta}(n)$ are uniquely determined since (31.12) defines $\eta_{q,\theta}(n-1)$ in terms of $\eta_{q,\theta}(n)$ and $\eta_{q,\theta}(q+1)$ is given.

> LEMMA 31.2. A sufficient condition that there exist
> a sequence $\zeta'(n)$ satisfying $\zeta'(n+1) = \zeta'(n)\overline{\xi(n+1)}$
> with the distances from $\zeta'(n)$ to the boundary $\overline{V'(n)}$
> forming a sequence essentially bounded away from 0 as
> required in Lemma 31.1 is the following: there exists
> an infinite set of negative integers N_k and a se-
> quence of elements $u_k \in V(N_k)$ with the property
> that for each sequence $\eta_{q,\theta}(n)$ the products
> $u_k\eta_{q,\theta}(N_k+1) \geq \varepsilon_q > 0$ for all k sufficiently
> large, and the ε_q form a sequence essentially
> bounded away from 0.

PROOF. Define $\zeta_k(n)$ as the $V(n)$-valued solution to $\zeta(n+1) = \zeta(n)\overline{\xi(n+1)}$ defined for $n \geq N_k$ with $\zeta_k(N_k) = u_k$. For k sufficiently large all the $\zeta_k(n)$ will be defined at $n = q$ and by the remark made earlier we have $\zeta_k(q)\eta_{q,\theta}(q+1) = \zeta_k(N_k)\eta_{q,\theta}(N_k+1) = u_k\eta_{q,\theta}(N_k+1) \geq \varepsilon_q$. But $\eta_{q,\theta}(q+1) = c(\theta)/c(\theta_O)$ and so we have

$$(31.13) \qquad \zeta_k(q)\,\frac{c(\theta)}{c(\theta_O)} \geq \varepsilon_q$$

for all sufficiently large k. Now it is possible to show that if $v = (v_1 v_2 \cdots v_r) \in \tilde{V}$ then the Euclidean distance of v to the boundary of the simplex $\tilde{V}$ is

$$\sqrt{\frac{r}{r-1}}\,v_j$$

where $v_j = \inf v_i$. On the other hand

$$(31.14) \qquad \inf_{\substack{\theta \in \theta(L(V)) \\ \theta \neq 0}} \left(v\,\frac{c(\theta)}{c(\theta_O)} \right) = \inf v_i = v_j$$

since $v\,c(\theta_O) = \Sigma\,v_i = 1$ and $vc(\theta)$ is a minimum when the non-zero components of $c(\theta)$ coincide with the minimum component of v. As a result of this we may use the quantity on the left of (31.14) as a measure

of the distance of v to the boundary, the latter quantity being defined
for $v \in \bar{V}$ as well. It follows that if we could find a single sequence
$\zeta'(n)$ satisfying the inequalities (31.13) and also satisfying
$\zeta'(n + 1) = \zeta'(n)\overline{\xi(n + 1)}$ then $\zeta'(n)$ would be the sequence sought after.
To find the $\zeta'(n)$ in question we take a subsequence of the k for which
$\zeta_k(0)$ converges, a subsequence of this for which $\zeta_k(-1)$ converges and
so forth. A diagonal subsequence then converges to the solution $\zeta'(n)$
desired. This proves the lemma.

As a result of this lemma our attention is focused on the ad-
joint sequences $\eta_{q,\theta}(n)$. A careful study will be made of these in the
next section.

§32. Adjoint Processes

32.1. <u>Conditional Distributions.</u> Let x be a random variable
defined on a sample space Ω and let $\mathscr{B}$ be a Borel field on Ω.

DEFINITION 32.1. Suppose x takes its values in a compact
separable space Λ. The condition distribution of x given $\mathscr{B}$ written
$E^*(x|\mathscr{B})$ is the random measure on Λ defined almost everywhere on Ω by

$$(32.1) \qquad E^*(x|\mathscr{B})(f) = E(f(x)|\mathscr{B}), \quad f \in C(\Lambda).$$

It is not difficult to show that $E^*(x|\mathscr{B})$ is itself "de-
fineable" almost everywhere using the fact that Λ is separable. That
is, although there are undenumerably many f in $C(\Lambda)$ and the right hand
side of (32.1) is defined for each f only up to a set of measure 0, it
is possible to assign a measure $\mu(\omega)$ to almost all points in Ω with
$\mu(\omega)(f) = E(f(x)|\mathscr{B})$. This $\mu(\omega)$ is a version of $E^*(X|\mathscr{B})$. We remark
that in this section we shall consider measure valued random variables
$\mu(\omega)$ which are assumed measurable in the weak sense: $\mu(f)(\omega)$ is
measurable for each f in $C(\Lambda)$. When we speak of a $C(\Lambda)$-valued random
variable we shall only require the notion for simple functions; i.e.,
where f_ω is a constant on each of a finite number of sets. However
what we shall say goes over for appropriate limits of such variables.

LEMMA 32.1. If $f_\omega \in C(\Lambda)$ is a $C(\Lambda)$-valued random
variable measurable with respect to $\mathscr{B}$ then
$E^*(x|\mathscr{B})(f_\omega) = E(f_\omega(x)|\mathscr{B})$ almost everywhere.

PROOF. Suppose first that $f_\omega = g$ for $\omega \in \Delta$ and $f_\omega = 0$ outside of Δ. Let x_Δ be the characteristic function of Δ. Then
$E^*(x|\mathscr{B})(f_\omega) = x_\Delta E^*(x|\mathscr{B})(g) = x_\Delta E(g(x)|\mathscr{B}) = E(x_\Delta g(x)|\mathscr{B}) = E(f_\omega(x)|\mathscr{B})$.
This proves the lemma for linear combinations of step functions and this
is all that we shall need.

Suppose that $Q : \Lambda \longrightarrow \Lambda'$ where Q is continuous. Then each measure μ on Λ is carried into a measure $Q\mu$ on Λ'. We now have

LEMMA 32.2. If the map $Q : \Lambda \longrightarrow \Lambda'$ depends upon ω but is constant on each of a finite number of subsets of Ω then

$$(32.2) \qquad Q \, E^*(x \,|\, \mathscr{B}) = E^*(Qx \,|\, \mathscr{B}) \quad ,$$

if Q is measurable with respect to $\mathscr{B}$.

PROOF. Let $f \in C(\Lambda')$ so that $f \circ Q \in C(\Lambda)$. We then have $Q \, E^*(x \,|\, \mathscr{B})(f) = E^*(x \,|\, \mathscr{B})(f \circ Q)$ and since $f \circ Q$ is measurable with respect to $\mathscr{B}$, $E^*(x \,|\, \mathscr{B})(f \circ Q) = E(f(Q(x)) \,|\, \mathscr{B}) = E^*(Qx \,|\, \mathscr{B})(f)$. This proves the lemma.

32.2. <u>The Fundamental Lemma</u>. Most of our results regarding the sequences $\eta_{q,\theta}(n)$ of §31.4 depend upon one lemma which we shall prove here after a preliminary definition and lemma.

DEFINITION 32.2. A set of fractional linear functionals $\{\ell_\alpha\}$ defined on cones V_α is <u>uniformly bounded</u> if there exists a $\delta > 0$ such that

$$(32.3) \qquad \delta < c(\theta_0(V_\alpha))\ell_\alpha < 1/\delta$$

for all α. (Alternatively, if $\ell_\alpha = h_\alpha / g_\alpha$ this requires that $\delta < \|h_\alpha\| / \|g_\alpha\| < 1/\delta$.)

The functionals ℓ_α may be unbounded when considered over the entire cone V_α; for boundedness in the sense of the foregoing definition only their behavior at the "central" vectors $c(\theta_0(V_\alpha))$ is relevant.

LEMMA 32.3. If Z is a standard inductive function of X where X is a finite state Markoff process then

$$(32.4) \qquad E(f(z_{-n-1}) \,|\, x_{-n}, \, x_{-n+1}, \, \ldots, \, x_0) = E(f(z_{-n-1}) \,|\, x_{-n})$$

for $f \in C(\Lambda)$ where Λ is the range of Z.

PROOF. It must be shown that $\bar{f}(x_{-n}, \, x_{-n+1}, \, \ldots, \, x_0) = E(f(z_{-n-1}) \,|\, x_{-n}, \, x_{-n+1}, \, \ldots, \, x_0)$ depends only on x_{-n}. We denote by $p(\pi, \, \pi')$ the transition probabilities of X. We now have

$$E(f(z_{-n-1})g(x_{-n}, \ x_{-n+1}, \ \ldots, \ x_0)) =$$

$$(32.5) \qquad E\{f(z_{-n-1})E(g(x_{-n}, \ x_{-n+1}, \ \ldots, \ x_0)|x_{-n-1}, \ z_{-n-1})\} =$$

$$E\{f(z_{-n-1})E(g(x_{-n}, \ x_{-n+1}, \ \ldots, \ x_0|x_{-n-1})\}$$

since Z is a st. ind. fn. of X. In terms of the transition probabilities we then have

$$E(f(z_{-n-1})g(x_{-n}, \ x_{-n+1}, \ \ldots, \ x_0)) =$$

(32.6)

$$E\Bigg\{ \sum_{i_1, i_2, \ldots, i_{n+1}} f(z_{-n-1})p(x_{-n-1}, \pi_{i_1})p(\pi_{i_1}, \pi_{i_2})\cdots p(\pi_{i_n}, \pi_{i_{n+1}})$$
$$g(\pi_{i_1}, \ldots, \pi_{i_{n+1}})\Bigg\} =$$

$$\sum_{i_1, i_2, \ldots, i_{n+1}} E(f(z_{-n-1})p(x_{-n-1}, \pi_{i_1}))p(\pi_{i_1}, \pi_{i_2})\cdots p(\pi_{i_n}, \pi_{i_{n+1}})$$
$$g(\pi_{i_1}, \ldots, \pi_{i_{n+1}}) \ .$$

On the other hand the left hand side of (32.6) may be written as

$$(32.7) \qquad E(\bar{f}(x_{-n}, x_{-n+1}, \ldots, x_0)g(x_{-n}, x_{-n+1}, \ldots, x_0)) =$$

$$\sum_{i_1, i_2, \ldots, i_{n+1}} p(\pi_{i_1})p(\pi_{i_1}, \pi_{i_2})\cdots p(\pi_{i_n}, \pi_{i_{n+1}})\bar{f}(\pi_{i_1}, \ldots, \pi_{i_{n+1}})g(\pi_{i_1}, \ldots, \pi_{i_{n+1}}).$$

We now let g range over all functions in Π^{n+1} and compare the right hand sides of (32.6) and (32.7). This gives

$$(32.8) \qquad E(f(z_{n-1})p(x_{-n-1}, \ \pi_{i_1})) = p(\pi_{i_1})\bar{f}(\pi_{i_1}, \ \ldots, \ \pi_{i_{n+1}}) \ .$$

This however shows that $\bar{f}$ depends only on π_{i_1} and this proves the lemma.

We can now prove

LEMMA 32.4. If there exists a non-degenerate projective inductive function of X (Definition 31.1) then there exists a set S in Ω_X^- with $P(S) = 1$ such that if $\xi \in S$ and $\eta(n)$ is a $\bigcup_{\pi \in \Pi} F(V_1(\pi))$-valued sequence

satisfying the adjoint equation $\eta(n) = \overline{\xi(n)}\eta(n + 1)$,
then $\{\eta(n)\}$ is uniformly bounded.

PROOF. Let $\Lambda = U_{\pi \in \Pi} V_1(\pi)$ and define the random measure μ_n
on Λ by

$$\mu_n = E^*(z_o | x_o, x_{-1}, \ldots, x_{-n}) \ ,$$

where Z is the non-degenerate solution to $z_{n+1} = z_n \overline{x_{n+1}}$. μ_n is a
random variable on $\Omega_{\overline{X}}$ and we note that the support of μ_n is in
$\overline{V_1(x_o)}$ since $z_o \in \overline{V_1(x_o)}$. Moreover with probability 1 we have
$\mu_n(\overline{V_1^O(x_o)}) = 1$ since, with probability 1, $z_o \in \overline{V_1^O(x_o)}$, Z being non-
degenerate. By (32.1) $\mu_n(f) = E^*(f(z_o) | x_o, x_{-1}, \ldots, x_{-n})$ and since
the latter converges almost everywhere by the martingale convergence
theorem we have

$$(32.9) \qquad \mu_n \longrightarrow \mu = E(z_o | x_o, x_{-1}, \ldots, x_{-n}, \ldots)$$

almost everywhere and μ will assign measure 1 to $\overline{V_1^O(x_o)}$ with probability 1. We take S to be the set of sample points in $\Omega_{\overline{X}}$ for which
both these statements are true.

Now we can also write

$$\mu_n = E^*(z_{-n-1} \ \bar{x}_{-n} \ \bar{x}_{-n+1} \ \cdots \ \bar{x}_o | x_o, x_{-1}, \ldots, x_{-n})$$

where we think of $\bar{x}_{-n} \ \bar{x}_{-n+1} \ \cdots \ \bar{x}_o$ as a transformation of $\overline{V_o(x_{-n})}$ to
$\overline{V_1(x_o)}$. By Lemma 32.2 this can be rewritten

$$(32.10) \qquad \mu_n = E^*(z_{-n-1} | x_o, x_{-1}, \ldots, x_{-n})\bar{x}_n \ \bar{x}_{-n+1} \ \cdots \ \bar{x}_o$$

where we retain the operator on the right for convenience of notation.
Consider now $E^*(z_{-n-1} | x_o, x_{-1}, \ldots, x_{-n})$. By (32.1) and Lemma 32.3
this measure depends only on x_{-n}

$$(32.11) \qquad E^*(z_{-n-1} | x_o, x_{-1}, \ldots, x_n) = \nu_{x_{-n}} \ .$$

Note that each $z_{-n-1} \in \overline{V_1^O(x_{-n-1})} = \overline{V_o^O(x_{-n})}$ with probability 1 since
Z is non-degenerate. As a result the measure ν_π must assign measure
1 to the interior $\overline{V_o^O(\pi)}$ as well.

We now have for $\xi \in S$

(32.12) $$v_\xi(-n)\overline{\xi(-n)}\,\overline{\xi(-n+1)}\,\ldots\,\xi(0) \longrightarrow \mu_\xi \quad.$$

Using this we shall prove that $\xi \in S$ satisfies the assertion of the lemma. Suppose therefore to the contrary that $\eta(n)$ satisfies the adjoint equation but is unbounded. We can then find an infinite sequence of integers $-n_q$ for which the $\xi(-n_q)$ are all the same, say $\xi(-n_q) = \pi$ (since there are only finitely many π), and such that

(32.13) $$c(\theta_0(V_0(\pi)))\eta(-n_q) \longrightarrow \infty \quad;$$

the case in which the $\eta(n)$ are unbounded from 0 can be reduced to this by taking $\eta'(n) = \eta(n)^{-1}$. We suppose that $\eta(n)$ is defined for $n \leq 1$. We think of $\eta(1) \in F(V_1(\xi(0)))$ as a mapping from $\overline{V_1^0(\xi(0))}$ to the half line $[0, \infty)$; on the boundary $\overline{V_1'(\xi(0))}$, $\eta(1)$ may take on the value ∞. Now $\overline{\xi(-n)}\,\overline{\xi(-n+1)}\,\ldots\,\overline{\xi(0)}$ was a mapping from $\overline{V_0(\xi(-n))}$ to $\overline{V_1(\xi(0))}$. By the hypothesis that there exists a non-degenerate inductive function of X, $\overline{\xi(-n)}\,\overline{\xi(-n+1)}\,\ldots\,\overline{\xi(0)}$ must take the interior $\overline{V_0^0(\xi(-n))}$ into the interior $\overline{V_1^0(\xi(0))}$. Since $v_{\xi(-n)}$ has its support in $\overline{V_0^0(\xi(-n))}$ the left side of (32.12) has its support in $\overline{V_1^0(\xi(0))}$ and so the map $\eta(1)$ may be applied to it. We may therefore write

(32.14) $$v_{\xi(-n)}\overline{\xi(-n)}\,\overline{\xi(-n+1)}\,\ldots\,\overline{\xi(0)}\,\eta(1) \longrightarrow \mu_\xi\eta(1) \quad.$$

However by the adjoint equation this becomes

(32.15) $$v_{\xi(-n)}\eta(-n) \longrightarrow \mu_\xi\eta(1)$$

and for $n = n_q$,

(32.16) $$v_\pi\eta(-n_q) \longrightarrow \mu_\xi\eta(1) \quad.$$

Returning to (32.13) let us suppose that $v \in \overline{V(\pi)}$ and that $\Gamma_{V_0(\pi)}(v, c(\theta_0(V_0(\pi))) \geq \varepsilon$. Here $c(\theta_0(V_0(\pi)))$ is to be identified with its image in $\overline{V(\pi)}$. Now if $\ell = u_1/u_2$ is a fractional linear functional on a cone V and $v_1, v_2 \in V$ then

$$v_1 - K_V(v_1, v_2)v_2 \in V, \quad v_2 - K_V(v_2, v_1)v_1 \in V$$

so that

$$(v_1u_1) - K_V(v_1, v_2)(v_2u_1) \geq 0, \quad (v_2u_2) - K_V(v_2, v_1)(v_1u_2) \geq 0$$

and

$$(32.17) \quad \frac{(v_1 u_1)}{(v_1 u_2)} \geq [K_V(v_1,\ v_2) v_2 u_1] \left[\frac{K_V(v_2, v_1)}{v_2 u_2} \right] = \Gamma_V(v_1,\ v_2) \frac{(v_2 u_1)}{(v_2 u_2)}$$

Accordingly $(v_1 \ell) \geq \Gamma_V(v_1,\ v_2)(v_2 \ell)$; similarly $(v_2 \ell) \geq \Gamma_V(v_1,\ v_2)$ $(v_1 \ell)$. We see then that (32.13) implies that for all v satisfying $\Gamma_{V_0(\pi)}(v,\ c(\theta_0(V_0(\pi)))) \geq \varepsilon$ we have $v\eta(-n_q) \longrightarrow \infty$. As a result, the portion of the measure ν_π having its support in $\Delta_\varepsilon = \{v : \Gamma_{V_0(\pi)}$ $(v,\ c(\theta_0(V_0(\pi)))) \geq \varepsilon\}$ is dissipated to ∞ by $\eta(-n_q)$. Hence if for some ε, Δ_ε actually carried part of the measure ν_π then (32.16) would not be possible since $\mu_\xi\eta(1)$ is a probability measure, i.e., has mass 1 on $(0,\ \infty)$. But $\nu_\pi(V_0^0(\pi)) = 1$ by the remark following (32.11), and therefore $\nu_\pi(\Delta_\varepsilon) \longrightarrow 1$ as $\varepsilon \longrightarrow 0$. This contradiction shows that $\eta(n)$ must be bounded and this completes the proof of the lemma.

We shall derive from this lemma several corollaries which depend upon the following definition.

DEFINITION 32.3. If Z is a non-degenerate inductive function of X then we say a process Y is <u>adjoint</u> to Z if the variables y_n take their values in $\overline{L(V_0(x_n))}$ or in $F(V_0(x_n))$ and if they satisfy

$$(32.18) \qquad\qquad\qquad y_n = \overline{x_n}\, y_{n+1}\ .$$

Y will be referred to as a <u>projective process</u> or an <u>F.L. process</u> according as $y_n \in \overline{L(V_0(x_n))}$ or $y_n \in F(V_0(x_n))$.

We have not specified here the topology on the sets $F(V_0(\pi))$; the reason is that, as we have pointed out, $F(V)$ is a subset of $\overline{L(V) \times L(V)}$ which is a compact set and an F.L. process may be thought of as taking its values in $\cup_\pi \overline{L(V_0(\pi)) \times L(V_0(\pi))}$.

For the subsequent discussion we shall need the following remarks. If $v \in V$ then $\bar{v}$ will denote the image of v in $\bar{V}$; $v/\|v\|$ will be the element of $\tilde{V}$ corresponding to v. If V_1 and V_2 are two cones $V_1 \times V_2$ denotes the cone of pairs $(v_1,\ v_2)$, $v_1 \in V_1$ and $v_2 \in V_2$ (we have already used $L(V) \times L(V)$). Note that $\|(v_1,\ v_2)\| = \|v_1\| + \|v_2\|$. The face in $V_1 \times V_2$ to which $(v_1,\ v_2)$ belongs is determined by the pair of faces to which v_1 and v_2 belong. **Consequently** we shall write either $\theta(v_1,\ v_2)$ or $(\theta(v_1),\ \theta(v_2))$. In imbedding $F(V)$ into $\overline{L(V) \times L(V)}$ we identify the fractional linear functional u_1/u_2 with the element $\overline{(u_1,\ u_2)}$. Note that this is not the same as $(\bar{u}_1,\ \bar{u}_2) \in$ $\overline{L(V)} \times \overline{L(V)}$. In general each element of $\overline{V_1 \times V_2}$ determines a pair of elements of $\bar{V}_1$ and $\bar{V}_2$ but not conversely.

COROLLARY 1. If Y is an F.L. process adjoint to the non-degenerate process Z then the range of Y is uniformly bounded in the sense of Definition 32.2.

PROOF. This follows directly from the lemma since the range of a process is determined by the range of any generic sequence of the process and for Y a generic sequence may be chosen to correspond to a sequence $\xi(n)$ in S, since $P(S) = 1$ (S is the set referred to in Lemma 32.4).

COROLLARY 2. Let (Y', Y'') be a composite process whose components are projective processes adjoint to Z and suppose that the variables y_n'' are non-zero with probability 1. Then there exists an F.L. process W adjoint to Z such that the variables w_n of W may be expressed almost everywhere in the form $w_n = \overline{(w_n', w_m'')}$ with $\overline{w_n'} = y_n'$ and $\overline{w_n''} = y_n''$. (w_n' and w_n'' themselves are not determined by w_n but $\overline{w_n'}$ and $\overline{w_n''}$ are.)

PROOF. Since $P(S) = 1$ we may find a sequence $(\xi(n), \eta'(n), \eta''(n))$ generic for (X, Y', Y'') with $\xi(n)$ belonging to S. We may suppose that $\eta'(n)$ and $\eta''(n)$ satisfy $\eta(n) = \overline{\xi(n)}\eta(n + 1)$. We may find a pair of vector valued sequences $\omega'(n)$ and $\omega''(n)$ such that $\overline{\omega'(n)} = \eta'(n)$ and $\overline{\omega''(n)} = \eta''(n)$ and since $\eta'(n)$ and $\eta''(n)$ satisfy $\eta(n) = \overline{\xi(n)}\eta(n + 1)$ it is possible to choose $\omega'(n)$ and $\omega''(n)$ such that they satisfy

$$(32.19) \qquad\qquad \omega(n) = \overline{\xi(n)}\, \omega(n + 1) \quad ;$$

this time $\overline{\xi(n)}$ is taken as an ordinary linear transformation. Now form the sequence

$$\varphi(n) = \left(\xi(n),\ \eta'(n),\ \eta''(n),\ \frac{\omega'(n)}{\|\omega'(n)\| + \|\omega''(n)\|},\ \frac{\omega''(n)}{\|\omega'(n)\| + \|\omega''(n)\|} \right)$$
$$(32.20)$$

whose last two components belong to the cone $L(V_o(\xi(n)))$. We cannot have $\|\omega''(n)\| = 0$ since $\overline{\omega''(n)} = \eta''(n)$ which is generic for Y'' and we assume that the y_n'' are non-zero. Since the last two components of $\rho(n)$ have norm ≤ 1 they lie in a compact subset of $L(V_o(\xi(n)))$. We may therefore form the limit processes of $\rho(n)$; let U be one of these. Then $U = (X, Y', Y'', W', W'')$. Note that it does not follow from the fact that $\overline{\omega'(n)} = \eta'(n)$ and $\overline{\omega''(n)} = \eta''(n)$ that $\overline{w'_n} = y'_n$ and $\overline{w''_n} = y''_n$ since the operation $v \longrightarrow \bar{v}$ is not a continuous one. In particular if $\|\omega'(n)\|$ were to become small in comparison with $\|\omega''(n)\|$ then in the limit process $w'_n = 0$ whereas the y'_n need not have been 0. However it follows from the choice of $\xi(n)$ that this cannot occur here. Namely the ratio $\omega'(n)/\omega''(n)$ evidently forms a solution to the adjoint equation over $\xi(n)$ and so by the definition of S, $\|\omega'(n)\|/\|\omega''(n)\|$ is bounded away from 0 and ∞. In particular the last two components in (32.20) are bounded away from the 0 element in $L(V_o(\xi(n)))$. Now for a subset of V bounded away from 0 the map $v \longrightarrow \bar{v}$ is continuous and hence we have $w'_n = y'_n$ and $w''_n = y''_n$. Finally since the ratio $\omega'(n)/\omega''(n)$ satisfies the adjoint equation over $\xi(n)$ it follows that w'_n/w''_n is an F.L. process adjoint to Z. It is possible to prove that U is an L-extension of (X, Y', Y'') so that U and hence W is ergodic as required. However this is not necessary since as we have seen a number of times the existence of an arbitrary process satisfying the conditions specified implies the existence of an ergodic process satisfying the same conditions.

If Y' and Y'' are two processes with y'_n and y''_n taking these values in the same projective space V_n, we shall say the $Y' \geq Y''$ if $\theta(y'_n) \geq \theta(y''_n)$ almost everywhere.

COROLLARY 3. Consider the complete solution Y^* to the equation

$$(32.18) \qquad\qquad y_n = \overline{x_n}\, y_{n+1}$$

where the y_n form a projective process. Then assuming there exists a non-degenerate projective inductive function of X among the components of Y^α of Y^* there is one that dominates all other components: $Y^o \geq Y^\alpha$ for all α.

PROOF. It suffices to show that there cannot exist two distinct maximal components in the sense of the foregoing ordering. Suppose then that Y' and Y'' are any two components of Y^*, we will show that there exists a component dominating both of these. To do so form the process $U = (X, Y', Y'', W', W'')$ of the preceding corollary and define Y by setting $y_n = \overline{w'_n + w''_n}$. Y is then a limit process of $\overline{\omega'(n) + \omega''(n)}$ and by (32.19) it follows that Y satisfies $y_n = \overline{x_n}\, y_{n+1}$. Since

$\theta(y_n) = \theta(w_n' + w_n'') \geq \theta(w_n') = \theta(y_n')$ since $y_n' = \overline{w_n'}$ it follows that $Y \geq Y'$. Similarly $Y \geq Y''$. As in the preceding corollary we may also suppose that U is ergodic. Hence the process (X, Y', Y'', Y) corresponds to a subprocess of Y^* and so Y determines a component of Y^* dominating both Y' and Y''. This proves the corollary.

We might point out that it is not always the case that the complete solution to equation (32.18) has a maximal component. For example, in the illustration of §31.1 if we take $\alpha_1 = \alpha_2 = 0$ and β_1 and β_2 are not both 1 then the only solutions to the equations $z_{n+1} = z_n \overline{x_{n+1}}$ (which is the same as (32.18) with n replaced by $-n$) are $z_n = 0$ and $z_n = \infty$, or in terms of the two dimensional cone V, the solutions are $z_n = \overline{(1, 0)}$ and $z_n = \overline{(0, 1)}$. There is no further solution and hence no solution dominating both of these.

> COROLLARY 4. Let $H = \{(\pi, \theta_1, \theta_2)\}$ with $\theta_1, \theta_2 \in \theta(V_o(\pi))$ such that there is a composite process (X, Y', Y'') whose components Y' and Y'' are projective processes adjoint to the non-degenerate process Z with $(\pi, \theta_1, \theta_2)$ in the range of the variables $(x_n, \theta(y_n'), \theta(y_n''))$. Suppose that $\xi(n)$ is any sample sequence of X and that $\eta(n) = \omega'(n)/\omega''(n)$ is an F.L.-solution (i.e., "fractional linear solution") of the adjoint equation $\eta(n) = \overline{\xi(n)}\,\eta(n+1)$. Then if $(\xi(0), \theta(\omega'(0)), \theta(\omega''(0))) \in H$, the sequence $\eta(n)$ will be uniformly bounded and there is a constant δ_o independent of $\xi(n)$ and $\eta(n)$ such that the bound δ of $\eta(n)$ (see (32.3)) satisfies

$$\delta \geq \delta_o \min\,\{K_{V'}(\omega'(0),c(\theta'))K_{V'}(c(\theta''),\omega''(0)),$$

(32.21)

$$K_{V'}(\omega''(0),c(\theta''))K_{V'}(c\theta'),\omega'(0))\}$$

> where $\pi = \xi(0)$, $\theta' = \theta(\omega'(0))$, $\theta'' = \theta(\omega''(0))$ and $V' = L(V_o(\pi))$.

PROOF. As is well known, the Markoff process X is also an "inverse Markoff process", i.e., it remains a Markoff process if the time scale is reversed. (If p_{ij} are the transition probabilities for the direct process X and p_i its stationary probabilities, then the inverse process has transition probabilities

$$p_{1j}^{*} = \frac{p_j}{p_1}\, p_{j1}$$

and the same stationary probabilities.) Now from the point of view of
the inverse process a solution to the equation $w_n = \varphi(x_n,\, w_{n+1})$ is an
inductive function of X. In particular if W is an F.L. process adjoint to Z so that $w_n = \overline{x_n}\, w_{n+1}$ then W is an inductive function of
the inverse process X, and since moreover $w_n \in \overline{L(V_o(x_n)) \times L(V_o(x_n))}$,
W is a projective inductive function. We may therefore apply the results
of §30 to this projective inductive function; in particular, by Theorem
30.1, W is a standard inductive function of the inverse process X.
This means that (X, W) is an inverse Markoff process and that (Lemma
30.1) $E(f(x_n)|x_{n+1},\, w_{n+1}) = E(f(x_n)|x_{n+1})$.

Now for each $\gamma \in H$, $\gamma = (\pi,\, \theta_1,\, \theta_2)$, there is a composite
process (X, Y', Y") whose components Y' and Y" are adjoint to Z,
such that $(y'_n,\, y''_n)$ pass through the faces $(\theta_1,\, \theta_2)$ with positive
probability. By Corollary 2 there is an F.L. process W adjoint to Z
whose components $(w'_n,\, w''_n)$ map onto $(y'_n,\, y''_n)$. Let us choose a particular one of the possible W and denote it W^γ. For the process (X, W^γ)
we find that with probability > 0, $(x_n,\, \theta(w'_n)\, \theta(w''_n)) = (\pi,\, \theta_1,\, \theta_2) = \gamma$.
In particular there will be a point $(\pi,\, \lambda'_o/\lambda''_o)$ in the range of (X, W^γ)
with $\theta(\lambda'_o) = \theta_1$, $\theta(\lambda''_o) = \theta_2$.

Now suppose that $(\pi_j,\, \lambda'/\lambda'')$ is any point in the range of
(X, W^γ) and suppose that $p_{1j} > 0$. Then $p_{j1}^{*} > 0$ and by the fact that
W^γ is a st. ind. fn. of the inverse process X, $P(x_n = \pi_1|x_{n+1} = \pi_j,$
$w^\gamma_{n+1} = \lambda'/\lambda'') = P(x_n = \pi_1|x_{n+1} = \pi_j) = p_{j1}^{*} > 0$. Since $w^\gamma_n = \overline{x_n}\, w^\gamma_{n+1}$ it
follows that the point $(\pi_1,\, \overline{\pi}_1(\lambda'/\lambda''))$ also belongs to the range of
(X, W^γ). Now since the sequence $\xi(n)$ of our corollary is assumed to
be a sample sequence of X all the transitions from $\xi(n)$ to $\xi(n + 1)$
have positive probability. By the foregoing it follows that if
$(\xi(n + 1),\, \lambda'/\lambda'')$ is in the range of (X, W^γ) then so is $(\xi(n),$
$\overline{\xi(n)}(\lambda'/\lambda''))$. In particular if $(\xi(0),\, \eta(0))$ is in the range of (X, W^γ)
and $\eta(n)$ satisfies the adjoint equation $\eta(n) = \overline{\xi(n)}\eta(n + 1)$ then
$(\xi(n),\, \eta(n))$ is in the range of (X, W^γ) for all n.

To prove the corollary we suppose that $\xi(n)$ and $\eta(n)$ are as
given with $(\xi(0),\, \theta(\omega'(0)),\, \theta(\omega''(0))) = (\pi,\, \theta',\, \theta'') \in H$. By what has
been shown we may choose a point $(\pi,\, \lambda'_o/\lambda''_o)$ in the range of (X, W^γ)
where $\gamma = (\pi,\, \theta',\, \theta'')$ such that $\theta(\lambda'_o) = \theta'$, $\theta(\lambda''_o) = \theta''$. Now form, in
the place of $\eta(n)$, the F.L. solution $\eta_o(n)$ to the equation
$\eta(n) = \overline{\xi(n)}\eta(n + 1)$ satisfying $\eta_o(0) = \lambda'_o/\lambda''_o$. By assumption,
$(\xi(0),\, \eta_o(0))$ is in the range of (X, W^γ) and hence all the $(\xi(n),\, \eta_o(n))$
are in the range of (X, W^γ). But by Corollary 1 the elements of the

range of any adjoint F.L. process W^γ are uniformly bounded and so we find that the sequence $\eta_0(n)$ is uniformly bounded. Moreover the bound depends only on γ since it is determined by the range of W^γ. Thus there is always <u>some</u> adjoint sequence $\eta_0(n)$ over $\xi(n)$ which is uniformly bounded and the bound may be taken as one of a finite set of numbers > 0. In particular, letting δ_1 be the smallest of these we have that the $\eta_0(n)$ are always bounded by δ_1.

To carry over this result to the original sequence $\eta(n)$ we observe the following: if $v \in V$ and

$$\frac{u_1'}{u_1''} \in F(V)$$

and

$$\frac{u_2'}{u_2''} \in F(V) \quad ,$$

then

$$(32.22) \qquad \left(v\,\frac{u_1'}{u_1''} \right) \geq K_{L(V)}(u_1', u_2')K_{L(V)}(u_2'', u_1'') \left(v\,\frac{u_2'}{u_2''} \right)$$

and

$$(32.23) \qquad \left(v\,\frac{u_1'}{u_1''} \right) \leq \frac{1}{K_{L(V)}(u_1'',u_2'')K_{L(V)}(u_2',u_1')} \left(v\,\frac{u_2'}{u_2''} \right)$$

To derive (32.22) we note that since $u_1' - K_{L(V)}(u_1', u_2')u_2' \in L(V)$ $(vu_1') - K_{L(V)}(u_1', u_2')(vu_2') \geq 0$ or $(vu_1') \geq K_{L(V)}(u_1', u_2')(vu_2')$. Similarly $(vu_2'') \geq K_{L(V)}(u_2'', u_2')(vu_2')$ and (32.22) follows by multiplying these two inequalities. (32.23) in turn follows from (32.22) by interchanging the subscripts 1 and 2.

To compare the bounds for the sequences $\eta_0(n)$ and $\eta(n)$ we must compare products of the form $(v\eta_0(n))$ and $(v\eta(n))$. Writing

$$\eta_0(n) = \frac{\omega_0'(n)}{\omega_0''(n)} \ , \ \ \eta(n) = \frac{\omega'(n)}{\omega''(n)}$$

we find that

$$\frac{1}{K(\omega''(n),\omega_O''(n))K(\omega_O'(n),\omega'(n))}\,(v\eta_O(n)) \geq (v\eta(n)) \geq$$

(32.24)

$$K(\omega'(n),\ \omega_O'(n))K(\omega_O''(n),\ \omega''(n))\,(v\eta_O(n))\quad,$$

where the K are taken with respect to the appropriate cones. However since $\omega'(n)$, $\omega''(n)$, $\omega_O'(n)$, $\omega_O''(n)$ all satisfy $\omega(n) = \overline{\xi(n)}\omega(n+1)$ and $K(\overline{\xi(n)}u_1,\ \xi(n)u_2) \geq K(u_1,\ u_2)$ (see the proof of Lemma 15.1 (iv)) it follows that for all n,

(32.25)
$$\frac{1}{\delta_1}\,(v\eta_O(n)) \geq (v\eta(n)) \geq \delta_1(v\eta_O(n))$$

where δ_1 is the smaller of $K_{V'}(\omega'(0),\ \omega_O'(0))K_{V'}(\omega_O''(0),\ \omega''(0)))$ $K_{V'}(\omega''(0),\ \omega_O''(0))K_{V'}(\omega_O'(0),\ \omega'(0))$ where $V' = L(V_O(\xi(0)))$. Thus since $\theta(\omega'(0)) = \theta' = \theta(\lambda_O') = \theta(\omega_O'(0))$ and $\theta(\omega''(0)) = \theta'' = \theta(\lambda_O'') = \theta(\omega_O''(0))$ none of the four quantities $K_{V'}(\cdot,\ \cdot)$ vanishes and so the original sequence $\eta(n)$ is uniformly bounded. Moreover we can compare the bounds of $\eta(n)$ and $\eta_O(n)$ using δ_1 as given. To obtain the final result of the corollary we use the fact that $K(u_1,\ u_2)K(u_2,\ u_3) \leq K(u_1,\ u_3)$ (see proof of Lemma 15.1) to replace, say, $K_{V'}(\omega'(0),\ \omega_O'(0))$ by $K_{V'}(\omega'(0),\ c(\theta'))$. This introduces a factor $K_{V'}(\omega_O'(0),\ c(\theta'))$ and since the $\omega_O'(0)$ are finite in number, namely, there is one for each $\gamma \in H$, this factor is absorbed in the constant δ_O. The same holds for the remaining terms. This proves the corollary.

32.3. <u>Restricted Solution to the Problem</u>. We are going to prove here that the assertion (A) made in §31.2 is valid if we make one added assumption. The assumption has to do with the maximal projective process adjoint to Z considered in Corollary 3 to Lemma 32.4. Namely, letting Y^O be this maximal projective process with variables $y_n^O \in \overline{L(V_O(x_n))}$, we assume that with positive probability y_n^O is in the interior of $\overline{L(V_O(x_n))}$. Strictly speaking we shall not prove (A) but rather the inductive step needed in proving (A). That is to say we assume that (A) holds whenever the maximal dimension of the $V(\pi)$ is $< d$ and we shall prove that it holds for the case of maximal dimension d. By Lemma 31.1 the induction step may be carried out if we can find corresponding to the given generic sequence $\xi(n)$ of X a sequence $\zeta'(n)$ which remains at a distance essentially bounded away from 0 from the boundary of $\overline{V_1(\xi(n))}$ and which satisfies the functional equation. By Lemma 31.2 the condition for this is that certain F.L. sequences $\eta_{q,\theta}$ satisfying the adjoint equation are bounded in a manner specified in the

lemma. Our object now is to prove that this is the case under the assumption
we now make regarding the maximal projective process Y^O adjoint to Z.

We shall require the following lemma which could have been
brought as a corollary to Theorem 29.2.

> LEMMA 32.5. Let X be a finite state Markoff process
> with range Π, let Λ be a finite set and ψ a
> mapping of $\Pi \times \Lambda$ into Λ. Then if $\xi(n)$ is a
> right-infinite sequence generic for X and Z
> satisfies $z_{n+1} = \psi(x_{n+1}, z_n)$ there is a sequence
> $\zeta(n)$ with $(\xi(n), \zeta(n))$ generic for (X, Z). Also
> if $\zeta(n)$ is any solution of $\zeta(n + 1) = \psi(\xi(n + 1),$
> $\zeta(n))$ then $\zeta^*(n) = (\xi(n), \zeta(n))$ is "eventually"
> regular in the sense that the truncated sequence
> $\zeta_k^*(n) = \zeta^*(n + k)$ is a regular sequence for some k.
> Moreover, there exists an essentially bounded se-
> quence of integers $k(v)$ (Definition 31.2) such that
> for any ψ-inductive function $\zeta(n)$ of the truncated
> sequence $\xi_v(n) = \xi(n + v)$, the composite sequence
> $(\xi_v(n), \zeta(n))$ must be regular after $k(v)$ terms.

PROOF. The set Λ being finite the map ψ must be compact.
Hence any ψ-inductive function Z of X is a standard inductive function
(Theorem 28.4) and (X, Z) is a $\Pi \times \Lambda$-valued Markoff process. The
first assertion of Theorem 29.2 also applies so that any solution Z has
a representative over $\xi(n)$. If $\xi(n)$ were left-infinite then Theorem
29.2 would apply in its entirety and $\zeta^*(n)$ would itself be regular. We
have already pointed out though that this does not hold for right-infinite
sequences $\xi(n)$ because the elements $\zeta^*(n)$ need not belong to the
range of some ergodic solution (X, Z). In case Λ is finite however it
is possible to show that $\zeta^*(n)$ eventually does lie in some ergodic
range (i.e., ergodic subset of $\Pi \times \Lambda$ for the transition probabilities
of the Markoff process (X, Z)). The proof of Theorem 29.2 shows that once
$\zeta^*(n)$ lies in an ergodic range the remainder of the sequence will be
regular. To see that the $\zeta^*(n)$ eventually lie in some ergodic subset
of $\Pi \times \Lambda$ we argue as follows. If it did not any limit process of $\zeta^*(n)$
would have its range outside any ergodic subset of $\Pi \times \Lambda$. But such a
limit process has the form (X, Z) with Z a ψ-inductive function of
X, and hence (X, Z) is a Markoff process with the transition probabilities determined by X. But a finite state Markoff process must have
an ergodic subset in its range. This shows that eventually $\zeta^*(n)$ lies
in an ergodic subset of $\Pi \times \Lambda$.

Only the last assertion of the lemma remains to be proved. To do so let us define $\psi^*(\pi)$ for $\pi \in \Pi$ as the map of $\Pi \times \Lambda$ into $\Pi \times \Lambda$ defined by $\psi^*(\pi)(\pi', \lambda') = (\pi, \psi(\pi, \lambda'))$. We shall show that for some $q > 0$ the element $\psi^*(\xi(q))\psi^*(\xi(q - 1)) \cdots \psi^*(\xi(1))\lambda^*$ belongs to an ergodic subset of $\Pi \times \Lambda$ for every λ^* in $\Pi \times \Lambda$. Since $\Pi \times \Lambda$ is finite it evidently suffices to know that a q exists for each λ^* in $\Pi \times \Lambda$. But this follows by what has already been shown since if for a particular λ^* no such q were to exist then the right-infinite sequence $\zeta^*(q) = \psi^*(\xi(q))\psi^*(\xi(q - 1)) \cdots \psi^*(\xi(1))\lambda^*$ would have the form $\zeta^*(q) = (\xi(q), \zeta(q))$ with $\zeta(q)$ a ψ-inductive function of $\xi(q)$ and no $\zeta^*(q)$ would belong to an ergodic subset of $\Pi \times \Lambda$. Thus there exists a q with the required property.

Since $\xi(n)$ is regular the finite block of terms $(\xi(1), \xi(2), \ldots, \xi(q))$ occurs in $\xi(n)$ with positive frequency. However a more precise statement may be made since $\xi(n)$ is generic for an ergodic and hence topologically ergodic process. We may therefore apply Theorem 7.1 to $\xi(n)$. Specifically let us define $\varphi(\pi_{i_1}, \pi_{i_2}, \ldots, \pi_{i_q})$ as 1 if $(\pi_{i_1}, \pi_{i_2}, \ldots, \pi_{i_q}) = (\xi(1), \xi(2), \ldots, \xi(q))$ and $\varphi(\pi_{i_1}, \pi_{i_2}, \ldots, \pi_{i_q})$ as 0 otherwise. Setting $\zeta(n) = \varphi(\xi(n + 1), \xi(n + 2), \ldots, \xi(n + q))$ and applying Theorem 7.1 we find that since $\sup_{n \in O}\zeta(n) = 1$ there exists for every ε an N such that some member of $(\zeta(n), \zeta(n + 1), \ldots, \zeta(n + N))$ equals 1 for all n outside a set of density $< \varepsilon$. This is equivalent to asserting that every (N+q)-tuple $(\xi(n + 1), \xi(n + 2), \ldots, \xi(n + N + q))$ contains a block of terms equal to $(\xi(1), \xi(2), \ldots, \xi(q))$, for n outside a set of density $< \varepsilon$. Hence for all n outside a set of density $< \varepsilon$, the entries $(\xi_\gamma(n), \zeta(n))$ must lie in an ergodic set for $n > N + q$ and hence the sequence is regular beyond this point. This proves the lemma.

Another preliminary lemma is the following:

LEMMA 32.6. Let $\xi(n)$ be a sample sequence for X and suppose that $\eta(n) = \omega'(n)/\omega''(n)$ is an F.L. solution to $\eta(n) = \overline{\xi(n)}\eta(n + 1)$ for $-\ell \leq n \leq 0$ with $\omega'(0) = c(\theta_1)$ and $\omega''(0) = c(\theta_2)$ where neither θ_1 nor θ_2 corresponds to the 0 face. There exists a constant $\sigma > 0$ depending only on the matrices $\{\bar{\pi}\}$ such that denoting $\theta(\omega'(-\ell))$ and $\theta(\omega''(-\ell))$ by θ' and θ'' and $L(V_0(\xi(-\ell))$ by V' we have

$$(32.26) \qquad \left. \begin{array}{l} K_{V'}(\omega'(-\ell),\ c(\theta')) \\[1.5ex] K_{V'}(c(\theta'),\ \omega'(-\ell)) \\[1.5ex] K_{V'}(\omega''(-\ell),\ c(\theta'')) \\[1.5ex] K_{V'}(c(\theta''),\ \omega''(-\ell)) \end{array} \right\} \quad \geq \sigma^{\ell}$$

PROOF. (32.26) states that the vectors $\omega'(n)$ and $\omega''(n)$ governed by the equation $\omega(n) = \overline{\xi(n)}\omega(n + 1)$ can increase or decrease at most exponentially. For example, since $c(\theta') = c(\theta(\omega'(-\ell)))$, the latter is the vector with all the non-zero components of $\omega'(-\ell)$ replaced by 1. Hence $K_{V'}(\omega'(-\ell),\ c(\theta'))$ is the smallest non-zero component of $\omega'(-\ell)$. Similarly $K_{V'}(c(\theta'),\ \omega'(-\ell))$ is the reciprocal of the largest component of $\omega'(-\ell)$. Thus we must show that there exists a $\sigma > 0$ such that if $\omega(n)$ satisfies $\omega(n) = \overline{\xi(n)}\omega(n + 1)$ and $\omega(0) = c(\theta)$ for some $\theta \neq \theta(0)$ then the largest component of $\omega(-\ell)$ is $\leq (1/\sigma)^{\ell}$ and the smallest non-zero component of $\omega(-\ell)$ is $> \sigma^{\ell}$. Now since the maximum term of $\omega(n)$ is bounded by the maximum row sum of $\overline{\xi(n)}$ multiplied by the maximum term of $\omega(n + 1)$ we see that the first condition will be satisfied if

$$\sigma < \frac{1}{\text{max. row sum of any } \overline{\pi}}$$

On the other hand the minimum non-zero term of $\omega(n)$ is obtained by some non-vanishing product in $\overline{\xi(n)}\omega(n + 1)$ and hence is greater than the minimum non-zero term of $\omega(n + 1)$ multiplied by the minimum non-zero term of $\overline{\xi(n)}$. Hence the second condition will be true if $\sigma < $ min. non-zero term of any $\overline{\pi}$. This proves the lemma.

With this we can supply the inductive step needed for the proof of (A). We state this as a lemma.

LEMMA 32.7. Let X, Z be as in (A), let $\xi(n)$ be a generic sequence for X and $V(n) = V_1(\xi(n))$, and assume that for the maximal projective process Y^O adjoint to Z the variables y_n^O take values interior to $L(V_O(x_n))$ with positive probability. Then there is a sequence ε_q essentially bounded away from 0 such that the $\eta_{q,\theta}$ defined in §31.4 satisfy

$$(32.27) \qquad c(\theta_O(V(n)))\eta_{q,\theta}(n + 1) > \varepsilon_q$$

for $n < n_q$. Thus the condition of Lemma 31.2 holds with the entire set of negative integers as $\{N_k\}$.

PROOF. We shall write $\eta_{q,\theta}(n) = \omega'_{q,\theta}(n)/\omega''_q(n)$ with $\omega'_{q,\theta}(q+1) = c(\theta) \in L(V(q))$ and $\omega''_q(q+1) = c(\theta_o)$, the vector of $L(V(q))$ with all components $= 1$. $\omega'_{q,\theta}$ and ω''_q both satisfy $\omega(n) = \overline{\xi(n)}\omega(n+1)$. We note that the denominator $\omega''_q(n)$ always dominates the numerator $\omega'_{q,\theta}(n)$ in the sense that $\omega''_q(n) - \omega'_{q,\theta}(n) \in L(V(n-1))$ since this is true for $n = q+1$, and so for every $v \in V(n-1), v\eta_{q,\theta}(n) \leq 1$. Consequently (32.27) says that $\eta_{q,\theta}(n)$ is eventually a uniformly bounded sequence of fractional linear functionals with the bounds ε_q essentially bounded away from 0. The proof of this runs along the following lines. If we knew that for some N, the triple $(\xi(N), \theta(\omega'_{q,\theta}(N)), \theta(\omega''_q(N)))$ is an element of H as defined in Corollary 4 to Lemma 32.4, then by that corollary it would follow that the sequence $\eta_{q,\theta}(n)$ for $n \leq N$ is uniformly bounded. We shall show that although this may not be the case there is another sequence $\bar{\eta}_{q,\theta} = \bar{\omega}'_{q,\theta}/\bar{\omega}''_q$ which is less that $\eta_{q,\theta}$ in the sense that $v\bar{\eta}_{q,\theta}(n) \leq v\eta_{q,\theta}(n)$ for which this is the case. It follows that the $\eta_{q,\theta}(n)$ are bounded and the bound exceeds that of the $\bar{\eta}_{q,\theta}(n)$. In order to prove that the ε_q form a sequence essentially bounded away from 0 we use the estimate given in (32.21). This shows that the bound depends upon the "distance" of $\omega'_{q,\theta}(N)$ and $\omega''_q(N)$ from the central elements in their respective faces. We refer to the $\kappa_{v'}(\omega'(0), c(\theta'))$ occurring in (32.21), with $\omega'(0)$ here replaced by $\omega'_{q,\theta}(N)$, etc. Now by Lemma 32.6 these "distances" depend upon the length of the interval from q to N and the problem reduces to showing that $|N - q|$ remains essentially bounded as a sequence in q. This as we shall see is a consequence of Lemma 32.5. We turn now to the details.

We have already made use of the fact that the process X is an inverse Markoff process, or, in other words, that the process $x'_n = x_{-n}$ is a Markoff process. A process W with w_n satisfying $w_n = \overline{x_n} w_{n+1}$ is an inductive function of X when a reversal of the "direction of time" is made. This was used in the proof of Corollary 4 to Lemma 32.4 in showing that for W a projective process adjoint to Z, (X, W) would be an inverse Markoff process. Similarly it follows that if the w_n are $\theta(L(V_o(x_n)))$-valued variables so that they have a finite range, then W is a compact inductive function of the inverse process X. In these cases the left-infinite sequence $\xi(n)$ generic for X acts as a right-infinite sequence generic for the inverse process. In particular, Lemma 32.5 applies. It follows that if we consider the solutions to

$$(32.28) \qquad\qquad \theta(n) = \overline{\xi(n)}\,\theta(n+1)$$

where $\theta(n) \in \theta(V_O(\xi(n))) = \theta(V_1(\xi(n-1))) = \theta(V(n-1))$ then they will
all eventually (i.e., for   n   sufficiently negative) be regular. More-
over, the composite sequence $(\xi(n), \theta^{(1)}(n), \ldots, \theta^{(r)}(n))$ consisting
of all the solutions to (32.28) will also eventually be regular since it
also has a finite range. If the $\theta(n)$ for which (32.28) holds are only
to be defined for $n \leq q + 1 \leq 0$,   (q + 1   to conform to the notation in
the definition of $\eta_{q,\theta}$)   then the solutions need not be continuable to
n = 0,   and so are not necessarily formed by truncating the solution defined for $n \leq 0$. Thus for all   q   the composite sequence consisting of
$\xi(n)$ and all the solutions up to   q   will become regular for $n \leq q - k(q)$,
but by the lemma referred to $k(q)$ will be essentially bounded in   q.
Now, to prove our lemma it will suffice to show that (37.27) holds for a
particular q with $\varepsilon_q > 0$ depending upon the $k(q)$ just defined. For
the $k(q)$ being essentially bounded it would follow easily that the ε_q
must be essentially bounded away from   0. Hence we shall take   q = - 1
and $k(-1) = k$ will represent the number of terms that have to be passed
before $\xi(n), \theta^{(1)}(n), \ldots, \theta^{(k)}(n))$ is regular. What we will show is
that $\eta_{-1,\theta}(n)$ is uniformly bounded and that a bound may be taken that
depends only on   k. It should be pointed out that some of the $\theta^{(j)}(n)$
that are distinct for   n = 0   may become identical for $n \leq - k$;   however beyond the point   n = - k,   no two $\theta^{(j)}(n)$ can take on the same
value if they are not already equal for   n = - k   since they would subsequently be equal by (32.28) and this would contradict the regularity of
the composite sequence.

We suppose that $(\xi(n), \theta^{(1)}(n), \ldots, \theta^{(r)}(n))$ for $n \leq - k$
is generic for the composite process $(X, T^{(1)}, \ldots, T^{(r)})$ with variables $(x_n, t_n^{(1)}, \ldots, t_n^{(r)})$.   Here, some of the $T^{(j)}$ may be identical
i.e., $t_n^{(j)} = t_n^{(1)})$ since the $\theta^{(j)}(n)$ may have become identical. We
note that by Lemma 32.5 every face valued solution to $t_n = \overline{x_n} \, t_{n+1}$
occurs among the $T^{(j)}$ since every such solution has a representative
over $\xi(n)$. In particular if $t_n^O = \theta(y_n^O)$ where Y^O is the maximal
projective solution to $y_n = \overline{x_n} \, y_{n+1}$,   then $T^O = T^{(1)}$ for some   i.
Now by our hypothesis regarding the maximal process Y^O it follows that
t_n^O is, with positive probability, the maximal face of the cone to which
it corresponds: $t_n^O = \theta_O(\mathbb{L}(V_O(x_n)))$. Hence if $\theta^{(1)}(n)$ represents t_n^O,
then for certain   n, $\theta^{(i)}(n) = \theta_O(V(n-1))$. Thus for these n,
$\theta^{(i)}(n) \geq \theta^{(j)}(n)$ for all   j. On the other hand, if $\theta^{(1)}(n)$ is the
solution that begins with $\theta^{(1)}(0) = \theta_O(V(-1))$ then $\theta^{(1)}(0) \geq \theta^{(j)}(0)$
for all   j   and so by (32.28), $\theta^{(1)}(n) \geq \theta^{(j)}(n)$ for all   j   and all
n. Thus eventually $\theta^{(i)}(n) = \theta^{(1)}(n)$. In other words, the $\theta^{(1)}(n)$
corresponding to the maximal Y^O may be taken as the solution whose value
at   n = 0   is $\theta_O(V(-1))$. We point out that this need not be so if we

discard the hypothesis that the maximal Y^O passes through the interior of some of the cones $L(V_o(x_n))$.

In general if $T^{(j)}$ is an arbitrary solution to $t_n = \overline{x_n}\, t_{n+1}$ there need not exist a projective process Y satisfying $y_n = \overline{x_n}\, y_{n+1}$ with $t_n^{(j)} = \theta(y_n)$. However we can always find some $T^{(\ell)}$ with $t_n^{(\ell)} \leq t_n^{(j)}$ such that $t_n^{(\ell)} = \theta(y_n)$ for some Y. The reason for this is the following. If $\theta^{(j)}(n)$ is generic for $t_n^{(j)}$ and we choose an arbitrary $\eta(0)$ with $\theta(\eta(0)) = \theta^{(j)}(0)$ then if we define $\eta(n)$ by $\eta(n) = \overline{\xi(n)}\,\eta(n+1)$ we obviously will have $\theta(\eta(n)) = \theta^{(j)}(n)$ for all n. If we take a limit process of $(\xi(n), \theta^{(j)}(n), \eta(n))$, denote it $(X, T^{(j)}, Y)$ then we will not necessarily have $\theta(y_n) = t_n^{(j)}$ since as has been pointed out θ is not a continuous map. However it is easy to show that $\theta(y_n) \leq t_n^{(j)}$ (this corresponds to the $\eta(n)$ accumulating toward the boundary of $\theta^{(j)}(n)$) so that $\theta(y_n)$ is the desired process.

We apply the foregoing to an analysis of the F.L. sequence $\eta_{-1,\theta}(n)$ for the case $q = -1$ that we are considering. We have $\eta_{-1,\theta}(n) = \omega'_{-1,\theta}(n)/\omega''_{-1}(n)$ where $\omega'_{-1,\theta}$ and ω''_{-1} are solutions of $\omega(n) = \xi(n)\,\omega(n+1)$ and $\omega'_{-1,\theta}(0) = c(\theta)$, $\omega''_{-1}(0) = c(\theta_o)$, θ, $\theta_o \in \theta(V(-1))$. The sequences $\theta(\omega'_{-1,\theta}(n))$ and $\theta(\omega''_{-1}(n))$ will be solutions to (32.28), and as such will be found among the $\theta^{(j)}(n)$. Moreover $\theta(\omega''_{-1}(n))$ will be a maximal solution since $\theta(\omega''_{-1}(0)) = \theta_o$, and by what has been said, for $n \leq -k$, it is generic for T^O where $t_n^O = \theta(y_n^O)$, Y^O being the maximal projective solution. As regards $\theta(\omega'_{-1,\theta}(n)) = \theta^{(j)}(n)$, it dominates (for $n \leq -k$) some solution $\theta^{(\ell)}(n)$ generic for $T^{(\ell)}$ with $t_n^{(\ell)} = \theta(y_n)$ for some Y.

We now contend that the element $\gamma = (\xi(-k), \theta^{(\ell)}(-k), \theta(\omega''_{-1}(-k)))$ belongs to H. For this it must be shown that γ lies in the range of $(x_n, \theta(y_n'), \theta(y_n''))$ for some (X, Y', Y'') with Y' and Y'' projective processes adjoint to Z. This follows from the foregoing since $(\xi(n), \theta^{(\ell)}(n), \theta(\omega''_{-1}(n)))$ is generic (for $n \leq -k$) for a process $(X, T^{(\ell)}, T^O)$ with $t_n^{(\ell)} = \theta(y_n)$ and $t_n^O = \theta(y_n^O)$. By Lemma 28.2 there is a process extending $(X, T^{(\ell)}, T^O)$, $(X, T^{(\ell)}, Y)$ and (X, T^O, Y^O) so that the composite process $(X, Y, Y^O, T^{(\ell)}, T^O)$ is defined and its variables are $(x_n, y_n, y_n^O, \theta(y_n), \theta(y_n^O))$. Hence the entries of the sequence $(\xi(n), \theta^{(\ell)}(n), \theta(\omega''_{-1}(n)))$ for $n \leq -k$ are in the range of $(x_n, \theta(y_n), \theta(y_n^O))$. (The necessity for combining the processes above arises from the requirement in the definition of H that X, Y' and Y'' be defined jointly and <u>a priori</u> the processes Y and Y^O have no connection.)

Now if $\theta(\omega'_{-1,\theta}(n))$ were equal to $\theta^{(\ell)}(n)$ rather than dominating it we would be through. For then Corollary 4 to Lemma 32.4 would apply to the F.L. solution $\eta_{-1,\theta}(n)$ for $n \leq -k$ with the result that $\eta_{-1,\theta}(n)$ would be uniformly bounded and with the bound depending upon the quantities $\kappa_{V'}(\omega'_{-1,\theta}(-k))$, $c(\theta(\omega'_{-1,\theta}(-k)))$, etc. By Lemma 32.6, since the $\omega'_{-1,\theta}(n)$ and $\omega''_{-1}(n)$ take on, for $n = 0$, the values $c(\theta)$ and $c(\theta_0)$, these quantities are bounded by a number depending only on k. This would show that the bound of $\eta_{-1,\theta}(n)$ depends only on k, and from this we could conclude that for all q, the $\eta_{q,\theta}(n)$ are uniformly bounded and that the bound depends only on $k(q)$.

However we do not have $\theta(\omega'_{-1,\theta}(n)) = \theta^{(\ell)}(n)$; nevertheless matters are not made any worse by this. For since $\theta(\omega'_{-1,\theta}(-k)) \geq \theta^{(\ell)}(-k)$ we can choose a vector in $\theta^{(\ell)}(-k)$ by replacing certain non-zero terms in $\omega'_{-1,\theta}(-k)$ with zeros. Call this vector $\omega'(-k)$. We then proceed to define $\omega'(n)$ for $n \leq -k$ by $\omega'(n) = \xi(n) \omega'(n + 1)$. Since $\theta(\omega'(-k)) = \theta^{(\ell)}(-k)$, Corollary 4 to Lemma 32.4 applies to the F.L. sequence $\omega'(n)/\omega''_{-1}(n)$. Moreover we clearly have $\eta_{-1,\theta}(n) = \omega'_{-1,\theta}(n)/\omega''_{-1}(n) \geq \omega'(n)/\omega''_{-1}(n)$ so that the bound for the $\omega'(n)/\omega''_{-1}(n)$ will also serve as a bound for $\eta_{-1,\theta}(n)$. We need therefore only show that the bound prescribed in (32.21) depends only on k. However the quantities $K_{V'}(\omega'(-k))$, $c(\theta(\omega'(-k)))$, etc. depend only upon the non-zero terms of $\omega'(-k)$ and since these are all components of $\omega'_{-1,\theta}(-k)$ as well, the $K_{V'}$ for $\omega'(-k)$ will not be smaller than the corresponding $K_{V'}$ for $\omega'_{-1,\theta}(-k)$. These however, together with the $K_{V'}$ for $\omega''_{-1}(-k)$, depend only on k, or rather, dominate a quantity depending only on k, by Lemma 32.6. This completes the proof of Lemma 32.7.

It might be helpful at this point to clarify the role of the assumption we have made here regarding the maximal process Y^0. In order to apply Corollary 4 of Lemma 32.4 to the F.L. sequence $\eta_{-1,\theta}(n)$ we must know that the numerator and denominator sequences of $\eta_{-1,\theta}(n)$ eventually pass through faces such that the triple $(\xi(n), \theta(\omega'_{-1,\theta}(n)), \theta(\omega''_{-1}(n)))$ is in H. This means that the faces $\theta(\omega'_{-1,1}(n))$ and $\theta(\omega''_{-1}(n))$ must correspond to certain projective processes. The purpose of our assumption regarding Y^0 was to ensure that this is the case for $\theta(\omega''_{-1}(n))$; this was pointed out in the course of the proof. As to $\theta(\omega'_{-1,\theta}(n))$ it suffices that the result be true for a <u>smaller numerator</u> $\omega'(n)$ since we are interested in bounding $\eta_{-1,\theta}(n)$ from below. However it would not suffice to have the result for a smaller denominator and for this reason it is necessary to know that $\theta(\omega''_{-1}(n))$ itself corresponds to some projective process. What our proof of the lemma has really shown is that if we form the sequences $\eta^0_{q,\theta}(n)$ with the same numerators as $\eta_{q,\theta}(n)$ but whose denominators $\omega^0_q(n)$ satisfy $\theta(\omega'_q(n)) = \theta^0(n)$ with $\theta^0(n)$ generic for T^0 where $t^0_n = \theta(y^0_n)$ and with $\omega^0_q(q + 1) = c(\theta')$ where $\theta' = \theta^0(q + 1)$ then the sequences $\eta^0_{q,\theta}(n)$ are uniformly bounded, etc. With the hypothesis made regarding Y^O it is possible to identify $\eta_{q,\theta}(n)$ with $\eta^O_{q,\theta}(n)$ and this implies the conclusion of the lemma.

32.4. _The Remaining Alternative_. The main result of this section is contained in the following theorem.

> THEOREM 32.1. If Z is a projective inductive func-
> tion of the finite state Markoff process X and
> $\xi(n)$ is generic for X then there exists a sequence
> $\zeta(n)$ with $(\xi(n),\ \zeta(n))$ generic for $(X,\ Z)$.

PROOF. We let d be the largest dimension of the faces to which the z_n belong with positive probability and we proceed by induction on d. The case $d = 1$ being trivial we are left with proving that if the theorem is true for smaller values of d it is true for the given one. We have already made several reductions of the problem; in particular we may assume that Z is a non-degenerate process and that all that need be shown is that the sequences $\eta_{q,\theta}(n)$ defined in §31.4 satisfy the condition of Lemma 31.2. Moreover by Lemma 32.7 we may suppose that the maximal projective process Y^O adjoint to Z (whose existence is guaranteed by Corollary 3 to Lemma 32.4) is such that the variables y^O_n always lie on the boundary of $\overline{L(V_O(x_n))}$, since the theorem has been proven in the alternative case.

We shall fix our attention on $\eta_{-1,\theta}(n)$; after we show that this sequence is uniformly bounded for n inside a fixed set $\{N_k\}$, it will be an easy matter to see how the bound for $\eta_{q,\theta}(n)$ in $\{N_k\}$ depends upon q. Now by the remarks at the close of §32.3 the result we would like for $\eta_{-1,\theta}(n)$ holds in any case for $\eta^O_{-1,\theta}(n)$ as defined there. To make use of this it is necessary to show that $\eta^O_{-1,\theta}(n)$ exists and for this we must prove the existence of the sequence $\omega^O_{-1}(n)$ with the properties: $\omega^O_{-1}(n) = \overline{\xi(n)}\ \omega^O_{-1}(n + 1)$, $\theta(\omega^O_{-1}(n)) = \theta^O(n)$ where $\theta^O(n)$ is generic for T^O where $t^O_n = \theta(y^O_n)$ (Y^O is as usual the maximal projective solution), and finally $\omega^O_{-1}(0) = c(\theta^O(0))$. For this it is only necessary to show that $\theta^O(n)$ exists with $(\xi(n),\ \theta^O(n))$ generic for $(X,\ T^O)$, for then $\theta^O(n) = \overline{\xi(n)}\ \theta^O(n + 1)$ so that if $\theta(\omega^O_{-1}(n)) = \theta^O(n)$ holds for $n = 0$ and $\omega^O_{-1}(n) = \overline{\xi(n)}\ \omega^O_{-1}(n + 1)$ then the relationship in question holds for all n. Consequently $\omega^O_{-1}(n)$ could be defined by the functional equation and the condition that $\omega^O_{-1}(0) = c(\theta^O(0))$. However the existence of $\theta^O(n)$ with $(\xi(n),\ \theta^O(n))$ generic for $(X,\ T^O)$ follows from Lemma 32.5, since T^O is an inductive function of the inverse process X and $\xi(n)$ appears as a right-infinite generic sequence for this process. Thus $\omega^O_{-1}(n)$ exists and defining

$\eta^{O}_{-1,\theta}(n)$ as the F.L. sequence with the same numerator as $\eta_{-1,\theta}(n)$ and denominator $\omega^{O}_{-1}(n)$, we may say that the sequence $\eta^{O}_{-1,\theta}(n)$ is uniformly bounded and that a bound may be determined in terms of the sequence $\xi(n)$ (i.e., it depends upon $k = k(-1) = $ number of terms before a specified sequence defined in terms of $\xi(n)$ becomes regular).

Now since $\eta_{-1,\theta}(n)$ and $\eta^{O}_{-1,\theta}(n)$ have the same numerator it follows that

$$(32.29) \qquad \eta^{O}(n) = \frac{\eta_{-1,\theta}(n)}{\eta^{O}_{-1,\theta}(n)} = \frac{\omega^{O}_{-1}(n)}{\omega''_{-1}(n)}$$

is an F.L. sequence; in fact, $\eta^{O}(n) = \eta_{-1,\theta'}(n)$ where $\theta' = \theta^{O}(0)$. Moreover by (32.29) it follows that if $\eta^{O}(n)$ is shown to be uniformly bounded for a sertain set of n, the same will be true of $\eta_{-1,\theta}(n)$ since $\eta^{O}_{-1,\theta}(n)$ is known to be uniformly bounded. Our theorem therefore reduces to showing that there is a set $\{N_{k}\}$ such that the $\eta^{O}(N_{k} + 1)$ (see Lemma 31.2) are uniformly bounded for k sufficiently large, and to seeing how the bound for these functionals depends upon $\xi(n)$ (in order to extend the result to $\eta_{-q,\theta}$). To simplify the notation we write $\eta^{O}(n) = \omega^{O}(n)/\omega(n)$; both numerator and denominator satisfy the adjoint equation over $\xi(n)$ and $\omega^{O}(0) = c(\theta^{O}(0))$, $\omega(0) = c(\theta_{O})$, i.e., $\omega(0)$ is the vector with all components equal to 1, whereas $\omega^{O}(0)$ has 1's only in the position corresponding to $\cdot\theta^{O}(0)$. Since $\omega^{O}(n)$ and $\omega(0)$ satisfy the adjoint equation we will have

$$\frac{\omega_{O}(-n)}{\omega(-n)} = \overline{\xi(-n)}\ \overline{\xi(-n+1)} \ \ldots\ \overline{\xi(-1)}\ \frac{\omega^{O}(0)}{\omega(0)}$$

and consequently to determine the behavior of $\omega^{O}(-n)/\omega(-n)$ it will be necessary to study the matrix products $\overline{\xi(-n)}\ \overline{\xi(-n+1)}\ \ldots\ \overline{\xi(-1)}$.

We do this be decomposing the matrices $\overline{\xi(-n)}$ by means of the sequence $\theta^{O}(n)$. That is, the matrix $\overline{\xi(-n)}$ transforms $L(V(n))$ into $L(V(n-1))$ and takes the face $\theta^{O}(n)$ into $\theta^{O}(n-1)$. Hence it takes the subspace $\overline{\theta^{O}(n)}$ (the closure of $\theta^{O}(n)$) into $\theta^{O}(n-1)$. We may therefore express $\overline{\xi(-n)}$ as

$$(32.30) \qquad \overline{\xi(-n)} = \begin{pmatrix} \xi_{11}(-n) & \xi_{12}(-n) \\ 0 & \xi_{22}(-n) \end{pmatrix}$$

where we have reordered the basis elements in $L(V(n))$ so that those corresponding to the subspace $\overline{\theta^{O}(n)}$ appear first; and similarly for $L(V(n-1))$. Given the matrices $\overline{\xi(-n)}$ the matrices $\xi_{ij}(-n)$ are then fully determined by the sequence $\theta^{O}(-n)$ (if we suppose that in reordering the basis elements a fixed procedure is followed). We note that the same procedure may be followed over any sample sequence $(\xi'(n),\ \theta'(n))$ of $(X,\ T^{O})$ and we thus obtain, as a subprocess of $(X,\ T^{O})$, a compound process $(x_{11}^{n},\ x_{12}^{n},\ x_{22}^{n})$ and moreover $(\xi_{11}(n),\ \xi_{12}(n),\ \xi_{22}(n))$ will be generic for the latter.

Using the decomposition (32.30) the product $\overline{\xi(-n)}\ \overline{\xi(-n+1)} \cdots \xi(-1)$ becomes

$$(32.31) \quad \begin{pmatrix} \xi_{11}(-n)\xi_{11}(-n+1)\cdots\xi_{11}(-1) & \displaystyle\sum_{j=1}^{n} \xi_{11}(-n)\cdots\xi_{12}(-j)\cdots\xi_{22}(-1) \\ 0 & \xi_{22}(-n)\xi_{22}(-n+1)\cdots\xi_{22}(-1) \end{pmatrix} .$$

If we write

$$\omega^{O}(0) = \begin{pmatrix} c' \\ 0 \end{pmatrix}, \quad \omega(0) = \begin{pmatrix} c' \\ c'' \end{pmatrix}$$

where c' and c'' consists of 1's we find that the problem of showing that $\omega^{O}(n)/\omega(n)$ is bounded amounts to showing that

$$\left\| \sum_{j=1}^{n} \xi_{11}(-n)\cdots\xi_{12}(-j)\cdots\xi_{22}(-1)c'' \right\|$$

and $\|\xi_{22}(-n)\xi_{22}(-n+1)\cdots\xi_{22}(-1)c''\|$ do not become arbitrarily large in comparison with $\|\xi_{11}(-n)\xi_{11}(-n+1)\cdots\xi_{11}(-1)c'\|$. If we define the norm of a matrix $\|M\|$ as the maximum of its row sums (for a matrix with non-negative entries) then this involves comparing

$$\left\| \sum_{j=1}^{n} \xi_{11}(-n)\cdots\xi_{12}(-j)\cdots\xi_{22}(-1) \right\|$$

and $\|\xi_{22}(-n)\xi_{22}(-n+1)\cdots\xi_{22}(-1)\|$ with $\|\xi_{11}(-n)\xi_{11}(-n+1)\cdots\xi_{11}(-1)\|$. If the $\xi_{ij}(n)$ were all 1-dimensional we would divide the first two by the third and what would be required would be to show that the series

$$(32.32) \qquad \sum_{k=1}^{\infty} \frac{\xi_{22}(-k)\,\xi_{22}(-k+1)\ldots\xi_{22}(-1)}{\xi_{11}(-k)\,\xi_{11}(-k+1)\ldots\xi_{11}(-1)}$$

converges. We point this out because our proof will be largely patterned after the 1-dimensional case and in particular will be based upon the convergence of a series of matrix products analogous to (32.32). (We might remark that the example of §31.1 for

$$\beta_1^{p_1}\beta_2^{p_2} < 1$$

leads to a situation of the kind under consideration here. The matrices $\xi_{ij}(n)$ are 1-dimensional and the $\bar{\pi}_1$ and $\bar{\pi}_2$ are then already given in the form (32.30).) We note that since we are operating on fractional linear functionals the matrix $\overline{\xi(-n)}$ and any scalar multiple thereof define the same operation. Our object in the next few paragraphs will be to show that the $\overline{\xi(n)}$ may be normalized in such a way that the $\xi_{11}(n)$ terms act essentially as identity operators. We shall then be left with showing that

$$\left\| \sum_{j=1}^{n} \xi_{12}(-j)\ldots\xi_{22}(-1) \right\|$$

and $\|\xi_{22}(-n)\xi_{22}(-n+1)\ldots\xi_{22}(-1)\|$ remain bounded.

The decomposition of the spaces $L(V(\pi))$ by means of faces $\theta^{O}(n)$ determines a corresponding decomposition of the dual spaces $V(\pi)$. Thus if $v\bar{\pi} = w$ for vectors v and w in $V_{O}(\pi)$ and $V_1(\pi)$ and $\bar{\pi}$ is decomposed into

$$\bar{\pi} = \begin{pmatrix} \pi_{11} & \pi_{12} \\ 0 & \pi_{22} \end{pmatrix}$$

then the relationship $v\bar{\pi} = w$ may be expressed by

$$(v^{(1)}v^{(2)}) \begin{pmatrix} \pi_{11} & \pi_{12} \\ 0 & \pi_{22} \end{pmatrix} = (w^{(1)}w^{(2)}) \ .$$

In particular, consider the given process Z satisfying $z_n \overline{x_{n+1}} = z_{n+1}$.
We may then take the z_n to be vector valued with $z_n \in V_1(x_n)$, this time satisfying $z_n \overline{x_{n+1}} \sim z_{n+1}$ with "$\sim$" signifying "equal to a scalar multiple of". This relationship then becomes

$$(32.33) \qquad (z_n^{(1)} z_n^{(2)}) \begin{pmatrix} x_{11}^{n+1} & x_{12}^{n+1} \\ 0 & x_{22}^{n+1} \end{pmatrix} \sim (z_{n+1}^{(1)} z_{n+1}^{(2)})$$

Now this implies that $z_n^{(1)} x_{11}^{n+1} \sim z_{n+1}^{(1)}$. It follows that the vectors $z_n^{(1)}$ determine a projective inductive function of (X, T^O) with a new assignment of cones and matrices. At this point we make use of the assumption that the variables y_n^O always belong to the boundary of $\overline{L(V_O(x_n))}$. As a result the decomposition of $L(V_O(x_n))$ affected by the $\theta(y_n^O)$ is always a proper one in the sense that $L(V_O(x_n))$ is decomposed into two non-trivial cones. It follows that the 0 matrix always occurs in the expression for $\overline{x_n}$. From this we see that the projective process Z' determined by the $z_n^{(1)}$ will have its maximal dimension $< d$, the maximal dimension of Z. Hence by our induction assumption, the assertion of our theorem holds for Z' and any generic sequence for (X, T^O) may be extended to a generic sequence for (X, T^O, Z'). In particular since $(\xi(n), \theta^O(n))$ is generic there will be a sequence $\zeta'(n)$ such that $(\xi(n), \theta^O(n), \zeta'(n))$ is generic for (X, T^O, Z'). From Z' we may obtain an equivalent vector-valued process $\tilde{Z}^{(1)}$ whose variables $\tilde{z}^{(1)}$ satisfy $\|\tilde{z}_n^{(1)}\| = 1$, and we suppose that $(\xi(n), \theta^O(n), \tilde{\zeta}^{(1)}(n))$ is the corresponding generic sequence for $(X, T^O, \tilde{Z}^{(1)})$. This requires only that $\zeta'(n)$ should never degenerate to 0 but this is excluded by the assumption that Z is non-degenerate, and by Lemma 30.7 which asserts that $\theta(\zeta'(n))$ is the face of $V_1(\xi(n)) = V_O(\xi(n + 1))$ determined by $\theta^O(n + 1)$. We now define, to take the place of the matrices $\overline{x_n}$, x_{11}^n, x_{12}^n, x_{22}^n, the matrices u_n, u_{11}^n, u_{12}^n, u_{22}^n by setting

$$(32.34) \qquad u_n = \frac{\overline{x_n}}{\|\tilde{z}_{n-1} x_{11}^n\|}$$

and letting u_{11}^n, u_{12}^n, u_{22}^n be the corresponding components of u_n:

$$(32.35) \qquad u_n = \begin{pmatrix} u_{11}^n & u_{12}^n \\ 0 & u_{22}^n \end{pmatrix} .$$

We find that

$$(32.36) \qquad \tilde{z}_n^{(1)} u_{11}^{n+1} = \frac{\|\tilde{z}_n^{(1)} x_{11}^{n+1}\|}{\|\tilde{z}^{(1)} x_{11}^{n+1}\|} = \tilde{z}_{n+1}^{(1)} ,$$

for, $\tilde{z}_n^{(1)} u_{11}^{n+1}$ is proportional to $\tilde{z}_{n+1}^{(1)}$ and since both have norm 1 we must have equality. Hence using the process $\tilde{Z}^{(1)}$ and the matrices u_n

the projective relationship $z_n^{(1)} x_{11}^{n+1} \sim z_{n+1}^{(1)}$ is replaced by an equality
of vectors. The usefulness of this normalization will become clear as we
proceed. We observe incidentally that the denominators $\|\tilde{z}_{n-1}^{(1)} x_{11}^n\|$ are
bounded away from 0. For it is obvious by the definition of $\cdot x_{11}^n$, x_{12}^n,
x_{13}^n that no row of x_{11}^n can vanish. Hence in the product $\tilde{z}_{n-1} x_{11}^n$
every component of $\tilde{z}_{n-1}$ will occur multiplied by a non-zero term of x_{11}^n.
Since the x_{11}^n are finite in number there is a lower bound to their non-
zero terms and since $\|\tilde{z}_{n-1}\| = 1$ there must be some component of $\tilde{z}_{n-1}$
exceeding $1/d$. Hence the denominators $\|\tilde{z}_{n-1} x_{11}^n\|$ are bounded from below
and the matrices u_n are bounded in norm. Thus the u_n, u_{11}^n, u_{12}^n, u_{22}^n
form processes and the sequences $u(n)$, $u_{11}(n)$, $u_{12}(n)$, $u_{22}(n)$ defined by
dividing the corresponding $\mathfrak{z}_{1j}(n)$ by $\|\zeta^{(1)}(n-1)\ \mathfrak{z}_{11}(n)\|$ are generic
for these processes.

    With these preliminaries out of the way we may proceed to the
main body of the proof. We now have to show that

$$\left\| \sum_{j=1}^{n} u_{11}(-n)\ldots u_{12}(-j)\ldots u_{22}(-1) \right\|$$

and $\|u_{22}(-n)u_{22}(-n+1)\ldots u_{22}(-1)\|$ do not become arbitrarily large in
comparison with $\|u_{11}(-n)u_{11}(-n+1)\ldots u_{11}(-1)\|$. The essential step in
showing this will be in proving that

$$\left\| \sum_{j=1}^{n} u_{22}(-j+1)u_{22}(-j+2)\ldots u_{22}(-1) \right\|$$

is bounded as $n \longrightarrow \infty$ and this will occupy us for the greater part of
our discussion.

    Set $U(-n) = u(-n)u(-n+1)\ldots u(-1)$ for $n \geq 1$ and let

$$(32.37) \qquad U(-n) = \begin{pmatrix} U_{11}(-n) & U_{12}(-n) \\ 0 & U_{22}(-n) \end{pmatrix}$$

correspond to the decomposition (32.35) of the u_n. At this point we make
use of the definition of Y^{o} as a maximal projective solution adjoint to
Z. Namely we will show that as a result of this the matrices $U_{12}(-n)$ are
such that no column of $U_{12}(-n)$ can vanish with the exception of a set of
n of density o. To show this suppose to the contrary that for a set of
n of positive upper density there is a column of $U_{12}(-n)$ that vanishes.

Then for a set of n of positive upper density the same column of $U_{12}(-n)$ vanishes (recall that the number of columns of $U_{12}(-n)$ is fixed since $U(-n)$ takes $V(-n-1)$ into the fixed space $V(0)$). Let us denote by I_n the indices $(1, 2, \ldots, i_n)$ corresponding to the subspace $\theta^0(n)$ in $V(n-1)$ and by J_n the remaining indices $i_n + 1, \ldots, r_n)$. Thus the vector $\omega^0(n)$ which may be written as a column vector (b_i) has $b_i \neq 0$ for $i \in I_n$ and $b_i = 0$ for $i \in J_n$. Suppose then that the column of $U_{12}(-n)$ corresponding to the index $j \in J_0$ vanishes for a set of n of positive upper density. It follows that if we form the product $\tau(-n) = U(-n)\tau(0)$ where $\tau(0)$ is the column vector with all components 0 but for the j-th component then the components of $\tau(-n)$ corresponding to indices in I_{-n} all vanish. On the other hand $\tau(-n)$ will never be the 0 vector. For this would mean that an entire column of $U(-n)$ vanishes; hence an entire column of $\bar{x}_{-n}\bar{x}_{-n+1}\cdots\bar{x}_{-1}$ would vanish with positive probability. But then

$$z_{-n-1}\bar{x}_{-n}\bar{x}_{-n+1}\cdots\bar{x}_{-1} = z_{-1}$$

would have a 0 component with positive probability contrary to our assumption that Z is non-degenerate. As a result the $\tau(-n)$ form a sequence of non-zero vectors satisfying $\tau(-n) = u(-n)\tau(-n+1)$ and such that $\theta(\tau(-n))$ is not $\leq \theta^0(-n)$, the face corresponding to the maximal projective solution Y^0. If we let $\overline{\tau(-n)}$ represent the image of $\tau(-n)$ in the projective space $\overline{L(V(-n-1))}$ then we have $\overline{\tau(-n)} = \overline{\xi(-n)}\ \overline{\tau(-n+1)}$. Now form a limit process of the sequence $(\xi(n), \overline{\tau(n)}, \theta^0(n))$. This has the form (X, Y, T^0) where Y is a projective process adjoint to Z (note that if $\tilde{\tau}(n) = \tau(n)/\|\tau(n)\|$, then since no column of $\xi(n)$ vanishes identically and the $\{\bar{\pi}\}$ are finite in number it follows that $\|\overline{\xi(n)}\ \tilde{\tau}(n+1)\|$ is bounded away from 0. We then apply the observation following Lemma 30.6 in §30.3 to deduce that Y is a projective process adjoint to Z.) However since for a set of n of positive upper density $\theta(\tau(-n))$ is not $\leq \theta^0(-n)$, it is possible to take the limit process in such a way that with positive probability, $\theta(y_n)$ is not $\leq t_n^0 = \theta(y_n^0)$. This result is incompatible with our assumption that Y^0 is maximal (by the usual argument of adjoining (X, Y^0) to (X, Y, T^0)).

From the foregoing it follows that there is some $U(-n)$ with no column of $U_{12}(-n)$ identically 0. As a result, if we form the product (of variable matrices)

$$(32.38) \qquad U_{-n} = u_{-n}u_{-n+1}\cdots u_{-1} = \begin{pmatrix} U_{11}^{-n} & U_{12}^{-n} \\ 0 & U_{22} \end{pmatrix}$$

then with positive probability the columns of U_{12}^{-n} will all be non-vanishing for some n.

We return to the relationship (32.33) involving the vectors $z_n = (z_n^{(1)} z_n^{(2)})$ which were normalized so that $\|z_n\| = 1$. Consider the process $\tilde{Z}$ with variables $\tilde{z}_n$ defined by

$$\tilde{z}_n = \frac{z_n}{\|z_n^{(1)}\|} \quad .$$

We may then write $\tilde{z}_n = (\tilde{z}_n^{(1)} \tilde{z}_n^{(2)})$ where the $\tilde{z}_n^{(1)}$ are the same as before. It might be mentioned that the range of $\tilde{Z}$ is unbounded in general since $\|z_n^{(1)}\|$ need not be bounded away from 0, but this will not affect our discussion in any way. Now consider the products $\tilde{z}_n u_{n+1}$. By (32.33) we may write $\tilde{z}_n u_{n+1} \sim \tilde{z}_{n+1}$. On the other hand the first component of the product is $\tilde{z}_n^{(1)} u_{11}^{n+1}$ and by (32.36) this equals $\tilde{z}_{n+1}^{(1)}$ so that we must have

$$(32.39) \qquad\qquad \tilde{z}_n u_{n+1} = \tilde{z}_{n+1} \quad .$$

By (32.38) we then have

$$\tilde{z}_{-n-1} U_{-n} = \tilde{z}_{-1} \quad .$$

Writing out the components we find

$$\tilde{z}_{-n-1}^{(1)} U_{11}^{-n} = \tilde{z}_{-1}^{(1)}$$

(32.40)

$$\tilde{z}_{-n-1}^{(1)} U_{12}^{-n} + \tilde{z}_{-n-1}^{(2)} U_{22}^{-n} = \tilde{z}_{-1}^{(2)} \quad .$$

From (32.40) we shall be able to conclude that for our generic point $\xi(n)$ the matrix sums

$$\sum_{n=1}^{\infty} U_{22}(-n) = \sum_{n=1}^{\infty} u_{22}(-n) u_{22}(-n+1) \cdots u_{22}(-1)$$

converges. (The convergence of a series of matrices with non-negative components is equivalent to the boundedness of the norms $\|\Sigma_{n=1}^{N} U_{22}^{(-n)}\|$.) We shall do this by investigating the (variable) products U_{22}^{-n}.

We recall that $\|M\|$, for a matrix M, is the maximum row sum of M; it is easily verified that $\|M_1 M_2\| \leq \|M_1\| \|M_2\|$. We shall use $\rho(v)$ to denote the least element (possibly 0) of a vector v. We then note that

$$(32.41) \qquad \|M\|\rho(v) \leq \|vM\| \leq \|v\|\|M\| \quad .$$

We also note that, as remarked before, the matrices u^{-n}_{1j} $1, j = 1, 2$ are all bounded. Also by definition it is seen that their least non-zero term is bounded away from 0. From this we may deduce that

$$(32.42) \qquad \|u^{-n}_{1j}M\| > \rho_1 \|M\|$$

for some fixed $\rho_1 > 0$ whenever the product in question is defined.

Our first step will be to prove the following: If

$$\lim_{n \to \infty} E\left(\frac{\log \|u^{-n}_{22}\cdots u^{-1}_{22}\|}{n}\right) = \beta \quad,$$

then

$$\frac{\log \|u^{-n}_{22}\cdots u^{-1}_{22}\|}{n} \longrightarrow \beta$$

in measure. Note that the limit in question always exists since $\log \|u^{-n-m}_{22} \cdots u^{-1}_{22}\| \leq \log \|u^{-n-m}_{22} \cdots u^{-n-1}_{22}\| + \log \|u^{-n}_{22} \cdots u^{-1}_{22}\|$; hence writing $a_n = E(\log \|u^{-n}_{22} \cdots u^{-1}_{22}\|)$ we find $a_{n+m} \leq a_n + a_m$ (by stationarity) and so a_n/n converges. Now by ergodicity, if m is sufficiently large and $\varepsilon > 0$

$$(32.43) \qquad \lim_{N \to \infty} \frac{1}{N+1} \sum_{j=0}^{N} \frac{\log \|u^{-m-j}_{22}\cdots u^{-1-j}_{22}\|}{m} < \beta + \varepsilon \,, \qquad a.e.$$

From this and the foregoing remarks it follows that

$$(32.44) \qquad \limsup_{N \to \infty} \frac{\log \|u^{-N}_{22}\cdots u^{-1}_{22}\|}{N} \leq \beta + \varepsilon$$

almost everywhere. Since $\varepsilon > 0$ is arbitrary we conclude

$$(32.45) \qquad \limsup_{N \to \infty} \frac{\log \|u^{-N}_{22}\cdots u^{-1}_{22}\|}{N} \leq \beta \quad .$$

Now if we have a bounded sequence of real-valued functions f_n with $E(f_n) \longrightarrow \beta$ and

$$\limsup_{n \to \infty} f_n \leq \beta$$

it is easy to show that $f_n \longrightarrow \beta$ in measure. For clearly $P(f_n \geq \beta + \varepsilon) \longrightarrow 0$ with $n \longrightarrow \infty$ since otherwise $f_n \geq \beta + \varepsilon$ infinitely often with positive probability. This together with the fact that the f_n are bounded and $E(f_n) \longrightarrow \beta$ implies that $P(f_n \leq \beta - \varepsilon) \longrightarrow 0$. This gives the desired result. As a consequence of this we may find a sequence n_k such that

$$(32.46) \qquad \frac{\log \|u_{22}^{-n_k} \ldots u_{22}^{-1}\|}{n_k} \longrightarrow \beta \qquad \text{a.e.}$$

     The next step will be to show that as a result of (32.40) we must have $\beta < 0$. Assume then that $\beta \geq 0$. From the second equation of (32.40) we find that $\tilde{z}_{-1}^{(2)} \geq \tilde{z}_{-n-1}^{(2)} U_{22}^{-n}$ (componentwise) so that

$$(32.47) \qquad K(\tilde{z}_{-1}^{(2)}, \; \tilde{z}_{-n-1}^{(2)} U_{22}^{-n}) \geq 1$$

where we have omitted the subscript V of K_V for convenience of notation. We find moreover that equality in (32.47) can take place only if some component of $\tilde{z}_{-n-1}^{(1)} U_{12}^{-n}$ vanishes. Now no component of $\tilde{z}_{-n-1}^{(1)}$ can vanish since Z is non-degenerate and so this can only happen if some column of U_{12}^{-n} vanishes. It follows that with positive probability

$$(32.48) \qquad K(\tilde{z}_{-1}^{(2)}, \; \tilde{z}_{-m-1}^{(2)} U_{22}^{-m}) > 1$$

for suitable m. This together with (32.47) implies

$$E\left(\log \frac{1}{K(\tilde{z}_{-1}^{(2)}, \tilde{z}_{-m-1}^{(2)} U_{22}^{-m})} \right) \leq -\alpha$$

for some $\alpha > 0$. By ergodicity we have

$$(32.49) \qquad \lim_{N \to \infty} \frac{1}{N} \sum_{j=1}^{N} \log \frac{1}{K(\tilde{z}_{-j}^{(2)}, \tilde{z}_{-j-m}^{(2)} u_{22}^{-j-m+1} \ldots u_{22}^{-j})} \leq -\alpha \; ,$$

almost everywhere. Now (§15.1) $K(v_1\sigma, v_2\sigma) \geq K(v_1, v_2)$ for a linear transformation σ so that (32.49) implies

$$(32.50) \quad \limsup_{N \to \infty} \frac{1}{N} \sum_{j=1}^{N} \log \frac{1}{K(\tilde{z}_{-j}^{(2)} u_{22}^{-j+1} \ldots u_{22}^{-1}, \ \tilde{z}_{-j-m}^{(2)} u_{22}^{-j-m+1} \ldots u_{22}^{-1})} \leq -\alpha \quad .$$

Also for two vectors in a cone V, $v_1 - K(v_1, v_2)v_2 \in V$ so that $\|v_1\| \geq K(v_1, v_2) \|v_2\|$ and

$$\frac{\|v_2\|}{\|v_1\|} \leq \frac{1}{K(v_1, v_2)}$$

Hence (32.50) gives

$$\limsup_{N \to \infty} \frac{1}{N} \sum_{j=1}^{N} \log \frac{\|\tilde{z}_{-j-m}^{(2)} u_{22}^{-j-m+1} \ldots u_{22}^{-1}\|}{\|\tilde{z}_{-j}^{(2)} u_{22}^{-j+1} \ldots u_{22}^{-1}\|} \leq -\alpha \quad .$$

or after performing the cancellations

$$(32.51) \qquad \limsup_{N \to \infty} \frac{1}{N} \sum_{i=1}^{m} \log \|\tilde{z}_{-N-1}^{(2)} u_{22}^{-N-1+1} \ldots u_{22}^{-1}\| \leq -\alpha \quad .$$

Now suppose that $\beta \geq 0$, i.e., that

$$\lim_{k \to \infty} n_k^{-1} \log \|u_{22}^{-n_k} \ldots u_{22}^{-1}\| \geq 0 \quad .$$

Then since the u_{22}^{-n} are bounded, it follows that

$$\lim_{k \to \infty} n_k^{-1} \log \|u_{22}^{-n_k+j} u_{22}^{-n_k+j+1} \ldots u_{22}^{-1}\| \geq 0$$

for any $j \geq 0$.

By (32.41) we have $\|\tilde{z}_{-N}^{(2)} u_{22}^{-N+1} \ldots u_{22}^{-1}\| \geq \rho(\tilde{z}_{-N}^{(2)})\|u_{22}^{-N+1} \ldots u_{22}^{-1}\|$, so that (32.51) implies

$$\limsup_{k \to \infty} n_k^{-1} \sum_{i=1}^{m} \log \rho(\tilde{z}_{-n_k+i}^{(2)}) \leq -\alpha$$

almost everywhere. Now this implies that the probability that for all but finitely many k

$$\sum_{i=1}^{m} \log \rho(\tilde{z}^{(2)}_{-n_k+1}) < -\frac{\alpha}{2} n_k$$

is 1 and this implies that the probability that all

$$\sum_{i=1}^{m} \log \rho(\tilde{z}^{(2)}_{-n_k+1}) \leq -\frac{\alpha}{2} n_k \qquad \text{for} \quad k \geq k_o$$

is positive if k_o is sufficiently large. We would then have

$$(32.53) \qquad P\left(\sum_{i=1}^{m} \log \rho(\tilde{z}^{(2)}_{-n_k+1}) < -\frac{\alpha}{2} n_k \right) > \delta$$

for some δ if $k \geq k_o$. But the $\tilde{z}^{(2)}_{-n}$ form a stationary sequence so that (32.53) implies that

$$P\left(\sum_{i=1}^{m} \log \rho(\tilde{z}^{(2)}_{i}) < -\frac{\alpha}{2} n_k \right) > \delta$$

and hence

$$P(\log \rho(\tilde{z}^{(2)}_{1}) = -\infty) \geq \frac{1}{m} P\left(\sum_{i=1}^{m} \log \rho(\tilde{z}^{(2)}_{i}) = -\infty \right) \geq \frac{\delta}{m} > 0 \quad .$$

But this is impossible since Z is non-degenerate so that the components of $\tilde{z}^{(2)}_{i}$ are non-vanishing with probability 1.

Hence $\beta < 0$, and therefore for some k

$$(32.54) \qquad E(\log \|u^{-k}_{22} \cdots u^{-1}_{22}\|) < 0 \quad .$$

From here it is not far to our goal. For, the sequence $(\xi(n), u_{22}(n))$ is generic for the process with variables (x_n, u^n_{22}) as we have seen. Hence (32.54) gives for the particular sequence $u_{22}(n)$:

$$(32.55) \qquad \lim_{N \to \infty} \frac{1}{N+1} \sum_{j=0}^{N} \log \|u_{22}(-k-j) \cdots u_{22}(-1-j)\| < 0 \quad .$$

As with equations (32.43) and (32.44) we conclude from this that

$$\limsup_{N \to \infty} \frac{\log \|u_{22}(-N) \cdots u_{22}(-1)\|}{N} < 0$$

or that

$$\limsup_{N \to \infty} \|u_{22}(-N) \cdots u_{22}(-1)\|^{\frac{1}{N}} < 1 \quad .$$

Finally we conclude from (32.56) that

$$\sum_{n=1}^{\infty} \|u_{22}(-n)\ldots u_{22}(-1)\| \quad,$$

and therefore

$$\sum_{n=1}^{\infty} u_{22}(-n)\ldots u_{22}(-1) \quad,$$

converges.

Having shown this we return to the original problem of showing that $\eta_0(n)$ is uniformly bounded for a certain set $\{N_k\}$ of n. The set $\{N_k\}$ will be given in terms of the sequence $\zeta^{(1)}_k(n)$ generic for $\tilde{Z}^{(1)}$. Namely we take $\{N_k\}$ to be the set of n for which $\rho(\tilde{\zeta}^{(1)}(n)) > \varepsilon_0$ where ε_0 is chosen sufficiently small so that the corresponding set of n is infinite. (In fact, the density of this set may be made arbitrarily close to 1 since the $\tilde{\zeta}^{(1)}(n)$ are generic for $\tilde{Z}^{(1)}$ and the variables of $\tilde{Z}^{(1)}$ have a vanishing component with probability 0.)

We recall that the proof of the boundedness of $\eta^0(n)$ was reduced to showing that

$$\frac{\left\| \displaystyle\sum_{j=1}^{n} u_{11}(-n)\ldots u_{12}(-j)\ldots u_{22}(-1) \right\|}{\|u_{11}(-n)u_{11}(-n+1)\ldots u_{11}(-1)\|} \quad \text{and}$$

(32.57)

$$\frac{\|u_{22}(-n)u_{22}(-n+1)\ldots u_{22}(-1)\|}{\|u_{11}(-n)u_{11}(-n+1)\ldots u_{11}(-1)\|}$$

are bounded for $-n \in \{N_k + 1\}$ (by (32.31) and the fact that since we are dealing with ratios, the $u_{ij}(-n)$ may be used in the place of $\xi_{ij}(-n)$). Clearly it suffices to show that the numerators are bounded and the denominators bounded away from 0. However by (32.41) we have

$$\|u_{11}(-n)u_{11}(-n+1)\ldots u_{11}(-1)\| \geq \frac{\|\tilde{\zeta}^{(1)}(-n-1)u_{11}(-n)\ldots u_{11}(-1)\|}{\|\tilde{\zeta}^{(1)}(-n-1)\|}$$

and by (32.36) and the condition that $\|\tilde{z}_n^{(1)}\| = 1$,

$$(32.58) \qquad \|u_{11}(-n)u_{11}(-n+1)\cdots u_{11}(-1)\| \geq \frac{\|\tilde{\zeta}^{(1)}(-1)\|}{\|\tilde{\zeta}^{(1)}(-n-1)\|} = 1 \, .$$

Hence the denominators are bounded. Now the numerator of the second fraction in (32.57) is not only bounded but tends to 0 by (32.56). Moreover by (32.41) we have

$$\left\| \sum_{j=1}^{n} u_{11}(-n)\cdots u_{12}(-j)\cdots u_{22}(-1) \right\| \leq$$

$$\frac{1}{\rho(\tilde{\zeta}^{(1)}(-n-1))} \left\| \sum_{j=1}^{n} \tilde{\zeta}^{(1)}(-n-1)u_{11}(-n)\cdots u_{12}(-j)\cdots u_{22}(-1) \right\|$$

and since $-n-1 \in \{N_k\}$ and by (32.36) again, the right hand side is less than

$$\frac{1}{\varepsilon_0} \left\| \sum_{j=1}^{n} \tilde{\zeta}^{(1)}(-j-1)u_{12}(-j)\cdots u_{22}(-1) \right\| \leq$$

$$(32.59) \qquad \frac{1}{\varepsilon_0} \left\| \sum_{j=1}^{n} u_{12}(-j)\cdots u_{22}(-1) \right\| \leq$$

$$\kappa_0 \left\| \sum_{j=1}^{n} u_{22}(-j+1)\cdots u_{22}(-1) \right\|$$

since $\|\tilde{\zeta}^{(-1)}(-j-1)\| = 1$ and the $\|u_{12}(-j)\|$ are bounded. Here κ_0 is a constant depending only on ε_0 and the original matrices $\bar{\pi}$ (since the bound of u_{ij} depends only on the original matrices). Since the infinite series

$$\sum_{j=1}^{\infty} u_{22}(-j+1)\cdots u_{22}(-1)$$

converges we find that the $\eta_0(N_k + 1)$ are uniformly bounded.

To complete the proof of the theorem we must see how the bound for $\eta_0(N_k + 1)$ depends upon the sequence $\xi(n)$, in order to determine how ε_q depends upon q. Now from the preceding results we find that ε_{-1}, the bound for the sequence $\eta^0(N_k + 1)$, may be expressed in terms of

$$\left\| \sum_{n=1}^{\infty} u_{22}(-n)\ldots u_{22}(-1) \right\| .$$

Our problem therefore is to show that

$$(32.60) \qquad \sigma_q = \left\| \sum_{n=1}^{\infty} u_{22}(-n+q+1)\ldots u_{22}(q) \right\|$$

is essentially bounded in $q \leq -1$. In order to do this we recall Theorem 7.3 which states that for an ergodic sequence the averages of its derived sequences are approached "almost uniformly". We apply this to the ergodic sequence $u_{22}(-n)$ and the derived sequence whose expectation is given in (32.54). Suppose the left hand side of (32.54) equal to $-\gamma$. Theorem 7.3 implies that for some sufficiently large N_0,

$$\frac{1}{N+1} \sum_{j=0}^{N} \log \|u_{22}(-k-j+q+1)\ldots u_{22}(-j+q)\| < -\frac{\gamma}{2}$$

for all $N > N_0$ if q is outside a set of upper density $< \varepsilon$. From this we may conclude that if N_0 is sufficiently large

$$(32.61) \qquad \|u_{22}(-N+q)\ldots u_{22}(q)\| < e^{-N\gamma/4k}$$

whenever $N > N_0$ and q is outside a set of upper density $> \varepsilon$. Using (32.61) we find that σ_q depends only upon N_0 and since a uniform N_0 may be chosen for all q outside a set of small density it follows that the σ_q are essentially bounded in q. This completes the proof of the theorem.

32.5. _Some Remarks_. Our first comment is that the results obtained in the last three sections apply not only to projective processes but also to ordinary vector-valued processes with matrix transformations. For this one need only observe that a vector v in R^m can be identified with the vector $(1, v)$ in R^{m+1} which in turn may be identified with the point $\overline{(1, v)}$ in the projective space associated with R^{m+1}. To this representation of vectors there corresponds the representation

$$\sigma \longrightarrow \begin{pmatrix} 1 & 0 \\ 0 & \sigma \end{pmatrix}$$

of matrices into matrices of higher dimension. By this means a given vector process Z satisfying $z_{n+1} = z_n \overline{x_{n+1}}$ in the sense of ordinary multiplication of vectors with matrices may be made to correspond to a projective process $\tilde{Z}$ whose variables $\tilde{z}_n$ satisfy $\tilde{z}_{n+1} = \tilde{z}_n \tilde{x}_{n+1}$ where the $\tilde{x}_{n+1}$ are projective transformations. Using this procedure we find that a generic X sequence may be extended to a generic sequence of (X, Z).

There is however an important difference between the vector valued solutions to

$$z_{n+1} = z_n \overline{x_{n+1}}$$

and the projective solutions. Namely, whereas the latter always exist, the former "in general" do not. If, for example X is taken as a random process whose variables are square matrices with positive entries then a vector valued process Z can exist only if each of the π_i in the range of X is a matrix with maximal eigenvalue 1, and if each of the products $\pi_{i_1} \pi_{i_2} \cdots \pi_{i_k}$ has the same property. Thus for dimension 1, the π_i would all have to equal 1. This may be seen as a consequence of Lemma 32.4 and its corollaries by considering the F.L. processes adjoint to the projective process corresponding to Z.

Our next comment is connected with the example of §31.1 where it was claimed that a non-trivial process Z exists if and only if

$$\beta_1^{p_1} \beta_2^{p_2} < 1 \quad .$$

We shall not go into all the details of this; however we point out that this inequality corresponds to the inequality $E(\log \|u_{22}^{-k} \cdots u_{22}^{-1}\|) < 0$, used in the proof of Theorem 32.1. The proof of our assertion accordingly becomes a corollary to the proof of Theorem 32.1. The case in which $\alpha_1 = \alpha_2 = 0$ is of some interest and we shall consider it in more detail. Here the operations $\overline{\pi}_1$, $\overline{\pi}_2$ become multiplication by β_1 and β_2 (at random) and our assertion states that the only way in which a Markoff process Z can exist with these as transitions is if $\beta_1 = \beta_2 = 1$. Now this may be proven by taking as the state space the line $(-\infty, \infty)$ and with $\bar{\pi}_1$, $\bar{\pi}_2$ corresponding to addition of $\log \beta_1$ and $\log \beta_2$. The stationary measure μ of the process must then satisfy

$$\int_{-\infty}^{\infty} f(t)d\mu(t) = p_1 \int_{-\infty}^{\infty} f(t + \log \beta_1)d\mu(t) +$$

$$+ p_2 \int_{-\infty}^{\infty} f(t + \log \beta_2)d\mu(t)$$

and taking $f(t) = e^{i\lambda t}$ shows that μ cannot be a probability measure.

There is however another way of showing this which is more in line with our present approach. We note that if $\alpha_1 = \alpha_2 = 0$ then the adjoint equation $y_n = \bar{x}_n \, y_{n+1}$ is the same as the original equation but with the inverse matrices: $y_{n+1} = y_n \, \bar{x}_n^{-1}$, and a change of indexing. Also there always exist two solutions to the adjoint equation corresponding to the vectors $\binom{1}{0}$ and $\binom{0}{1}$. Now if there exists a non-degenerate solution to the original equation then by Corollaries 1 and 2 to Lemma 32.4 there exists a bounded F.L. process with numerator and denominator corresponding to these vectors. Over any sequence of X, the F.L. solution would have the form

$$\eta(-n) = \binom{1}{0} \Big/ \begin{pmatrix} 0 \\ \beta_1^{-\nu_1} & \beta_2^{-\nu_2} \end{pmatrix} \quad .$$

For this to be bounded we must have $\beta_1^{-\nu_1} \beta_2^{-\nu_2}$ bounded away from 0 and ∞ for all sample sequences and this is clearly only possible if $\beta_1 = \beta_2 = 1$.

By the same type of reasoning it is possible to prove:

If X is a finite state Markoff process and $y_{n+1} - y_n = f(x_{n+1})$ where the $f(x_n)$ are numerical values associated with the variables x_n, then the y_n form a subprocess of X and $y_n = F(x_n)$ for some function F.

This is done by applying the results of this section to the matrices

$$\overline{x_n} = \begin{pmatrix} 1 & 0 \\ 0 & \exp f(x_n) \end{pmatrix} .$$

The last result, we remark, might also be obtained by an analysis of the stationary measure of the (necessarily) Markoff process (x_n, y_n).

§33. Application to Statistical Predictability

33.1. <u>Resumé</u>. The notion of statistical predictability was based upon the idea of a continuously predictable cover $\tilde{X}$ of a process X. $\tilde{X}$ was first defined abstractly as the minimal continuously predictable process extending X. Strictly speaking $\tilde{X}$ was the minimal process with the property that its one-sided version is continuously predictable to all of X (not necessarily $\tilde{X}$); we found however that in case X is finitely-valued then $\tilde{X}$ is itself c.p. We then defined statistical predictability as the condition that X statistically determines $\tilde{X}$ at a preassigned point in Ω_X^-. What this amounted to was that the point in question be generic for X and that it have exactly one generic extension to $\tilde{X}$. We might point out that $\tilde{X}$ is by definition an L-extension of X; otherwise (Lemma 26.1) we could not expect that two distinct generic points of $\tilde{X}$ restrict to distinct generic points of X.

The abstractly defined c.p. cover of X could be given a more precise form if X is finitely-valued and a still more precise form if X is a finitely-valued subprocess of an m-Markoff process (or just Markoff process), i.e., if X is a finite dimensional process. In either case $\tilde{X}$ had the form of a composite process (X, Z) where Z and X are related by a functional equation that just about displays Z as an inductive function of X. When X is a finite dimensional process we have Theorem 24.2 to the effect that the range of X may be identified with linear operations on a cone V with the range of Z a subset $V_1 \subset V$ and

$$(33.1) \qquad\qquad < z_{n-1} x_n, \ \theta > z_n = z_{n-1} x_n$$

where the operators x_n act on the right and θ is a positive functional on V. The subset V_1 is (as suggested by (33.1)) the set of vectors for which $< v, \theta > = 1$. The cone V is not necessarily a cone in the sense used in this chapter but is the quotient space of such a cone V by a certain equivalence relation (or subspace of the corresponding Euclidean space (§24)).

If we identify the elements in V_1 with the elements of the projective space $\overline{V}$ associated with V then (33.1) generally has the form $z_n = z_{n-1}x_n$ where x_n now represents a projective transformation. We say "generally" because it may happen that $z_{n-1}x_n = 0$ in which case (33.1) does not imply that $z_n = 0$ whereas $z_n = z_{n-1}x_n$ would imply this. There will therefore be two tasks confronting us before we can apply the results of this chapter to statistical predictability. We must first generalize these results to the more general case of cones that are quotient spaces of the type of cone considered here. Secondly we must show that $z_{n-1}x_n$ is 0 only with probability 0 so that (33.1) may be replaced by the projective equation.

First, however, a few words to clarify the relevance of our foregoing results to the problem under discussion. All of our results in this chapter had to do with the case the X was a finite state Markoff process. In the present case X is a finite dimensional process, i.e., $x_n = \varphi(y_n)$ where Y is a finite state Markoff process. However if we can write $z_n = z_{n-1}x_n$ then we can also write $z_n = z_{n-1}\overline{y_n}$ with $\overline{y_n} = \varphi(y_n)$. Consequently Z is a projective inductive function of the <u>Markoff process</u> Y. Theorem 32.1 would then imply that any generic point of Y has a generic extension to (Y, Z). This implies that any generic point of X <u>derived from a generic point of</u> Y has a generic extension to (X, Z). Thus at least half of the condition for statistical predictability would be fulfilled at a sequence derived from a generic Markoff sequence. As we have pointed out in §22 this does not include every generic point of the subprocess.

33.2. <u>Application of Theorem 32.1</u>. It would not have been difficult to carry through the entire discussion of §§30-32 for general finite-dimensional, finitely generated cones. In our case we may take advantage of the fact that the cone V has the form V'/H where V' is a cone of the type considered in §§30-32 and H is a subspace of the Euclidean space. Moreover the linear transformations x_n of V are induced by linear transformations of V' (see §24). As a result of this one can show that a solution of $z_{n+1} = z_n x_{n+1}$ has the form $z_n = h(z_n')$ where h is the canonical map of $V' \longrightarrow V'/H$ and the z_n' also satisfy $z_{n+1}' = z_n'x_{n+1}$. In particular it follows from this that Theorem 32.1 holds also for this case; we simply extend first to Z' and then restrict this extension to Z.

To show that the process Z' exists with the properties described we simply take a generic right-infinite sequence $(\xi(n), \zeta(n))$ for the process (X, Z). If $\zeta'(1)$ is any element of the projective space of V' with $h(\zeta'(1)) = \zeta(1)$ and $\zeta'(n)$ is defined inductively by $\zeta'(n + 1) = \zeta'(n)\,\xi(n + 1)$ then $h(\zeta'(n)) = \zeta(n)$ for all n. Hence a limit process

(X, Z) of $(\xi(n), \zeta'(n))$ fulfills the requirements with the exception that (X, Z') need not be ergodic. However by the usual argument, if some process exists at all with the required properties there exists an ergodic process with these properties.

Our next step is to show that (33.1) actually does define a projective inductive function of X (and hence of Y). That is, we must show that the factor $< z_{n-1}x_n,\ \theta >$ vanishes with probability 0, or, what amounts to the same thing, that $z_{n-1}x_n$ vanishes with probability 0. This is a consequence of

LEMMA 33.1. If X is a finitely-valued process with range $\Lambda = \{a_1, a_2, \ldots, a_r\}$ then

$$(33.2) \qquad \sum_{i=1}^{r} [x_0 = a_i]\, E([x_0 = a_i] \mid TA_X^-)$$

is non-zero with probability 1 (§23.1).

PROOF. Let x be the characteristic function of the set at which (33.2) vanishes; we shall show that $E(x) = 0$. From

$$\sum_{i=1}^{r} E(x[x_0 = a_i])\, E([x_0 = a_i] \mid TA_X^-) = 0$$

we obtain

$$(33.3) \qquad \sum_{i=1}^{r} E\{E(x[x_0 = a_i] \mid TA_X^-)\, E([x_0 = a_i] \mid TA_X^-)\} = 0 \quad .$$

Now since $x \leq 1$ we have $E([x_0 = a_i] \mid TA_X^-) \geq E(x[x_0 = a_i] \mid TA_X^-) \geq 0$ so that (33.3) implies

$$\sum_{i=1}^{r} E\{(E(x[x_0 = a_i] \mid TA_X^-))^2\} = 0$$

which implies that $E(x[x_0 = a_i] \mid TA_X^-) = 0$ for each a_i and summing over a_i: $E(x \mid TA_X^-) = 0$ whence $E(x) = 0$.

The meaning of the assertion of the lemma is that with probability 1 the prediction measure assigns positive probability to the value that

actually does occur next. This is seen if we replace TA_X^- by $TA_{\tilde{X}}^-$ for
the c.p. process $\tilde{X}$ in which case the conditional expectations correspond
to evaluations with prediction measures. Using this lemma we can show that
$< z_{n-1} x_n,\ \theta >$ vanishes with probability 0. For the factor $<z_{n-1} x_n,\ \theta>$
of (33.1) corresponds to the factor $\mu_{T\tilde{\omega}}([x_1 = a_1])$ of (23.1) at a point
for which $x_0 = a_1$. As in Lemma 23.1 this factor becomes $E([x_0 = a_1]\ \big|\ TA_X^-)$
where $x_0 = a_1$ (compare with (23.5)). But by the lemma
$E[x_0 = a_1]E([x_0 = a_1]\ \big|\ TA_X^-)$ is non-zero with probability 1 so that each
component $[x_0 = a_1]E([x_0 = a_1]\ \big|\ TA_X^-)$ must be non-zero for $x_0 = a_1$.
As a result we find that the process Z in question is a projective inductive function of X (if we perform a harmless change of variables so
that the z_n take their values in the projective space of V rather than
in V_1). From this it follows that:

If $\xi(n)$ is a finitely-valued derived sequence of a Markoff
sequence $\eta(n)$, then $\xi(n)$ has a generic extension to the c.p. cover
$\tilde{X}$ of the process $X(\xi)$.

For $\tilde{X} = (X,\ Z)$ and Z is a projective inductive function of
X and so it is a projective inductive function of the Markoff process
$Y = X(\eta)$ (where we assume as we may by Lemma 11.3 that $\xi(n)$ depends
only on $\eta(n) : \xi(n) = \varphi(\eta(n))$. Since X is a subprocess of $(X,\ Z)$ and
Y there exists a common extension of $(X,\ Z)$ and Y by Lemma 28.2 and
thus the composite process $(Y,\ Z)$ will be well-defined. By Theorem 32.1
the generic point η extends to a generic point $(\eta,\ \zeta)$ of $(Y,\ Z)$ and
it follows that $(\xi,\ \zeta)$ will be generic for $(X,\ Z) = \tilde{X}$.

33.3. <u>Uniqueness</u>. For the statistical predictability of the
sequence $\xi(n)$ above there is one more condition that needs to be verified.
Namely we must show that $\xi(n)$ has only one generic extension to $\tilde{X}$.
Suppose then that $(\xi(n),\ \zeta(n))$ is generic for $\tilde{X} = (X,\ Z)$ where Z is
taken as a projective process. It is first of all necessary to prove that
$\zeta(n)$ satisfies $\zeta(n + 1) = \zeta(n)\xi(n + 1)$. As we have remarked earlier
this is not obvious because the operations $\xi(n)$ are not continuous on the
projective space of V. Now according to Lemma 30.6, $\zeta(n)$ would satisfy
the functional equation if $\xi(n)$ were generic for a Markoff sequence.
Thus if we knew that $(\eta(n),\ \zeta(n))$ is generic for $(Y,\ Z)$ our assertion
would follow. Although this is not evident we may still proceed as in the
proof of Lemma 30.6. As in the proof of that lemma we conclude that
$\zeta(n + 1) = \zeta(n)\xi(n + 1)$ holds whenever $\zeta(n)$ is not in the null space of
the operator $\xi(n + 1)$. Our next step is to establish that $(\eta(n),\ \zeta(n))$
is stochastic and its limit process $X(\eta,\ \zeta) = (Y,\ Z)$. For this we recall
that $(X,\ Z) = \tilde{X}$ is an L-extension of X. Hence the process $(Y,\ Z)$
extending $(X,\ Z)$ and Y must be an L-extension of Y. It follows that

the process (Y, Z) is uniquely determined by the fact that it extends (X, Z) and Y (and the fact that it is the least such). Since a limit process of $(\eta(n), \zeta(n))$ has these properties it follows that it is unique and identical to (Y, Z). We may conclude from this that for infinitely many n, $(\eta(n), \theta(\zeta(n)))$ is in the range of (Y, W) with W the process whose variables are given by $w_n = \theta(z_n)$. This is done as in the proof of Lemma 30.6, noting that it is sufficient that $(\eta(n), \zeta(n))$ be a stochastic sequence representing (Y, Z), for it is only necessary to know that the $(\eta(n), \zeta(n))$ come arbitrarily close to a point in the range of (Y, Z). Since Y is a Markoff process and W is finitely-valued it follows that (Y, W) is a Markoff process with transition probabilities determined by those of Y. We deduce from this that if $(\eta(n), \theta(\zeta(n)))$ is in the range of (Y, W) for $n = n_1$ and if $\zeta(n + 1) = \zeta(n)\xi(n + 1) = \zeta(n)\varphi(\eta(n + 1))$ for $n_1 \leq n < n_2$ then $(\eta(n_2), \theta(\zeta(n_2)))$ is again in the range of (Y, W). From this it follows that $\eta(n_2)$ cannot be in the null space of $\xi(n_2 + 1) = \varphi(\eta(n_2 + 1))$ since $(\eta(n_2 + 1), 0)$ cannot be in the range of (Y, W). Hence $\zeta(n + 1) = \zeta(n)\xi(n + 1)$ for $n = n_2$ as well. Thus the functional relationship holds for all $n \geq n_1$ where n_1 is one of the infinite set of n for which $(\eta(n), \theta(\zeta(n)))$ is in the range of (Y, W). It follows that the functional relationship holds for all n.

Finally we shall need one more lemma.

LEMMA 33.2. Let X' be an L-extension of X and suppose that the variables of X' take their values in a metric space Λ'. If two generic points $\xi_1'(n)$ and $\xi_2'(n)$ of X' are extensions of the same sample point of X, then the sequence $d(\xi_1'(n), \xi_2'(n))$ is a null sequence (Definition 4.2).

PROOF. It is not hard to see by Definition 4.2 that it suffices to show that for any $f \in C(\Lambda')$ the sequence $\rho(n) = |f(\xi_1'(n)) - f(\xi_2'(n))|$ forms a null sequence. This in turn will follow if it is shown that

$$(33.4) \qquad \limsup_{N \to \infty} \frac{1}{N + 1} \sum_{n=0}^{N} \rho(-n)^2 = 0 \quad .$$

Now the sequences $f(\xi_1'(-n))$ and $f(\xi_2'(-n))$ may be expressed as $\psi(T^n(\xi_1'))$ and $\psi(T^n\xi_2')$ where $\psi = f(x_0')$ is a function on $\Omega_{X'}^-$. Since X' is an L-extension of X, $\psi \in L(A_X^-) \subset L^2(A_X^-)$. Consequently we may find a sequence of functions $\psi_k \in A_X^-$ with $\psi_k \longrightarrow \psi$ in $L^2(A_X^-)$. Now

$\psi_k(T^n\xi_1!) = \psi_k(T^n\xi_2!)$ since $\psi_k \in A_X^-$ and so

$$\rho(-n)^2 = |\psi(T^n\xi_1!) - \psi(T^n\xi_2!)|^2 \leq$$

$$2\,|\psi(T^n\xi_1!) - \psi_k(T^n\xi_1!)|^2 + 2\,|\psi_k(T^n\xi_2!) - \psi(T^n\xi_2!)|^2$$

(using $|\alpha + \beta|^2 \leq 2|\alpha|^2 + 2|\beta|^2$). This gives

$$\lim_{N \to \infty} \sup \frac{1}{N+1} \sum_{n=o}^{N} \rho(-n)^2 \leq 2\,E(|\psi - \psi_k|^2) + 2\,E(|\psi - \psi_k|^2)$$

so that letting $k \longrightarrow \infty$ gives the desired result.

From this and the foregoing remarks we can show that the sequence $\xi(n)$ can have at most one generic extension to $\tilde{X}$. For suppose that $(\xi(n), \zeta_1(n))$ and $(\xi(n), \zeta_2(n))$ are two such extensions. Since $\tilde{X}$ is an L-extension of X it follows from the preceding lemma that $d(\zeta_1(n), \zeta_2(n))$ is a null sequence where d is a metric on the projective space of V. We also know that $\zeta_1(n)$ and $\zeta_2(n)$ both satisfy $\zeta(n+1) = \zeta(n)\xi(n+1)$. We may therefore apply Lemma 30.8 (which extends readily to the case of arbitrary finitely generated cones) and we find that $\zeta_1(n)$ and $\zeta_2(n)$ are identical. With this we have completed the proof of

THEOREM 33.1. If $\xi(n)$ is a finitely-valued derived
sequence of a finite-valued Markoff sequence then
$\xi(n)$ is statistically predictable.

BIBLIOGRAPHY

[1]   Bohr, H., "Kleinere Beiträge zur Theorie der Fastperiodische
      Functionen II," _Danake Videnskabernes Selskab_, 10, (1931).

[2]   van der Corput, J. G., "Diophantische Ungliechungen I, Zur
      Gleichverteilung Modulo Eins," _Acta Mathematica_, 56, (1931).

[3]   Doob, J. L., _Stochastic Processes_, New York, (1953).

[4]   Favard, J., "Sur les équations différentielles linéaires à
      coefficients presque périodiques," _Acta Mathematica_, 51, (1928).

[5]   Helson, H., and Lowdenslager, D., "Prediction theory and Fourier
      Series in Several Variables," _Acta Mathematica_, 99, (1958).

[6]   Hewitt, E. and Savage, L. J., "Symmetric measures on cartesian
      products," _Trans. Amer. Math. Society_, 80, (1955).

[7]   Hopf, E., _Ergodentheorie_, Berlin, (1937).

[8]   Kolmogoroff, A. N., "Stationary sequences in Hilbert space,"
       (Russian), _Bull. Math. Univ. Moscow_, 2, No. 6 (1941), 40 pp.
       (English translation by Natasha Artin.)

[9]   Littlewood, J. E., "On the problem of n bodies," _Meddel. Lunds_
       Univ., Mat. Sem. Suppl. M. Riesz, (1952).

[10]   Loomis, L. H., _An Introduction to Abstract Harmonic Analysis_,
       New York, (1953).

[11]   Masani, P., "Sur les processus vectoriels minimaux de rang maximal,"
       Comptes Rendus, 246, 1958, p. 2215; "Sur la prévision linéaires
       d'un processus vectoriel à densité spectrale non bornée," _Comptes_
       Rendus, 246, 1958, p. 2337.

[12]   Segal, I. E., "Abstract probability spaces and a theorem of
       Kolmogoroff," _Amer. Jour. of Math._, 76 (1954).

[13]   Weyl, H., "Gleichverteilung die Zahlen Mod 1," _Math. Ann._, 77,
       (1916).

[14]   Wiener, N., _Extrapolation, Interpolation, and Smoothing of Stationary_
       Time Sequences, New York, (1949).

[15]   Wiener, N., and Masani, P., "The prediction theory of multivariate
       stochastic process, I, The regularity condition," _Acta Mathematica_,
       98, (1957). "II The linear predictor," _Acta Mathematica_, 99, (1958).

[16]   Wiener, N., and Wintner, A., "Harmonic analysis and ergodic theory,"
       Amer. Jour. of Math., 63, (1941).

PRINCETON MATHEMATICAL SERIES

Edited by Marston Morse and A. W. Tucker

PRINCETON UNIVERSITY PRESS

PRINCETON, NEW JERSEY

Printed in the United States
4102